THE PHYSICS OF ASTROPHYSICS
Volume I : RADIATION

THE PHYSICS OF ASTROPHYSICS

Volume I

RADIATION

FRANK H. SHU

PROFESSOR OF ASTRONOMY
UNIVERSITY OF CALIFORNIA, BERKELEY

UNIVERSITY SCIENCE BOOKS
Mill Valley, California

University Science Books
20 Edgehill Road
Mill Valley, CA 94941
Fax: (415) 383-3167

Production manager: Mary Miller
Copy editor: Aidan Kelly
TEX formatter: Ed Sznyter
Text and jacket designer: Robert Ishi
Indexer: BevAnne Ross
Printer and binder: The Maple-Vail Book Manufacturing Group

Library of Congress Catalog Number: 91-65168

ISBN 0-935702-64-4

Printed in the United States of America
10 9 8 7 6 5 4 3 2 1

To Helen

Contents

Preface

Modern astronomers need to know a lot of physics. Unfortunately, the typical curriculum in most graduate astronomy departments leaves little time for the student to learn this material through formal course work. Ideally, while in graduate school, a prospective astronomer with a thorough preparation in undergraduate physics and mathematics would take a two-semester graduate sequence in quantum mechanics (leading up to Dirac's equation), a two-semester sequence in classical electrodynamics (including multipole radiation and the theory of special relativity), a semester of statistical mechanics (including Gibbs's ensembles and some applications in quantum statistical mechanics), a semester or more of gas dynamics and plasma physics (including shockwave theory and magnetohydrodynamics), and a year or more of advanced mathematical methods (including the theory of complex variables, ordinary and partial differential equations, numerical analysis, Fourier analysis, and various asymptotic approximation techniques). Unfortunately, such a program would take the budding professional astronomer through the first two years of graduate school without time for a single course in astronomy!

To cope with this dilemma, teachers of graduate courses in astronomy have traditionally included the relevant physics and mathematics as part of the background lecture material. This compromise, however, is very inefficient (e.g., there is a lot of repetition of subject matter such as radiative transfer or atomic and molecular physics) during an era when a veritable explosion of phenomenological knowledge increasingly demands more lecture time for exposure to the frontiers of the field.

At many universities this challenge has given birth to one or more basic courses in the physics prerequisite to many different subject areas in astronomy. I have taught such courses at the State University of New York at Stony Brook (1968–1973) and the University of California at Berkeley (1973–present). Innovations urged by J. Arons have caused the format at Berkeley to evolve from a single-semester course on "astrophyical processes" to the present format of a year-long sequence—the first semester of which deals with "radiation processes" and the second with "gas dynamics and magnetohydrodynamics."

This two-volume text on *The Physics of Astrophysics* grew from my lecture notes for the reorganized two-semester sequence. It is aimed at first-year graduate students and well-prepared seniors in astronomy and physics. The first volume deals with *Radiation*, the second with *Gas Dynamics*. The excellent text by G. Rybicki and A. Lightman on *Radiative Processes in Astrophysics* (Wiley) was my model for the first half of the sequence, but my presentation of the basic material pays more attention to low-energy phenomena (e.g., radio astronomy) and statistical astrophysics (e.g., rate equations).

Although these two volumes were written to form a coherent whole, they can be decoupled for use in separate courses. In particular, when I teach the two-course sequence, the first is not a prerequisite for the second. In writing both volumes, I have been guided by the following pedagogical assumptions and philosophy.

A. Although my discussion emphasizes processes rather than objects, so that the topics are arranged in a sequence formed more by the tradition of physics than by that of astronomy, I usually try to motivate the development by using concrete examples from astrophysics. When faced with a choice between abstract principles or practical applications, I always opt for the practical approach.

B. I have tried to make explanations detailed enough to indicate the important points in the reasoning, but I refrain from displaying every step of a derivation, in order not to divert the attention of the reader from more serious matters. The student may wish to read straight through each chapter to get the central thrust of the physical ideas, and then—as an aid to long-term memory and detailed understanding—return with paper and pencil to work out the missing steps in the formal mathematics.

C. For beginning students, who may not know in which subfields they wish to specialize, I believe it better to cover a lot of ground coherently than to delve deeply into any particular subject. Astronomers of the future will need tools that allow them to explore in many different directions.

D. Along the same lines, I prefer to assign long problems that require a sustained attack on a practical astronomical situation than to contrive short problems that require only a few simple steps to demonstrate a limited objective. I hope I have supplied enough hints along the way so that the student does not become frustrated by getting stuck in the midst of an extended calculation.

This first volume nominally has as its subject matter the emission, absorption, and scattering of radiation by matter, but it includes, in fact, many related topics as well: radiative transfer, statistical physics, classical electrodynamics, atomic and molecular structure. To use the book effectively, the student should have had: some atomic physics and kinetic theory, with exposure to the concepts of the mean-free path and the equilibrium thermodynamic distributions of matter and radiation; a course in quantum mechanics in which the Schrödinger equation is solved for the structure of the hydrogen atom, but not necessarily including the effects of electron spin; electrodynamics with Maxwell's equations expressed in differential form; special relativity with some exposure to the notion of four-vectors and Lorentz transformations; and classical mechanics at a level that uses the Lagrangian and Hamiltonian formulations of the subject. Gaps in a few of these subject areas will not prove fatal, but the student should then be prepared to use heavily the reference material listed at the beginning of each chapter.

This book adopts boldface symbols, e.g., $\mathbf{A}$, for vectors in ordinary three-space; in addition, we use the notation $\mathbf{P}$ and $\mathbf{Q}$ to represent pressure and quadrupole tensors (of dimension 3×3). We denote quantities having four-components by arrows, e.g., $\vec{u}$ for the four-velocity of special relativity. Double-arrowed characters mean 4×4 tensors or matrices, e.g., $\overleftrightarrow{\mathbf{R}}$ for the scattering matrix involving the radiative transfer of the Stokes parameters. To avoid overcomplicating the print face, we forego the arrow convention for variables that refer to dimensions in an internal space, e.g., Dirac matrices and spinors.

Vector calculus in three spatial dimensions constitutes a mathematical tool used freely throughout the book. Apart from the standard theorems of Green, Stokes, etc., the student should be conversant with the identity involving the triple vector product: $\mathbf{A} \times (\mathbf{B} \times \mathbf{C}) = (\mathbf{A} \cdot \mathbf{C})\mathbf{B} - (\mathbf{A} \cdot \mathbf{B})\mathbf{C}$, and the mnemonic that the minus sign accompanies the dot product that involves the vector in the middle. I also assume that he or she knows the Cartesian-tensor analogs for this expression: $\epsilon_{ikm}\epsilon_{mj\ell} = \delta_{ij}\delta_{k\ell} - \delta_{i\ell}\delta_{kj}$, but I do not use the latter formula very much. The reader should also know that the dot and the cross can be interchanged in the triple scalar product: $\mathbf{A} \cdot (\mathbf{B} \times \mathbf{C}) = (\mathbf{A} \times \mathbf{B}) \cdot \mathbf{C}$. Finally, I assume that the student knows how to modify these expressions if one or more of the quantities $\mathbf{A}$, $\mathbf{B}$, $\mathbf{C}$ represents the del operator ∇, and that the direct product of two vectors, as expressed by the dyadic $\mathbf{AB}$, yields a second-rank tensor.

References to journal articles and review papers occur in the text where they appear (usually for the first and last time). Citations of books, to which the reader may want to refer in depth, are given in full in the bibliography.

PART I

RADIATIVE TRANSFER AND STATISTICAL MECHANICS

1

Specific Intensity and
the Equation of Radiative Transfer

Reference: Chandrasekhar, *Radiative Transfer*, Chapter 1.

Almost everything we know about the astronomical universe derives from the laborious gathering of light from faint celestial sources. The information encoded in electromagnetic radiation from the cosmos cannot be deciphered unless we know how to read the message. This book contains the fundamentals of the physics needed to analyze electromagnetic radiation from astronomical sources. As outlined in the Table of Contents, this volume separates into three related parts. Our discussion in Part I begins with the different concepts involved in a statistical treatment of the transport of radiation: *rays* when freely propagating and *waves* or *photons* when interacting with matter. We assume that the reader has some familiarity with the validity of the classical description of radiation as electromagnetic waves if many photons are involved (Part II). We also assume that he or she knows that a quantum treatment is warranted if the interaction with matter takes place one photon at a time (Part III).

DEFINITION OF SPECIFIC INTENSITY

Our definition of specific intensity follows from consideration of Figure 1.1. Let dE be the amount of radiant energy which crosses in time dt the area dA with unit normal $\hat{\mathbf{n}}$ in a direction within solid angle $d\Omega$ centered about $\hat{\mathbf{k}}$ and with photon frequency between ν and $\nu + d\nu$. The monochromatic specific intensity I_ν is then defined by the equation:

$$\text{energy}: \quad dE = I_\nu(\hat{\mathbf{k}}, \mathbf{x}, t)\hat{\mathbf{k}} \cdot \hat{\mathbf{n}} \, dA \, d\Omega \, d\nu \, dt. \tag{1.1}$$

The quantity I_ν has the cgs units, $[I_\nu] = \text{erg cm}^{-2}\,\text{s}^{-1}\,\text{Hz}^{-1}$ steradian^{-1}. The rationale for this definition comes from the conservation of I_ν in the absence of interactions with matter.

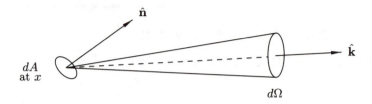

FIGURE 1.1
Definition of specific intensity $I_\nu(\hat{\mathbf{k}}, \mathbf{x}, t)$.

Proof: Consider the passage of radiant energy from element of area dA to dA' as depicted in Figure 1.2.

$$dE' = I_\nu(\hat{\mathbf{k}}, \mathbf{x}', t')\hat{\mathbf{k}} \cdot \hat{\mathbf{n}}' \, dA' \, d\Omega' \, d\nu \, dt, \tag{1.2}$$

where

$$d\Omega = \hat{\mathbf{k}} \cdot \hat{\mathbf{n}}' \, dA'/s^2, \tag{1.3}$$

$$d\Omega' = \hat{\mathbf{k}} \cdot \hat{\mathbf{n}} \, dA/s^2, \tag{1.4}$$

with s being the distance along the ray path between the two elements of area,

$$\mathbf{x}' = \mathbf{x} + s\hat{\mathbf{k}}, \tag{1.5}$$

and t' is the delayed time allowing for a finite speed of propagation c,

$$t' \equiv t + s/c. \tag{1.6}$$

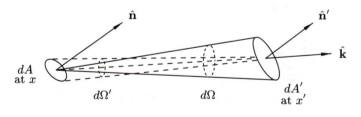

FIGURE 1.2
Proof of constancy of I_ν in a vacuum.

In the absence of interaction with matter, $dE' = dE$; substitution of equations (1.3) and (1.4) into equations (1.1) and (1.2) yields the desired result,

$$I_\nu(\hat{\mathbf{k}}, \mathbf{x}', t') = I_\nu(\hat{\mathbf{k}}, \mathbf{x}, t), \tag{1.7}$$

for propagation in a vacuum. When interactions do take place, we replace the above result with a transport equation of the form,

$$\frac{\partial I_\nu}{\partial t} + c\hat{\mathbf{k}} \cdot \boldsymbol{\nabla} I_\nu = \text{sources} - \text{sinks}. \tag{1.8}$$

We shall shortly expand on what we mean by the right-hand side of equation (1.8), but in the interim, we find it useful to relate I_ν to some quantities that may be more familiar to physicists.

RELATIONSHIP TO PHOTON DISTRIBUTION FUNCTION AND OCCUPATION NUMBER

The photon distribution function $F_\alpha(\mathbf{x}, \mathbf{p}, t)$ is defined so that $F_\alpha(\mathbf{x}, \mathbf{p}, t)$ $d^3x\, d^3p$ = number of photons of spin state α at time t with position within d^3x centered on $\mathbf{x}$ and with momentum within d^3p centered on $\mathbf{p}$. The index α has only two values, 1 and 2, even though spin $s = 1$ has three possible projections along the direction of motion, $m_s = -1, 0, +1$, because the state $m_s = 0$ has no physical meaning for a particle traveling at the speed of light.

The vector momentum of a photon satisfies

$$\mathbf{p} = \hbar\mathbf{k} = (h\nu/c)\hat{\mathbf{k}}. \tag{1.9}$$

The radiant energy contained by all photons occupying the elemental phase-space volume $d^3x\, d^3p$ is

$$dE = \sum_{\alpha=1}^{2} h\nu F_\alpha(\mathbf{x}, \mathbf{p}, t)\, d^3x\, d^3p. \tag{1.10}$$

A beam of photons traveling for time dt in the direction $\hat{\mathbf{k}}$ through an element of area dA whose normal equals $\hat{\mathbf{n}}$ occupies a spatial volume

$$d^3x = (c\, dt)(\hat{\mathbf{k}} \cdot \hat{\mathbf{n}})\, dA.$$

If their momenta have a small spread of solid angle $d\Omega$ and magnitude dp, the corresponding element of momentum volume equals

$$d^3p = p^2\, d\Omega\, dp = \left(\frac{h^3\nu^2}{c^3}\right) d\Omega\, d\nu.$$

Comparison of equations (1.1) and (1.10) now yields the correspondence

$$I_\nu = \sum_{\alpha=1}^{2} \left(\frac{h^4 \nu^3}{c^2} \right) F_\alpha(\mathbf{x}, \mathbf{p}, t). \tag{1.11}$$

In quantum statistical mechanics, h^3 represents a fundamental unit of phase-space volume. We define the occupation number for each (manifested) photon spin state α as $\mathcal{N}_\alpha \equiv h^3 F_\alpha$. Equation (1.11) can now be written

$$I_\nu = \sum_{\alpha=1}^{2} \left(\frac{h \nu^3}{c^2} \right) \mathcal{N}_\alpha(\mathbf{x}, \mathbf{p}, t). \tag{1.12}$$

PROPERTIES UNDER THERMODYNAMIC EQUILIBRIUM

Toward the end of the nineteenth century Kirchhoff showed that (blackbody) radiation in thermodynamic equilibrium with matter has a distribution that depends on only the temperature T and on no other material properties of the enclosure.

Proof: Suppose I_ν differed for two enclosures at the same T over some frequency interval $\delta\nu$. Connect the enclosures with a filter that passes only radiation of that frequency interval. Radiant energy would now flow from one enclosure to another, making it possible to extract useful work from two reservoirs at the same temperature T, in violation of the second law of thermodynamics. Thus our assumption must be in error, and the monochromatic specific intensity of a radiation field in thermodynamic equilibrium with matter must be a universal function of T; in modern notation,

$$\text{Kirchhoff}: \quad I_\nu = B_\nu(T). \tag{1.13}$$

Notice that blackbody radiation is isotropic and unpolarized.

Kirchhoff set a goal for future generations of finding the correct functional form of $B_\nu(T)$. Planck found the appropriate formula—first by interpolating (in entropy) between two forms (Rayleigh-Jeans and Wien laws) and next by coming up with a statistical argument using quantized oscillators to emit and absorb the radiation field. Both derivations gave

$$\text{Planck}: \quad B_\nu = \frac{2h\nu^3/c^2}{e^{h\nu/kT} - 1}. \tag{1.14}$$

Bose found a quantum-statistical method to treat an ideal gas of photons directly, and Einstein generalized Bose's treatment to indistinguishable material particles. Applied to a perfect gas of non-mutually interacting bosons (particles with integer spins), quantum statistical mechanics yields for the occupation number of each spin state at energy level ϵ:

$$\text{Bose-Einstein:} \quad \mathcal{N}_\alpha = \frac{1}{e^{(\epsilon - \mu)/kT} - 1}, \tag{1.15}$$

where μ is the chemical potential (Lagrange multiplier to assure number conservation). When photons are absorbed and emitted by matter (rather than simply scattered), as they must be to come into thermodynamic equilibrium with matter, number conservation becomes an irrelevant constraint. Thus the chemical potential μ of photons can be taken to be zero. Applied to photons, furthermore, $\epsilon = h\nu$; therefore the occupation number of blackbody photons of each spin state α at temperature T is given by

$$\text{Bose:} \quad \mathcal{N}_\alpha = \frac{1}{e^{h\nu/kT} - 1}. \tag{1.16}$$

Substitution of the thermodynamic distribution (1.16) into equation (1.12), with $I_\nu = B_\nu(T)$, rederives equation (1.14).

Three aspects of the preceding discussion deserve comment. First, photons should behave like particles in the limit of high energies. Indeed, when $h\nu \gg kT$, equation (1.16) has the approximate form

$$\mathcal{N}_\alpha \approx e^{-h\nu/kT}$$

that Boltzmann found appropriate for material distributions.

Second, photons should behave like classical waves in the limit of large occupation numbers. From equation (1.16), we see that $\mathcal{N}_\alpha \gg 1$ when $h\nu \ll kT$, since in this limit, we may expand $e^{h\nu/kT} = 1 + h\nu/kT + \cdots$ to obtain

$$\mathcal{N}_\alpha \approx kT/h\nu \gg 1.$$

This case corresponds to the Rayleigh-Jeans limit, wherein the energy content, $h\nu\mathcal{N}_\alpha$, of photons of frequency ν within a phase-space volume h^3 is given by the equipartition theorem as kT ($kT/2$ each for vibrations of **E** and **B**).

Third, equation (1.15) holds for material particles only if they are bosons. For fermions (particles of half-integer spin) satisfying the Pauli exclusion principle, quantum statistical mechanics yields

$$\text{Fermi-Dirac:} \quad \mathcal{N}_\alpha = \frac{1}{e^{(\epsilon - \mu)/kT} + 1}. \tag{1.17}$$

Notice the $+1$ which replaces the -1 in the denominator. The $+1$ assures that no more than one fermion of a given spin state (or, more generally, of a given internal quantum state) can occupy a given translational state (of phase-space volume h^3). The difference between $+1$ for an ideal Fermi-Dirac gas and -1 for an ideal Bose-Einstein gas can be stated as follows:

fermions avoid being in a place where there are other fermions like themselves (with $\mathcal{N}_\alpha$ smaller than the classical value), whereas bosons like to be where there are other bosons like themselves (with $\mathcal{N}_\alpha$ larger than the classical value). As we shall see in a later chapter, this tendency is the basis for the phenomenon of the *stimulated emission* of photons (spin 1). In contrast, neutrinos (spin 1/2) are fermions—with, as far as we can tell, zero rest mass and zero chemical potential when they are copiously emitted and absorbed by hot dense matter. In thermodynamic equilibrium, neutrinos of energy $h\nu$ would have occupation number

$$\text{Neutrinos:} \quad \mathcal{N}_\alpha = \frac{1}{e^{h\nu/kT} + 1}, \tag{1.18}$$

and instead of suffering stimulated emission, they would exhibit the phenomenon of *suppressed emission.*

EQUATION OF RADIATIVE TRANSFER

Astronomers like to write the transport equation (1.8) per unit length rather than per unit time; so the equation of radiative transfer usually reads

$$\frac{1}{c}\frac{\partial I_\nu}{\partial t} + \hat{\mathbf{k}} \cdot \boldsymbol{\nabla} I_\nu = \frac{1}{4\pi}\rho j_\nu - \rho\kappa_\nu^{\text{abs}} I_\nu - \rho\kappa_\nu^{\text{sca}} I_\nu + \rho\kappa_\nu^{\text{sca}} \oint \phi_\nu(\hat{\mathbf{k}}, \hat{\mathbf{k}}') I_\nu(\hat{\mathbf{k}}')\, d\Omega', \tag{1.19}$$

where ρ is the mass density per unit volume of the gas, j_ν is its emissivity per unit mass, κ_ν^{abs} is its total absorption opacity (absorption cross-section per unit mass), κ_ν^{sca} is its total scattering opacity, and $\phi_\nu(\hat{\mathbf{k}}, \hat{\mathbf{k}}')$ is the scattering probability density (from $\hat{\mathbf{k}}'$ to $\hat{\mathbf{k}}$). The last satisfies the normalization and reversibility constraints:

$$\oint \phi_\nu(\hat{\mathbf{k}}, \hat{\mathbf{k}}')\, d\Omega' = 1 = \oint \phi_\nu(\hat{\mathbf{k}}, \hat{\mathbf{k}}')\, d\Omega. \tag{1.20}$$

On the right-hand side of equation (1.19), ρj_ν represents the source for radiation that comes from true emission (the term is reduced by a factor of 4π to make it per steradian); $-\rho\kappa_\nu^{\text{abs}} I_\nu$ represents the amount of light removed from the beam, per unit length of photon travel, by the effects of true absorption; $-\rho\kappa_\nu^{\text{sca}} I_\nu$ represents the analogous amount scattered out of the relevant beam; and the last term represents the integral contribution scattered into the beam from any other line of sight.

Later in this text we will practice calculating j_ν, κ_ν^{abs}, etc., for specific microscopic radiation processes. For now, we will assume that such calculations exist, and we regard equation (1.19) as one to solve for I_ν when j_ν, κ_ν^{abs}, etc. are given. In point of fact, not only do interactions of photons with matter change the radiation field, but, in general, they will also affect the physical state of the matter. That is, real interactions are often severely and nonlinearly coupled, and this fact constitutes the principal difficulty in computing their values. In any case, before we can proceed with the simpler problem of solving formally for I_ν given j_ν, κ_ν^{abs}, etc., we need to carefully decompose the qualitatively differing microscopic contributions to j_ν and κ_ν^{abs}.

CORRECTION FOR STIMULATED EMISSION

According to Einstein, atoms and molecules (which emit and absorb photons one at a time) have two components to their emissivity j_ν:

$$j_\nu = j_\nu^{\text{spontaneous}} + j_\nu^{\text{induced}}, \tag{1.21}$$

where

$$j_\nu^{\text{induced}} \propto I_\nu. \tag{1.22}$$

Notice that j_ν^{induced} is not isotropic (it will depend on the propagation direction $\hat{\mathbf{k}}$ if I_ν does), so that stimulated emission occurs coherently and in the same direction as the carrier photons. In contrast, $j_\nu^{\text{spontaneous}}$ can usually (but not always) be taken as isotropic in the local frame of rest of the gas. Only for a static medium (bulk or fluid velocity = 0, although individual particles can have random thermal motions) will the local rest frame of the matter correspond to the laboratory frame. When fluid motions exist (especially when relativistic motions are present), a proper treatment of radiative transfer becomes much more complicated than it is for the static case; hence, for most of this text, we will assume that the effects of fluid motion are negligible, except when we explicitly state otherwise.

In any case, because of the proportionality in equation (1.22), astronomers conventionally incorporate $\rho j_\nu^{\text{induced}}/4\pi$ as part of the term $-\rho\kappa_\nu^{abs}I_\nu$, and they speak of the resulting κ_ν^{abs} as "true absorption corrected for stimulated emission." We will illustrate how to do this in detail for some important microscopic examples in future chapters. For now, we will use j_ν though we really mean $j_\nu^{\text{spontaneous}}$. The importance of carrying through the preceding decomposition is that we can then assume that j_ν and κ_ν^{abs} are independent of $\hat{\mathbf{k}}$ (for a static isotropic medium).

What about "stimulated scattering"? Is there such a thing? Yes, although most textbooks on radiative transfer ignore any discussion of the effect, because for *coherent scattering* (propagation redirection but no change of photon frequency—i.e., momentum exchange but no energy exchange), the effects of stimulated scattering exactly cancel out in the source and sink terms. (For nearly coherent scattering, they would nearly cancel out.) We shall return to this topic in a later chapter, but for now we will adopt the usual treatment and leave equation (1.19) as it stands. Moreover, for many applications, we may make a further simplification, and assume

$$\oint \phi_\nu(\hat{\mathbf{k}}, \hat{\mathbf{k}}')(\hat{\mathbf{k}}\, d\Omega \text{ or } \hat{\mathbf{k}}' d\Omega') = 0, \tag{1.23}$$

since electrons, atoms, molecules, and even grains scatter equally in the forward and backward directions when $\lambda = c/\nu$ is greater than the "size" of the scattering particle. Scattering problems where the phase function ϕ_ν does not possess forward-backward symmetry—or, even worse, where the scattering does not occur coherently—add computational complications but no conceptual difficulties.

2

Moment Equations and Conduction Approximation

Reference: Schwarzschild, *Structure and Evolution of the Stars*, Chapter 2.

Solution of the transfer equation (1.19), which represents a complicated integro-partial-differential equation in six phase-space variables (three components of $\mathbf{x}$, two angles $\hat{\mathbf{k}}$, and ν) plus time, would give us a lot of information that we often do not need. For many purposes, including a first step in numerical iterative methods to solve equation (1.19), we may be satisfied with a contracted description which integrates out the dependence on photon propagation direction $\hat{\mathbf{k}}$, but retains the information on the distribution with space $\mathbf{x}$ (and time t) and photon frequency ν. In preparation for such a contracted description, we first define some angular moments of I_ν:

$$\begin{pmatrix} cE_\nu \\ \mathbf{F}_\nu \\ c\mathbf{P}_\nu \end{pmatrix} = \oint \begin{pmatrix} 1 \\ \hat{\mathbf{k}} \\ \hat{\mathbf{k}}\hat{\mathbf{k}} \end{pmatrix} I_\nu \, d\Omega. \tag{2.1}$$

Some texts (e.g., Mihalas 1978) denote successive angular moments, when divided by 4π, as J_ν, $\mathbf{H}_\nu$, $\mathbf{K}_\nu$, etc., but we prefer the notation of equation (2.1) to remind ourselves that E_ν is the monochromatic energy density in the radiation field, $\mathbf{F}_\nu$ is the monochromatic energy flux, and $\mathbf{P}_\nu$ is the monochromatic pressure tensor. The integrands for E_ν and $\mathbf{F}_\nu$ in (2.1) are obvious from the fundamental definition of I_ν, equation (1.1), but what about $\mathbf{P}_\nu$?

We recall from electrodynamics or fluid mechanics that the different components of a stress tensor refer to the rate of transfer of momentum across surfaces with certain orientations. The rate of momentum transfer per photon of momentum $\mathbf{p}$ across a surface of unit normal $\hat{\mathbf{k}}$ equals $c\hat{\mathbf{k}}\mathbf{p}$; therefore the total rate of momentum transfer, the radiation stress tensor $\mathbf{P}_{\mathrm{rad}}$, by all photons in the distribution, is locally given by

$$\mathbf{P}_{\mathrm{rad}} = \sum_{\alpha=1}^{2} \int c\hat{\mathbf{k}}\mathbf{p}F_\alpha d^3p = \sum_{\alpha=1}^{2} \int \hat{\mathbf{k}}\hat{\mathbf{k}}F_\alpha \frac{h^4\nu^3}{c^3} \, d\Omega \, d\nu, \tag{2.2}$$

11

where we have used $d^3p = (h^3\nu^2/c^3)d\Omega d\nu$. Invoking equation (1.11), we obtain the identification

$$\mathbf{P}_{\text{rad}} = \frac{1}{c} \int \hat{\mathbf{k}}\hat{\mathbf{k}} I_\nu \, d\Omega d\nu, \qquad (2.3)$$

which corresponds to the third row of equation (2.1) if we equate $\mathbf{P}_{\text{rad}}$ with the integral of $\mathbf{P}_\nu$ over all frequencies ν. More generally, we denote the radiation energy density, flux, and pressure tensor by

$$\begin{pmatrix} E_{\text{rad}} \\ \mathbf{F}_{\text{rad}} \\ \mathbf{P}_{\text{rad}} \end{pmatrix} = \int_0^\infty \begin{pmatrix} E_\nu \\ \mathbf{F}_\nu \\ \mathbf{P}_\nu \end{pmatrix} d\nu. \qquad (2.4)$$

CONSERVATION RELATIONS FOR THE RADIATION FIELD

Given the preceding, we now begin our derivation of the moment equations. Begin by noting that $\hat{\mathbf{k}}$ commutes with ∇ and $\partial/\partial t$. If we multiply equation (1.19) by 1 and integrate over all solid angles of photon propagation, we obtain

$$\frac{\partial E_\nu}{\partial t} + \nabla \cdot \mathbf{F}_\nu = \rho(j_\nu - c\kappa_\nu^{\text{abs}} E_\nu), \qquad (2.5)$$

where we have assumed that j_ν and κ_ν^{abs} have no dependence on $\hat{\mathbf{k}}$, and where we have used equation (1.20) to cancel the contributions from the scattering terms. Equation (2.5) has the generic form of a conservation equation,

$$\frac{\partial}{\partial t}(\text{density of quantity}) + \nabla \cdot (\mathbf{flux} \text{ of quantity}) = \text{sources} - \text{sinks}. \quad (2.6)$$

Here the quantity is the energy of radiation of frequency ν, and the zeroth moment of the radiative transfer equation therefore states simply that the time-rate of change of photon energy equals the difference between what is emitted and what is absorbed by matter; i.e., energy is conserved in the matter-radiation system. Coherent scattering does not contribute to this budget, because no energy is exchanged between matter and radiation, photon by photon, if we make the approximation that all scatterings preserve $h\nu$.

Notice that equation (2.5) plays the natural role as a scalar partial differential equation (PDE) in $\mathbf{x}$ and t to solve for E_ν. Unfortunately, even if j_ν and κ_ν^{abs} were known, equation (2.5) contains three components of another variable, $\mathbf{F}_\nu$, that we have no way of obtaining without solving the original equation (1.19), and then integrating $\hat{\mathbf{k}} I_\nu$—see equation (2.1). Alternatively, we could try to obtain an equation for $\mathbf{F}_\nu$ by multiplying

equation (1.19) by $\hat{\mathbf{k}}$ and integrating over all solid angles. By inspection again, this produces the result

$$\frac{1}{c}\frac{\partial \mathbf{F}_\nu}{\partial t} + c\boldsymbol{\nabla} \cdot \mathbf{P}_\nu = -\rho(\kappa_\nu^{\text{abs}} + \kappa_\nu^{\text{sca}})\mathbf{F}_\nu. \tag{2.7}$$

To derive equation (2.7) we have used the forward-backward symmetry assumption, equation (1.23), to eliminate the contribution of the photons scattered into our beam. Physically, if such photons are as likely to be scattered forward as backward, then they can contribute nothing to the momentum exchange. In contrast, the photons scattered out of a beam into all other directions, the term represented by $-\rho\kappa_\nu^{\text{sca}}\mathbf{F}_\nu$, can yield a net momentum transfer that is opposite to the net flow of radiative energy $\mathbf{F}_\nu$. The same goes for the absorption of photons, which involves momentum exchange, followed by *isotropic* remission, which does not. The quantity

$$\kappa_\nu \equiv \kappa_\nu^{\text{abs}} + \kappa_\nu^{\text{sca}} \tag{2.8}$$

appearing in equation (2.7) is called the total opacity (corrected for stimulated emission but not for stimulated scattering), or simply, the opacity. In any case, if we now recall that the net momentum flow of photons is $1/c$ times the net energy flow, we see that the first moment equation (2.7) represents the law of conservation of momentum, just as the zeroth moment equation (2.5) represents the law of conservation of energy.

Notice also the following general rule: what appears as a source for the radiation field must reappear as a sink for the matter, and vice versa. In other words, when we add the energy or momentum equations for matter and radiation, the internal exchanges between them must not survive as net terms on the right-hand sides of the resulting sum. In particular, if momentum is removed at a certain rate from the radiation field, the back reaction must appear as a force for the matter. Thus the radiative force per unit area per unit length, i.e., the force per unit volume, acting on the gas must equal

$$\mathbf{f}_{\text{rad}} = \frac{\rho}{c}\int_0^\infty \kappa_\nu \mathbf{F}_\nu \, d\nu. \tag{2.9}$$

THE NEED FOR A CLOSURE CONDITION

Having dispensed with these digressions, let us return to the original issue. Equation (2.7) gives, as expected, an equation for $\mathbf{F}_\nu$; unfortunately, the set composed of the scalar equation (2.5) for E_ν and the vector equation (2.7) for $\mathbf{F}_\nu$ still does not form a closed set, because we have gained *five* new variables in the independent components of the symmetric tensor $\mathbf{P}_\nu$. The number 5 enters because the third row in equation (2.1) shows explicitly that $\mathbf{P}_\nu$ is a symmetric tensor; so three off-diagonal elements are redundant. The last fact explains why we do not need to specify which index is being summed in the "dot" product $\nabla \cdot \mathbf{P}_\nu$. Moreover, since $\hat{\mathbf{k}} \cdot \hat{\mathbf{k}} = 1$ for a unit vector, the trace of $\mathbf{P}_\nu$ equals E_ν and cannot vary independently of the latter variable; this removes one more degree of freedom from the nine entries of the tensor represented as a 3×3 matrix. Thus the appearance of $\mathbf{P}_\nu$ introduces $9 - 3 - 1 = 5$ new independent quantities.

We see, therefore, the emergence of a general pattern. Without making some physical assumption, we find that no finite number of the moment equations will ever form a closed set; new variables always appear, making the number of variables larger than the set of equations. Taking the $(\ell - 1)$-th moment of the transfer equation generally introduces $2\ell + 1$ new quantities associated with the ℓ-th moment variable. Only the *infinite* set of all moment equations (multiplication by 1, $\hat{\mathbf{k}}$, $\hat{\mathbf{k}}\hat{\mathbf{k}}$, ... and integrating over $d\Omega$) would contain the same amount of information as the original kinetic equation (1.19). An analogy holds with taking the Fourier-series expansion of a function. No finite number of Fourier coefficients contains the same amount of information as the original function; only the infinite set of Fourier coefficients can reproduce all of the bumps and wiggles. Since we need here to specify *two* angles of photon propagation (ϑ, φ), the actual decomposition would correspond to one in *spherical harmonics*, $Y_{\ell m}(\vartheta, \varphi)$. Thus the scalar E_ν relates to the single coefficient of the Y_{00} term; the vector $\mathbf{F}_\nu$ relates to the three coefficients of the Y_{1m} term, with $m = -1, 0, +1$; the symmetric tensor $\mathbf{P}_\nu$, apart from the trace, $\mathrm{tr}[\mathbf{P}_\nu] = E_\nu$, relates to the five coefficients of the Y_{2m} term, with $m = -2, -1, 0, +1, +2$; etc.

This analogy provides us, however, with some hope. If small bumps and wiggles do not interest us, we don't need all the Fourier (or spherical harmonic) coefficients; the lowest-order coefficients contain the gross behavior of the function. In other words, if we don't need to know all the details of a distribution function, the lowest-order moments specify its important properties (average value, degree of anisotropy, spread about mean, etc.). Thus, for many purposes, if we could use only equations (2.5) and (2.7) to compute E_ν, $\mathbf{F}_\nu$, and $\mathbf{P}_\nu$, even in some approximate manner, we might be quite content. To carry out such a program, we need to find or postulate a supplementary (tensor) equation that relates these three variables, i.e., that gives the high-order moments in terms of the lower-order ones. This constitutes the problem of finding a *closure relation*. In principle, we could search for a closure condition at a higher-order set of equations than just equations (2.5) and (2.7); in practice, people have generally limited searches to this level, for good reasons. Physically, equations (2.5) and (2.7) correspond to *conservation laws*, suggesting that they are special members of a much larger and less-interesting set of possible moment equations. Being related to familiar physical quantities, E_ν, $\mathbf{F}_\nu$, and $\mathbf{P}_\nu$ allow us to use physical intuition to find a useful approximate relation among them much more easily than if we thought in the abstract about some other set of higher-order moments. For many people, as it is, tensors that exert stress already border on the unfamiliar and bring on feelings of strain. Finally, when approximate methods work well, they usually do so after just a few terms, and if they don't work well, taking more terms usually doesn't improve matters very rapidly!

Under what conditions can we find good closure conditions? At the extremes. If a region is optically thin (photon mean-free path long compared to the macroscopic distances of interest), we can consider photons to fly in straight lines, basically unimpeded by the presence of matter. We shall consider in Chapter 5 the corresponding closure relation that might work for such situations. On the other hand, if a region is optically thick (photon mean-free path short compared to the macroscopic scales of interest), repeated scatterings, emissions, and absorptions must produce a nearly isotropic radiation field. We consider this situation in its simplest context in the next section.

THE RADIATION CONDUCTION APPROXIMATION

Consider the situation in the interior of a star like the Sun. The mean-free path $\ell_\nu \equiv 1/\rho\kappa_\nu$ for a typical photon in the Sun's interior might amount to 0.5 cm, with roughly equal contributions from absorption and scattering. This is roughly 10^{11} times smaller than the radius of the Sun, $R_\odot = 7 \times 10^{10}$ cm; i.e., the "optical depth" from the surface of the Sun to a typical point in its interior is $\sim 10^{11}$, making it the closest thing to a perfect thermodynamic enclosure known in the solar system. For the interior of the Sun, not only must the radiation field I_ν be nearly completely isotropic, but it must also be nearly Planckian,

$$I_\nu \approx B_\nu(T), \qquad (2.10)$$

where T is the local matter temperature. If we use this approximation to calculate E_ν and $\mathbf{P}_\nu$ from equation (2.1), we obtain

$$E_\nu \approx 4\pi B_\nu(T)/c, \qquad \mathbf{P}_\nu \approx [4\pi B_\nu(T)/3c]\mathbf{I}, \qquad (2.11)$$

where $\mathbf{I}$ is the unit tensor that has the Kronecker delta δ_{ij} as its components in Cartesian form. In other words, $\mathbf{P}_\nu = P_\nu\mathbf{I}$, where P_ν is a monochromatic scalar pressure that satisfies the familiar relationship $P_\nu = E_\nu/3$ valid for a relativistic isotropic gas. Usually this relationship is stated in the frequency-integrated form: $P_{\mathrm{rad}} = E_{\mathrm{rad}}/3$, where P_{rad} and E_{rad} actually have the dimensions of pressure and energy density, i.e., erg cm^{-3}.

Notice that we do not use equation (2.10) to evaluate $\mathbf{F}_\nu$ because only the small *nonisotropic* part of I_ν can lead to a nonvanishing energy flux.

In other words, photons going in opposite directions cancel in their contribution to the integral of $\hat{\mathbf{k}}I_\nu$ if we approximate $I_\nu(\hat{\mathbf{k}}, \mathbf{x})$ to be independent of $\hat{\mathbf{k}}$.

To see how accurate are the approximations given in equation (2.11), we can arrive at them by a different line of reasoning. Under conditions of close thermodynamic coupling between matter and radiation, we expect the emissivity j_ν to be given by Kirchhoff's law:

$$j_\nu = 4\pi\kappa_\nu^{\text{abs}}B_\nu(T). \tag{2.12}$$

For a blackbody, which absorbs everything that falls on it (κ_ν having the same uniformly high value for all ν), the specific emissivity would be exactly proportional to $B_\nu(T)$; for more realistic materials in thermodynamic equilibrium with matter at temperature T, $B_\nu(T)$ is weighted by the factor κ_ν^{abs} that measures the ability of the material to absorb (and therefore to emit) at frequency ν. As we shall prove below, this makes the rate of local radiative emission j_ν a large term in equation (2.5), which can be balanced only by an almost exactly equal amount of local radiative absorption, represented by the term $c\kappa_\nu^{\text{abs}}E_\nu$, thereby recovering the blackbody relation $E_\nu = 4\pi B_\nu(T)/c$.

To begin, notice that, in order of magnitude, $\partial E_\nu/\partial t \sim E_\nu/t_\odot$, where $t_\odot$ is the evolutionary timescale of the interior of the Sun (several billion years). Thus its ratio to $c\rho\kappa_\nu^{\text{abs}}E_\nu$ must be as the mean-free path $\ell_\nu = 1/\rho\kappa_\nu$, typically 0.5 cm, is to $ct_\odot$, several billion light-years. This ratio, $\sim 10^{-28}$, is negligibly small by anyone's standards; so we may safely ignore the term $\partial E_\nu/\partial t$ in equation (2.5) when dealing with the Sun's interior. In a similar manner, we can argue that the term $(1/c)\partial\mathbf{F}_\nu/\partial t$ in equation (2.7) is smaller than the term $\rho\kappa_\nu\mathbf{F}_\nu$ by the same factor. Thus $\rho\kappa_\nu\mathbf{F}_\nu$ can be balanced only by the remaining term $c\boldsymbol{\nabla}\cdot\mathbf{P}_\nu$, with a magnitude of $\sim cE_\nu/R_\odot$, since $\mathbf{P}_\nu$ and E_ν have comparable orders of magnitude. This demonstrates that $|\mathbf{F}_\nu|$ is $\sim (\ell_\nu/R_\odot)$ times smaller than cE_ν, which equals the monochromatic energy transport that photons could carry if they could only fly instead of having to random walk.

What about the term $\boldsymbol{\nabla}\cdot\mathbf{F}_\nu$ in equation (2.5)? It must be of order $|\mathbf{F}_\nu|/R_\odot \sim (\ell_\nu/R_\odot^2)cE_\nu$, which is smaller than the term $c\rho\kappa_\nu^{\text{abs}}E_\nu$ in equation (2.5) by the factor $(\ell_\nu/R_\odot)^2 \sim 10^{-22}$. We have now demonstrated the desired result: that the only way to satisfy equation (2.5) results in a very near equality between E_ν and $j_\nu/c\kappa_\nu^{\text{abs}} = 4\pi B_\nu(T)/c$.

En route to this conclusion, we have also shown that equation (2.7) is well approximated by [see equation (2.11)]

$$\mathbf{F}_\nu = -\frac{c}{\rho\kappa_\nu}\boldsymbol{\nabla}\cdot\mathbf{P}_\nu = -\frac{4\pi}{3\rho\kappa_\nu}\frac{\partial B_\nu}{\partial T}\boldsymbol{\nabla}T. \tag{2.13}$$

Integrating equation (2.13) over all ν yields

$$\mathbf{F}_{\text{rad}} = -\frac{4\pi}{3\rho\kappa_{\text{R}}}(\boldsymbol{\nabla}T)\int_0^\infty \frac{\partial B_\nu}{\partial T}\, d\nu, \tag{2.14}$$

where we have defined the *Rosseland mean opacity* κ_{R} by the weighted transmission average:

$$\frac{1}{\kappa_{\text{R}}} \equiv \frac{\int_0^\infty (1/\kappa_\nu)(\partial B_\nu/\partial T)\, d\nu}{\int_0^\infty (\partial B_\nu/\partial T)\, d\nu}. \tag{2.15}$$

Since

$$\int_0^\infty \frac{\partial B_\nu}{\partial T}\, d\nu = \frac{d}{dT}\int_0^\infty B_\nu(T)\, d\nu = \frac{d}{dT}\left(\frac{caT^4}{4\pi}\right), \tag{2.16}$$

we may now write equation (2.14) as

$$\mathbf{F}_{\text{rad}} = -\frac{c}{3\rho\kappa_{\text{R}}}\boldsymbol{\nabla}(aT^4). \tag{2.17}$$

Equation (2.17) is often called the radiation conduction equation, because it implies that the radiative flux is proportional to minus the gradient of the temperature T ("Fourier's law"). Indeed, recognizing aT^4 as the energy density of the blackbody radiation, we see that equation (2.17) has the general form for diffusive fluxes ("Fick's law"):

diffusive **flux** $= -\mathcal{D}\boldsymbol{\nabla}$(density of quantity being diffused),

where $\mathcal{D}$ is the diffusivity. Indeed, this comparison allows us to identify the radiative diffusivity as having the characteristic formula,

$$\mathcal{D}_{\text{rad}} = \frac{1}{3}c\ell,$$

where $\ell \equiv 1/\rho\kappa_{\text{R}}$ is the (Rosseland) mean-free path of the diffusing particles (photons). A "random walk" slows down the free-flight speed c by a typical factor of $\ell/R_\odot$, so that the time $R_\odot^2/\mathcal{D}_{\text{rad}}$ for photons to diffuse to the surface of the Sun is roughly $3R_\odot/\ell$ times longer than the free-flight time $R_\odot/c$ of about 2 s. This process prevents the Sun from releasing its considerable internal reservoir of photons in one powerful blast, but instead regulates it to the stately observed luminosity of $L_\odot = 3.86 \times 10^{33}$ erg s^{-1}. In any case, apart from being useful for rough order-of-magnitude arguments, the accurate equation (2.17) constitutes one of the fundamental equations underlying the whole theory of stellar structure and evolution.

To complete this section, we notice that the radiative force per unit volume acting on the matter can be obtained by substituting equation (2.13) into equation (2.9):

$$\mathbf{f}_{\text{rad}} = -(\boldsymbol{\nabla}T)\frac{4\pi}{3c}\int_0^\infty \frac{\partial B_\nu}{\partial T}\, d\nu = -\frac{1}{3}\boldsymbol{\nabla}(aT^4), \tag{2.18}$$

where we use equation (2.16). We recognize $aT^4/3$ as the radiation pressure P_{rad} of blackbody radiation of temperature T; so equation (2.9) reduces, in the limit of large optical thickness, to the familiar expression,

$$\mathbf{f}_{\mathrm{rad}} = -\boldsymbol{\nabla} P_{\mathrm{rad}}. \tag{2.19}$$

Finally, note that none of the formulae derived in this section assumed spherical symmetry. They apply to stars severely distorted by rotation or tidal forces (in a close binary) as well as to rounder examples, like the Sun.

3

Solution of the Problem of Radiative Transfer

Reference: Mihalas, *Stellar Atmospheres*, Chapter 3.

In this chapter, we consider the solution of problems of radiative transfer where too little optical depth exists to make radiation conduction a good approximation. We begin by giving a general discussion that permits arbitrary spatial variations in all three dimensions.

FORMAL SOLUTION OF THE EQUATION OF TRANSFER

If light travels across an object in an interval short compared with its evolutionary time (see Chapter 2), we may ignore time-dependence in the corresponding problem of radiation transport. Equation (1.19) then can be written

$$\hat{\mathbf{k}} \cdot \nabla I_\nu + \rho \kappa_\nu I_\nu = \rho \left(\frac{j_\nu}{4\pi} + \kappa_\nu^{\mathrm{sca}} \Phi_\nu \right), \qquad (3.1)$$

where $\kappa_\nu = \kappa_\nu^{\mathrm{abs}} + \kappa_\nu^{\mathrm{sca}}$ is the total opacity and Φ_ν is the intensity weighted by the angular phase function for scattering:

$$\Phi_\nu(\hat{\mathbf{k}}, \mathbf{x}) \equiv \oint \phi_\nu(\hat{\mathbf{k}}, \hat{\mathbf{k}}') I_\nu(\hat{\mathbf{k}}', \mathbf{x}) \, d\Omega'. \qquad (3.2)$$

For isotropic scattering, $\phi_\nu = 1/4\pi$, and Φ_ν becomes the mean intensity J_ν.

By introducing the *source function*,

$$S_\nu(\hat{\mathbf{k}}, \mathbf{x}) \equiv \frac{1}{\kappa_\nu} \left(\frac{j_\nu}{4\pi} + \kappa_\nu^{\mathrm{sca}} \Phi_\nu \right), \qquad (3.3)$$

and by defining the ray-path derivative $\hat{\mathbf{k}} \cdot \nabla$ to be d/ds, we may write equation (3.1) as

$$\frac{dI_\nu}{ds} + \rho \kappa_\nu I_\nu = \rho \kappa_\nu S_\nu. \qquad (3.4)$$

19

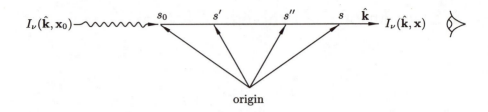

origin

FIGURE 3.1
Geometry for ray-path integration for I_ν.

Introducing an optical depth variable

$$\tau_\nu \equiv \int_{s_0}^s \rho\kappa_\nu ds, \tag{3.5}$$

where s_0 represents the back boundary of our object, we see that the left-hand side of equation (3.4) possesses an integrating factor

$$e^{-\tau_\nu} \frac{d}{ds} \left(e^{\tau_\nu} I_\nu \right) = \rho\kappa_\nu S_\nu. \tag{3.6}$$

Recognizing $\rho\kappa_\nu ds$ as $d\tau_\nu$ and integrating in s from s_0 to s (in τ_ν from 0 to τ_ν), we obtain

$$I_\nu(\hat{\mathbf{k}}, \tau_\nu)e^{\tau_\nu} - I_\nu(\hat{\mathbf{k}}, 0) = \int_0^{\tau_\nu} S_\nu e^{t_\nu} dt_\nu.$$

Introducing a new dummy variable $\tau_\nu' = \tau_\nu - t_\nu$, we obtain, upon transposing terms,

$$I_\nu(\hat{\mathbf{k}}, \tau_\nu) = I_\nu(\hat{\mathbf{k}}, 0)e^{-\tau_\nu} + \int_0^{\tau_\nu} S_\nu(\hat{\mathbf{k}}, \tau_\nu')e^{-\tau_\nu'} d\tau_\nu'. \tag{3.7}$$

Equation (3.7) looks deceptively simple. The source function (3.3) depends, in reality, not on the monochromatic optical depth τ_ν, but on the spatial position $\mathbf{x}$. The trouble with using τ_ν' as an integration variable in place of source position $\mathbf{x}'$ is that the former has a different value at different frequencies ν and different propagation directions $\hat{\mathbf{k}}$ for each field (observer) position $\mathbf{x}$. Except for special circumstances, if we are interested in more than one frequency and one propagation direction per field point, we would be well advised to perform the integration in physical space and with a more explicit notation (see Figure 3.1):

$$I_\nu(\hat{\mathbf{k}}, \mathbf{x}) = I_\nu(\hat{\mathbf{k}}, \mathbf{x}_0) \exp[-\tau_\nu(\hat{\mathbf{k}}, \mathbf{x}, \mathbf{x}_0)]$$
$$+ \int_{s_0}^{s} S_\nu(\hat{\mathbf{k}}, \mathbf{x}') \exp[-\tau_\nu(\hat{\mathbf{k}}, \mathbf{x}, \mathbf{x}')]\rho(\mathbf{x}')\kappa_\nu(\mathbf{x}') \, ds', \quad (3.8)$$

where

$$\tau_\nu(\hat{\mathbf{k}}, \mathbf{x}, \mathbf{x}') \equiv \int_{s'}^{s} \rho(\mathbf{x}'')\kappa_\nu(\mathbf{x}'') \, ds'' \qquad (3.9)$$

represents the optical depth along a ray path from $\mathbf{x}'$ to $\mathbf{x}$, with

$$\mathbf{x}' = \mathbf{x} - (s - s')\hat{\mathbf{k}}, \quad \mathbf{x}'' = \mathbf{x} - (s - s'')\hat{\mathbf{k}}, \quad \mathbf{x}_0 = \mathbf{x} - (s - s_0)\hat{\mathbf{k}} \quad (3.10)$$

denoting backward positions along the ray path at distances $(s-s')$, $(s-s'')$, and $(s - s_0)$ from $\mathbf{x}$. In this notation, $\mathbf{x}_0$ represents the incident point at which a ray from a background source first enters the region of interest (see Figure 3.1).

Equation (3.8) states that, apart from an exponential attenuation of an initial value, I_ν results from the superposition of all sources (emitted and scattered light) along the line of sight, weighted by the extinction factor $e^{-\tau_\nu'}$ from the source position $\mathbf{x}'$ to the field (observation) point $\mathbf{x}$. To verify that equation (3.8) satisfies equation (3.1) in general, notice that $\hat{\mathbf{k}} \cdot \nabla$ operating on the attenuated initial term $[\propto I_\nu(\hat{\mathbf{k}}, \mathbf{x}_0)]$ exactly cancels $-\rho\kappa_\nu$ times the same term; so this term can be ignored in what follows. When $\hat{\mathbf{k}} \cdot \nabla$ operates on the upper limit s of the integral, we get $\hat{\mathbf{k}} \cdot \nabla s$ times the integrand evaluated at $s' = s$, i.e., $\hat{\mathbf{k}} \cdot \hat{\mathbf{k}} = 1$ times $\rho(j_\nu/4\pi + \kappa_\nu^{\mathrm{sca}}\Phi_\nu)$ evaluated at $\mathbf{x}' = \mathbf{x}$. On the other hand, when $\hat{\mathbf{k}} \cdot \nabla$ operates on the integrand, we get $-\hat{\mathbf{k}} \cdot \nabla\tau_\nu$ times the original integral, i.e., $-\rho\kappa_\nu$ times the original integral. Thus the solution (3.8) satisfies equation (3.1) (Q.E.D.).

RADIATIVE EQUILIBRIUM

In order to use equation (3.8), we must be able to specify j_ν and Φ_ν. The latter, unfortunately, depends on an integral [equation (3.2)] of I_ν. Even j_ν will usually turn out to depend on the solution I_ν. In particular, for *static* distributions of matter near a strong source of radiation (e.g., a star), the condition of steady state requires that radiation neither add nor subtract a net amount of energy from the gas:

$$\rho \int_0^\infty (j_\nu - c\kappa_\nu^{\mathrm{abs}} E_\nu) \, d\nu = 0, \qquad (3.11)$$

where E_ν is the monochromatic energy density of the ambient radiation field [see equation (2.1)] and numerically equals $4\pi J_\nu/c$. Equation (3.11) places an integral constraint at each point $\mathbf{x}$ in space on the possible variation of j_ν and I_ν. Equivalently, we may integrate equation (2.5) over all frequencies ν and apply equation (3.11) to obtain the constraint of radiative equilibrium as [see the frequency-integrated and time-independent version of equation (2.5)]:

$$\nabla \cdot \mathbf{F}_{\mathrm{rad}} = 0. \tag{3.12}$$

LOCAL THERMODYNAMIC EQUILIBRIUM

To fix ideas let us consider a simple but important example. Suppose all the level populations of the internal and external degrees of freedom that can contribute to electromagnetic radiation are characterized by their thermodynamic values at a common temperature T. In such a situation, the matter is said to be in *local thermodynamic equilibrium* (LTE). The assumption of LTE differs from TE (complete thermodynamic equilibrium) in that the ambient radiation field need not be Planckian at the same T. Thus matter in LTE with $I_\nu \neq B_\nu(T)$ constitutes a first generalization beyond the "radiation conduction" approximation considered in the last chapter. The assumption of LTE constitutes a good approximation if collisional processes among particles either dominate competing photoprocesses or are in equilibrium with them at a common matter and radiation temperature. In all other situations they may fail badly, especially for the internal degrees of freedom. For rarefied matter in the spaces between stars and planets, non-LTE effects generally come into play, and we then need a full microscopic treatment of how atomic and molecular levels are populated and depopulated.

When LTE does apply, the emissivity j_ν has a thermal value given by Kirchhoff's law:

$$j_\nu = 4\pi \kappa_\nu^{\mathrm{abs}} B_\nu(T). \tag{3.13}$$

Specifying j_ν now boils down to specifying $T(\mathbf{x})$. [We assume that $\kappa_\nu(\mathbf{x})$ is known if $\rho(\mathbf{x})$ and $T(\mathbf{x})$ are given.] For situations in which radiation provides the dominant means for heating and cooling the gas, we cannot really specify $T(\mathbf{x})$ in advance, but must find it as part of the overall problem.

If we substitute equation (3.13) into equation (3.11), we require that the distribution of matter temperature $T(\mathbf{x})$ satisfies the integral constraint of *radiative equilibrium* (first formulated for stellar photospheres by Karl Schwarzschild in 1906):

$$4\pi \int_0^\infty \kappa_\nu^{\mathrm{abs}} B_\nu(T)\, d\nu = c \int_0^\infty \kappa_\nu^{\mathrm{abs}} E_\nu\, d\nu. \tag{3.14}$$

The left-hand side is some function of T (and possibly $\mathbf{x}$ through $\kappa_\nu^{\mathrm{abs}}$); the right, some function of $\mathbf{x}$. Thus the two sides fix $T(\mathbf{x})$ if $E_\nu(\mathbf{x})$ is known, i.e., if we have the solution for I_ν. [Given E_ν, we can find the value for

T at each $\mathbf{x}$ from equation (3.14), for example, by Newton's method for extracting roots.]

But we don't have the solution (3.8) until we specify $T(\mathbf{x})$; thus, even without the coupling due to (coherent) scattering, equation (3.8) often amounts to a complicated nonlinear integral equation for I_ν. Exceptions occur if agents (e.g., cosmic rays in the interstellar medium) other than photons maintain the matter temperature T. Also, even if radiative processes dominate the heating and cooling, we might attack equation (3.8) as an integral equation for most of the photons, i.e., for the continuum, and then use the resulting self-consistent temperature distribution and continuum radiation field as a background for computing radiative transfer in the lines.

In this chapter, we will concern ourselves primarily with a broad overview. Various methods exist to solve the associated integral equation; the most direct involves an iterative technique, as follows. Make an initial guess for $J_\nu(\mathbf{x}) = cE_\nu/4\pi$; approximate Φ_ν by J_ν (okay, if the scattering is nearly isotropic); and solve equation (3.14) to obtain $T(\mathbf{x})$. With equation (3.13) now giving j_ν, integrate (3.8) to obtain $I_\nu(\hat{\mathbf{k}}, \mathbf{x})$. Substitute this value into the first row of equation (2.1) to compute a new E_ν and into equation (3.2) to obtain a new Φ_ν. Continue to iterate in this manner until satisfactory convergence is achieved.

In practice, straight iterative schemes have, at best, a 50–50 chance of converging (see Problem Set 1), and other techniques (e.g., "complete linearization"; see Mihalas 1978) are better for extracting precise numerical solutions, at least for problems involving a restricted number of spatial dimensions and angular propagation directions (e.g., one each). We leave further discussion of such methods to texts on stellar atmospheres theory, and focus here on intuitive discussions of the physical nature of the solution.

EMISSION AND ABSORPTION LINES, LIMB DARKENING

Suppose T and Φ_ν are given to us somehow (consistent with radiative equilibrium for a stellar atmosphere, more arbitrary for the interstellar medium or some other environment). What does the resulting radiation field look like in the continuum and in lines? Under LTE conditions, the source function (3.3) becomes

$$S_\nu = (1 - A_\nu)B_\nu(T) + A_\nu\Phi_\nu, \tag{3.15}$$

where

$$A_\nu \equiv \kappa_\nu^{\text{sca}}/\kappa_\nu \tag{3.16}$$

is the scattering albedo. Consider now the problem of the formation of LTE spectral lines under optically thick and thin conditions. For simplicity, let us ignore the effects of scattering at first, so that $S_\nu = B_\nu(T)$, a smooth function of ν.

When the medium is optically thin at all relevant frequencies (e.g., the upper chromosphere and corona of the Sun; emission nebulae in our own Galaxy and in external galaxies) and there is no background source $I_\nu(\hat{\mathbf{k}}, \mathbf{x}, \mathbf{x}_0) = 0$, we may approximate the exponential $\exp[-\tau_\nu(\hat{\mathbf{k}}, \mathbf{x}, \mathbf{x}')]$ in equation (3.8) by unity. Moreover, we expect κ_ν^{abs} to be larger at the frequency of a resonance line than at neighboring spectral intervals. The indicated integration in equation (3.8) therefore gives a larger I_ν at the line frequency than at neighboring frequencies for every line of sight, *independent of the details of the density and temperature distributions ρ and T*. In other words, both angularly resolved and angularly unresolved observations would yield emission brighter in lines than in the continuum. An *emission-line spectrum* appears in this situation, because gas has more emissive power where it has more absorptive power, and if the medium is optically thin, every emitted photon flying in the proper direction will reach the observer.

An upper limit exists, however, to how bright thermal emission lines can become with increasing optical depth. For simplicity, suppose the emitting medium has a uniform temperature T. If $\kappa_\nu^{\text{sca}} = 0$, the source function $S_\nu = B_\nu(T)$ is independent of position if T is. With no background source, equation (3.7) now has the integral,

$$I_\nu = B_\nu(T)(1 - e^{-\tau_\nu}), \qquad (3.17)$$

where τ_ν is the total optical depth at frequency ν across the source along our particular line of sight. Since $(1 - e^{-\tau_\nu})$ cannot exceed 1, equation (3.17) states that a thermal body at a uniform temperature T cannot yield a specific intensity, in lines or a continuum, that is higher than a blackbody of the same temperature. (Non-LTE effects can provide counterexamples, the most celebrated being emission from masers.) For small optical depths, $(1 - e^{-\tau_\nu}) \approx \tau_\nu$, and the emission, normalized relative to a blackbody, is larger at frequencies where the optical depth is larger, i.e., in lines (see Figure 3.2).

The advantage that lines have over the continuum saturates when τ_ν begins to approach and exceed unity in both. Moreover, if the medium is (at least partially) optically thick in the continuum, we expect the temperature T to decrease outward from the source toward the observer. Recall that the discussion in Chapter 2 of radiation conduction shows the energy flux to be directed down a temperature gradient; conversely, a matter-temperature gradient will develop whenever the radiation flux in optically thick regions is nonzero. When we observe at the frequency of a line, the higher optical depth causes us to look less deeply into the object than at neighboring frequencies. Thus, at a resonant frequency, we see to less-hot layers than in the continuum, and equation (3.8) would then yield dark spectral lines

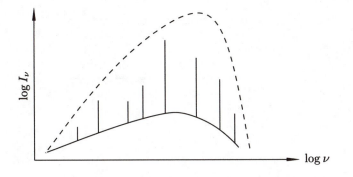

FIGURE 3.2
Limitation of emission lines and continuum to blackbody value in region of
uniform temperature T and where LTE applies.

superimposed on a brighter continuum (Figure 3.3). In other words, this
situation—e.g., the photosphere of the Sun and other stars—tends to produce an *absorption-line spectrum*.

   The effect described in the preceding has an extreme example where
there is a bright background source, of specific intensity $I_\nu(0)$, in addition
to an emitting and absorbing medium in LTE at a uniform temperature T.
If $\kappa_\nu^{\mathrm{sca}} = 0$, equation (3.7) implies for this case

$$I_\nu = I_\nu(0)e^{-\tau_\nu} + B_\nu(T)(1 - e^{-\tau_\nu}), \tag{3.18}$$

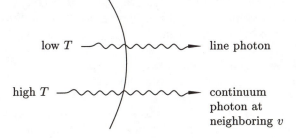

FIGURE 3.3
Radiation emergent from an atmosphere where the temperature increases
inward tends to produce an absorption-line spectrum.

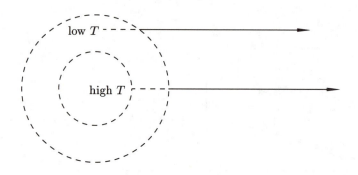

FIGURE 3.4
The physical situation behind limb darkening.

an equation much used in radio studies of the interstellar medium. If the background is brighter than the emitting and absorbing foreground, the positive combination $[I_\nu(0) - B_\nu(T)]e^{-\tau_\nu}$ will decrease with increasing τ_ν; i.e., lines will appear darker than the surrounding continuum. Historically, Arthur Schuster advanced in 1902–1905 the "reversing layer" type of analysis represented by equation (3.18) to explain the dark Fraunhofer lines of the solar spectrum. (He considered a foreground scattering layer rather than an absorbing one.) Modern stellar-atmosphere theory has generally abandoned the naive concept of two discrete layers at different temperatures in favor of more continuous variations, but models of comparable sophistication have yet to make their way into the literature of other fields (e.g., the interstellar medium, quasars, etc.).

Applied to the continuum, the same type of reasoning explains the phenomenon of *limb darkening* (see Figure 3.4). When we look at the limb (edge) of the Sun, we see—for a given optical depth—to less-hot layers than we do when we look toward the center. Thus photographs of the Sun in "white light" show its limb to be darker than its central face. In Chapter 4 we will give a quantitative estimate of the magnitude of this effect (first discussed by Karl Schwarzschild in 1906).

Since coherent scattering conserves the total number of photons at each frequency, we must reason more carefully when we come to "absorption" lines formed by the scattering of light by material distributed near a continuum source. Such scattering occurs more strongly in the line than in the neighboring continuum, so we naively expect the formation of a spectral feature. If, however, coherent scattering constitutes the only agent present, one observer would be able to see an "absorption" line only at the expense of other observers seeing an "emission" line. For a system in which the scattering medium is distributed spherically symmetrically with respect to the continuum source, all observers located on the celestial sphere are equivalent; so no net effect can occur! Nevertheless, spectral lines would remain absent only as long as the region remains unresolved. For a resolved source, e.g., the shell of a planetary nebula, the continuous background appears much brighter when an observer looks directly toward the central source than when he or she looks at the limb of nebula. Thus, as astronomers have recently demonstrated for actual observations of a planetary nebula, when scattering in a spectral line takes place, a line of sight toward the central star produces an absorption feature; a line of sight toward the limb of the nebula shows an emission feature. Averaged over the whole source, however, the spectral feature disappears!

How do dark lines form by scattering, then, in the atmosphere of an unresolved star? Line photons traveling out of the star in the direction of the observer can scatter out of the line of sight as well as scatter toward the deeper and denser layers of the atmosphere, where collisions among emitting and absorbing atoms can thermalize the radiation field (redistribute the frequencies into a more continuous spread). A compensating effect does not arise for inward-going photons scattering into our line of sight, since the presence of a boundary introduces a basic asymmetry to the problem (even if the atmosphere had an isothermal structure). Thus, in general, dark lines in stellar atmospheres will form by a combination of true absorption and true scattering.

4

Plane-Parallel Atmospheres

Reference: Kourganoff, *Basic Methods in Transfer Problems*, Chapters 1–4.

In this chapter, we consider approximate solutions for the problem of radiative transfer in the continuum, with a particular goal of extracting the implied *emergent spectral energy distribution* (distribution with ν) and *law of limb darkening* (variation with angle). We restrict ourselves to an idealized plane-parallel geometry and to the case where the assumptions of LTE and radiative equilibrium both hold. We shall not discuss methods for the exact solution of the problems so posed, but choose to emphasize approximate ones that lay bare the relevant physics and obtain 90 percent of the correct answer at 10 percent of the calculational effort. For more accurate (but also computationally more intensive) methods, consult Kourganoff (1963) or Chandrasekhar (1960).

PLANE-PARALLEL LTE ATMOSPHERE

If the transition between optically thick and optically thin regions occurs in a very narrow layer, calculations are simpler. For many stellar atmospheres, this transition layer (the photosphere) spans typically only $10^{-3}R_*$, where the stellar radius R_* is defined by the location where $F_{\rm rad} = \sigma T^4$ holds in an integration that proceeds outward from the deep interior. The resulting temperature, at which the star radiates *as if it were a blackbody* (in total flux), is called the *effective temperature* $T_{\rm eff}$. When variations of r can be ignored, radiative equilibrium [see equation (3.12)] requires $F_{\rm rad} = \text{constant} \equiv \sigma T_{\rm eff}^4$. Furthermore, for a slab geometry,

$$\hat{\mathbf{k}} \cdot \boldsymbol{\nabla} = -\mu \frac{d}{dz}, \tag{4.1}$$

where μ is the cosine of the angle ϑ of a ray path with respect to the vertical, and z is a depth variable (increasing *inward*) equal to zero at the

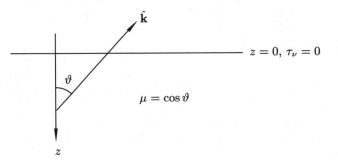

FIGURE 4.1
Geometry for plane-parallel atmosphere.

"top" of the atmosphere (where optical depth zero is reached). Figure 4.1
illustrates the geometry.

Such a configuration allows us to separate the angle dependence and
the position dependence of the path interval ds in equation (3.1),

$$ds = -dz/\mu, \qquad (4.2)$$

so that the angle variable μ enters effectively as a parameter only in the
equation of transfer, rather than as something whose variations we need to
consider when taking derivatives or performing integrals along ray paths.
(If the radius of curvature r were important, a dispacement ds along a
nonradial ray path would change the angle ϑ between $\hat{\mathbf{k}}$ and $\hat{\mathbf{r}}$, with the
complications that we will discuss in a later chapter.)

We assume that the atmosphere can be taken to possess infinite total
optical depth. Recall that e^{-100} is already effectively zero, not to mention
$\exp(-10^{11})$. If we further assume isotropic scattering and LTE (a reasonable approximation for transitions involving the continuum), the source
function (3.15) becomes

$$S_\nu = (1 - A_\nu)B_\nu(T) + A_\nu J_\nu, \qquad (4.3)$$

where A_ν is the scattering albedo $\kappa_\nu^{\mathrm{sca}}/\kappa_\nu$ and J_ν is the mean specific
intensity. The specific intensity emergent from the top of the atmosphere
has the expression [see equation (3.7) with upper and lower limits of the
integral reversed]

$$I_\nu(\mu, \tau_\nu = 0) = \int_0^\infty S_\nu e^{-\tau_\nu/\mu} \frac{d\tau_\nu}{\mu}, \qquad (4.4)$$

where $\mu > 0$ for outgoing rays and τ_ν is now the optical depth measured in the *vertical* direction (rather than along a ray path),

$$\tau_\nu \equiv \int_0^z \rho\kappa_\nu \, dz. \tag{4.5}$$

Previously, e.g., in equations (3.8) and (3.7), we let τ_ν denote what we now call the *slant optical depth* τ_ν/μ. If we take care to distinguish between outgoing rays ($\mu > 0$) and ingoing rays ($\mu < 0$), we can find expressions similar to equation (4.4) for the specific intensity at interior points.

GREY OPACITIES

A particularly clean problem arises for the idealized case when the opacities are grey, i.e., when κ_ν^{abs} and κ_ν^{sca} are independent of ν (roughly true in the solar atmosphere). If the scattering albedo $A_\nu \equiv A$ does not depend on ν, the integration of the source function (4.3) over all frequencies yields the identification:

$$S = (1 - A)B(T) + AJ, \tag{4.6}$$

where we have denoted the integral of $B_\nu(T)$ over all ν as $B(T) = caT^4/4\pi$. On the other hand, with $\kappa_\nu \equiv \kappa$ independent of ν, radiative equilibrium, equation (3.14) with $E = 4\pi J/c$, requires $J = B$ for $A \neq 1$. The result that the frequency-integrated mean specific intensity equals the local blackbody value (for total emission to balance total absorption when the opacity is grey) does not imply, as we shall see later, that the *monochromatic* specific intensity I_ν has either the spectral composition or the angular distribution of a true Planckian $B_\nu(T)$. Moreover, for $A = 1$ (conservative isotropic scattering), radiative energy conservation enforces only $F_{rad} =$ constant, without requiring $S = J$ to also equal B. For any other value of A, equation (4.6) does require

$$S = J = B. \tag{4.7}$$

Except for the issue whether the mean intensity J is related to the local matter temperature [as expressed through $B(T)$], we see from the preceding discussion that many of the distinctions between absorption and scattering vanish when the opacities are grey and the scattering is isotropic. Indeed, Chandrasekhar (1960, p. 293) explicitly shows that the grey problems of pure absorption ($A = 0$) and pure isotropic scattering ($A = 1$) have identical angular distributions and optical depth variations for the integrated light $I(\mu, \tau)$; consequently, we may take any linear combination of the two ($0 < A < 1$) and come to the same conclusion.

In any case, with $\tau_\nu \equiv \tau$ independent of ν, equation (4.4) may now be integrated over all frequencies to yield

$$I(\mu, 0) = \int_0^\infty S(\tau) e^{-\tau/\mu} \frac{d\tau}{\mu}. \tag{4.8}$$

With the same assumptions, the frequency-integrated counterpart of equation (2.7) becomes

$$c \frac{d}{d\tau}(P_{\text{rad}}) = F_{\text{rad}}, \tag{4.9}$$

where $d\tau \equiv \rho\kappa dz$ increases inward toward the center of the star. Equation (4.9) has the solution

$$cP_{\text{rad}} = F_{\text{rad}}(\tau + \tau_0), \tag{4.10}$$

where τ_0 is an integration constant.

EDDINGTON APPROXIMATION

So far our equations are exact, to the extent that the physical assumptions (slab geometry, isotropic scattering, grey opacities) are valid. To make further progress, we adopt Eddington's closure approximation:

$$P_{\text{rad}} \approx \frac{1}{3} E_{\text{rad}} = \frac{4\pi J}{3c}. \tag{4.11}$$

This relationship holds under more general conditions than completely isotropic radiation; for example, it remains valid if the radiation field is separately isotropic for the outward and inward directions. Combined with equation (4.10) and $S = J$, equation (4.11) leads to the identification that the source function is a linear function of optical depth,

$$S(\tau) = \frac{3}{4\pi} F_{\text{rad}}(\tau + \tau_0), \tag{4.12}$$

for any value of A.

Substitution of equation (4.12) into equation (4.8) now yields (with the substitution of factorials for complete Γ functions):

$$I(\mu, 0) = \frac{3}{4\pi} F_{\text{rad}}(\mu + \tau_0). \tag{4.13}$$

To determine the integration constant τ_0, we require that the outwardly emergent intensity carry the flux F_{rad}, i.e.,

$$F_{\text{rad}} = 2\pi \int_0^1 \mu I(\mu, 0)\, d\mu. \tag{4.14}$$

If we substitute equation (4.13) into the right-hand side of equation (4.14), a little manipulation yields the identification

$$\tau_0 = 2/3. \tag{4.15}$$

Equation (4.13) now gives the well-known limb-darkening result (for any combination of thermal emission or isotropic scattering) that the emergent specific intensity,

$$I(\mu, 0) = I(1, 0)\frac{3}{5}\left(\mu + \frac{2}{3}\right), \tag{4.16}$$

has $2/5$ the value at the limb ($\mu = 0$) that it does at the center ($\mu = 1$). (Figure out for yourself the equivalence between rays traveling from different points in a thin spherical shell toward a given observer and rays traveling from the same point in a plane-parallel atmosphere toward different observers.) The exact solution by Chandrasekhar and by Hopf of this problem yields $I(0,0)/I(1,0) = 0.3439$, not bad agreement with the result obtained by adopting Eddington's approximation, when we consider the anisotropy that this ratio implies for the radiation field at the boundary of the atmosphere.

When $A \neq 1$ (so that some true absorption and emission occur to allow energy interchange between matter and radiation), equations (4.7), (4.10), and (4.11), with $F_{\rm rad} = \sigma T_{\rm eff}^4$ (and $\sigma = ca/4$), imply a temperature variation with depth given by the famous law

$$T^4 = \frac{3}{4}T_{\rm eff}^4\left(\tau + \frac{2}{3}\right). \tag{4.17}$$

Equation (4.17) represents a special property of the present problem that fails to be valid more generally, namely, the existence of a T-τ relationship that holds for the problem of radiative transfer *independently of the mechanical structure of the atmosphere* (the distribution of ρ and T as functions of z). In any case, equation (4.17) implies that the gas temperature equals the effective temperature at a continuum optical depth of $2/3$; whereas the boundary temperature T_0 at $\tau = 0$ equals $2^{-1/4}T_{\rm eff} = 0.8409T_{\rm eff}$. The exact result is $T_0 = 0.8112T_{\rm eff}$.

EMERGENT SPECTRAL ENERGY DISTRIBUTION

For an LTE grey stellar atmosphere in which we neglect the effects of scattering ($A = 0$), the emergent monochromatic intensity is given by equation (4.4) when τ_ν is set equal to τ and S_ν is set equal to $B_\nu(T)$:

$$I_\nu(\mu, 0) = \int_0^\infty B_\nu[T(\tau)]e^{-\tau/\mu}\frac{d\tau}{\mu}, \tag{4.18}$$

where $T(\tau)$ is given in the Eddington approximation by equation (4.17). For no star other than the Sun do we have access to (angularly resolved) measurements of $I_\nu(\mu, 0)$. For other stars, we must be content with the *emergent spectral energy distribution*:

$$F_\nu(0) \equiv 2\pi \int_0^1 \mu I_\nu(\mu, 0) \, d\mu. \tag{4.19}$$

Substituting equation (4.18) into equation (4.19), we reverse the order of integration and obtain

$$F_\nu(0) = 2\pi \int_0^\infty E_2(\tau) B_\nu[T(\tau)] \, d\tau, \tag{4.20}$$

where $E_n(\tau)$ is the n-th exponential integral,

$$E_n(\tau) \equiv \int_1^\infty x^{-n} e^{-\tau x} \, dx, \tag{4.21}$$

with convenient numerical approximations given in Abramowitz and Stegun 1965.

As an empirical fact (see Problem Set 1), the observed energy distributions for actual stars (e.g., the Sun) with a given T_{eff} agree qualitatively but not quantitatively with the theoretical calculations for an LTE grey atmosphere. Starting around the 1920s and continuing into the 1940s, astronomers (principally Milne, Stromgren, and Munch) began to use this discrepancy to try to infer the frequency variation of the unknown source for the continuum opacity κ_ν in the Sun (and other late-type stars). The theoretical tool they used assumed that if the model grey opacity κ were chosen equal to the Rosseland mean of the true opacity (assumed to be purely absorptive), then the T-τ relation given by equation (4.17) would continue to be a good approximation. Recall that in the underlying optically thick regions, the radiative flux can be expressed as

$$F_{\text{rad}} = \frac{c}{3\rho\kappa_{\text{R}}} \frac{d}{dz}(aT^4). \tag{4.22}$$

If we define $d\tau \equiv \rho\kappa_{\text{R}} \, dz$ and set $F_{\text{rad}} = caT_{\text{eff}}^4/4$, we may integrate equation (4.22) to obtain

$$T^4 = \frac{3}{4}T_{\text{eff}}^4(\tau + \tau_0), \tag{4.23}$$

which agrees with equation (4.17) if we identify the integration constant τ_0 as $2/3$.

The astronomers now assumed that the unknown κ_ν could be expressed as a function $\alpha(\nu)$ of ν alone (with no dependence on z) times κ_{R}, i.e., that $\tau_\nu = \alpha(\nu)\tau$ in equation (4.4). By varying $\alpha(\nu)$, they could try to match the observed energy distributions of the actual stars. They discovered that the frequency dependence $\alpha(\nu)$ of the unknown continuum opacity in the Sun and other late-type stars had to increase from a minimum at about $16,000\,\text{Å}$, reach a maximum around $8,000\,\text{Å}$, and decline again toward shorter wavelengths. No then-known terrestrial material had these properties. There arose, then, a serious "opacity puzzle" in that, as late as 1940, no one knew what source of radiation determines the color of the Sun, and ultimately acts as the source of all hues on Earth!

Rupert Wildt made the suggestion that solved the puzzle. He proposed that the negative ion H^- (atomic hydrogen with a second electron loosely attached to it, with a binding energy of $0.754\,\text{eV}$) provides the continuum opacity of the Sun and most other late-type stars. Both *bound-free* transitions (with $h\nu > 0.754\,\text{eV}$, i.e., $\lambda < 16,500\,\text{Å}$) and *free-free* transitions (with any $h\nu$) contribute to the process. The free-free process is somewhat unusual in comparison with the normal example that we will study in Chapter 15 in that the free (ionized) state of the H^- ion is a neutral H atom plus an electron. We would not expect the interaction of a neutral atom (even one with an asymmetric distribution of electronic charge) and an electron to be able to emit (or absorb) much radiation, but the large abundance of neutral H atoms in the solar atmosphere makes up for the deficiency. The free electrons, as well as those that attach themselves to the H^-, come mainly from easily ionized metallic species in the solar atmosphere, e.g., sodium.

Nevertheless, Wildt's suggestion overtaxed the ability of quantum physicists to make quantitative calculations, because none of the standard perturbation techniques for multi-electron atoms (see Chapter 27) yields very good results for the H^- ion. In 1945 the greatest astrophysicist of our time, Chandrasekhar (and a year later with Breen), undertook a delicate quantum-mechanical calculation of this unusual species with a single bound state, and found that its absorption cross section for bound-free and free-free transitions, combined with the likely abundance of the H^- ion in the solar atmosphere (see Chapter 7), satisfies all of the observational requirements (see Figure 4.2.) It was a glorious moment for radiative-transfer theory.

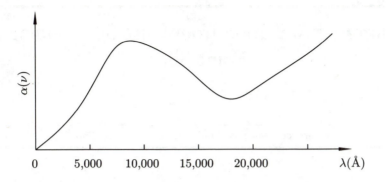

FIGURE 4.2
Frequency dependence of H^- bound-free and free-free opacity at conditions
typical of the solar atmosphere.

5

Infrared Radiation from Dust Surrounding a Point Source

Reference: Mihalas, *Stellar Atmospheres*, pp. 250–258.

In Chapter 4 we discussed how knowledge of the temperature distribution with mean optical depth allows us to infer the frequency dependence of the opacity from the emergent spectral energy distribution of an observed object. The reverse should also hold; if we know the frequency-dependent opacity, we ought to be able to extract information concerning the mechanical structure (e.g., density and temperature distribution) from the spectral energy distribution of an observed object. This program has achieved some success when applied to infrared radiation from dusty winds emanating from evolved stars (e.g., T.W. Jones and K.M. Merrill, 1976, *Ap. J.*, **209**, 509) and from infalling envelopes of gas and dust onto protostars (e.g., F.C. Adams and F.H. Shu, 1986, *Ap. J.*, **308**, 836). The nonrelativistic motion of the matter does not enter into the problem of transfer of radiation in the continuum.

A complication enters because of the presence of two sources of photons: (1) a small central source (we speak of a star, but the center could contain a disk as well, or the problem could be scaled up to a dusty, active galactic nucleus), (2) re-emission from dust grains in an extended envelope that absorb the light from the central source and reprocess it into the infrared. The complication proves to be minor, because in the infrared the wavelength of light (many microns) much exceeds the size of the grains (typically, 0.1 micron). Thus the grains behave like Rayleigh particles, and the scattering opacity $\kappa_\nu^{\mathrm{sca}} \propto \nu^4$ can be neglected in comparison with the absorption opacity $\kappa_\nu^{\mathrm{abs}} \propto \nu$ or ν^2 at small ν. In the approximation that dust grains are purely absorptive, we may unambiguously split off the directed radiation field from the central source, approximated as a point, from the diffuse radiation field arising from thermal emission by the dust.

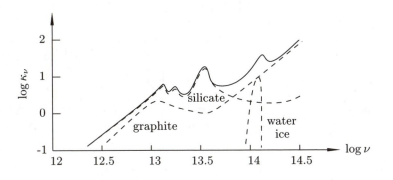

FIGURE 5.1
Opacity for mixture of silicate, graphite, and water-ice dust grains.

DUST OPACITY

Knowledge of interstellar dust opacities has improved substantially in recent years (see, e.g., B.T. Draine and H.M. Lee, 1984, *Ap. J.*, **285**, 89.). Figure 5.1 plots the dust cross section per unit mass, including the mass of the gas (which amounts to 99 percent of the total). Note that $\log \nu = 13.5$ corresponds to $\lambda = 10 \, \mu m$. A model which adopts a mixture of silicate and graphite grains, coated perhaps with a mantle of ice (mostly water), suffices to explain the shape of the continuous opacity associated with interstellar dust, as well as the main absorption features at 3.1, 10, and 20 μm. (There is also much evidence that the interstellar medium contains both very small dust grains and large organic molecules, i.e., polycyclic aromatic hydrocarbons, whose behaviors—particularly in the near-infrared and ultraviolet—are not described well along the classical lines assumed here; but see the astrophysical literature for these complications.)

The different constituents—water, silicate minerals, and graphite particles—evaporate (under protostellar conditions) at temperatures of $T_{ice} \approx$ 150 K, $T_{si} \approx 1,500$ K, and $T_{gr} \approx 2,000$ K. Thus we can take κ_ν to be a sum of Heaviside step functions,

$$\kappa_\nu = \kappa_\nu^{gr}\Theta(T_{gr} - T) + \kappa_\nu^{si}\Theta(T_{si} - T) + \kappa_\nu^{ice}\Theta(T_{ice} - T), \qquad (5.1)$$

where $\Theta(x) = 1$ for $x > 0$ and $\Theta(x) = 0$ for $x < 0$, and where the spatial variation of the dust temperature T remains to be determined.

DECOMPOSITION OF RADIATION FIELD INTO DIRECTED AND DIFFUSE PARTS

To begin, we divide the total specific intensity into two parts,

$$I_\nu = I_\nu^* + \mathcal{I}_\nu. \tag{5.2}$$

If the size of the central source is small, we may approximate its radiation field as having an angular dependence of a delta function peaked in the radial direction $\hat{\mathbf{r}}$:

$$I_\nu^*(\hat{\mathbf{k}}, \mathbf{r}) = F_\nu^*(\mathbf{r})e^{-\tau_\nu^*(\mathbf{r})}\delta(\hat{\mathbf{k}} - \hat{\mathbf{r}}), \tag{5.3}$$

where $\hat{\mathbf{r}}F_\nu^*(\mathbf{r})$ is the unattenuated monochromatic energy flux $\propto 1/r^2$ from the central source. If the central source were a spherically symmetric star (not necessary for this part of our problem), then $F_\nu^* = L_\nu^*/4\pi r^2$, with L_ν^* being the emergent spectral luminosity distribution of the star. In equation (5.3), τ_ν^* is the optical depth at point $\mathbf{r}$ in the dust shell to the central source,

$$\tau_\nu^*(\mathbf{r}) \equiv \int_0^r \rho(\hat{\mathbf{r}}r')\kappa_\nu(\hat{\mathbf{r}}r')\,dr', \tag{5.4}$$

where we make no distinction here between κ_ν and κ_ν^{abs}.

If we ignore scattered stellar light (which contains, in any case, little of the total energy budget for deeply embedded sources), the diffuse radiation field $\mathcal{I}_\nu$ arises from thermal emission from the dust and satisfies the equation of transfer:

$$\hat{\mathbf{k}} \cdot \boldsymbol{\nabla}\mathcal{I}_\nu = \rho\kappa_\nu[B_\nu(T) - \mathcal{I}_\nu]. \tag{5.5}$$

Equation (5.5) has the usual formal solution, equation (3.8), with S_ν given by $B_\nu(T)$. This formal solution has no utility until we can specify the temperature distribution $T(\mathbf{r})$. In what follows, we assume the density distribution $\rho(\mathbf{r})$ either to be known from a separate dynamical calculation, or to be derived by empirical fitting of the observed data points.

Radiative balance of emitted and absorbed radiation requires

$$4\pi \int_0^\infty d\nu \kappa_\nu B_\nu(T) = \int_0^\infty d\nu \kappa_\nu \oint I_\nu \,d\Omega. \tag{5.6}$$

If we substitute equation (5.2) into equation (5.6), we obtain the desired equation for T:

$$\kappa_P a T^4 = \kappa_E \mathcal{E} + \frac{1}{c}\int_0^\infty \kappa_\nu F_\nu^*(\mathbf{r}')e^{-\tau_\nu^*(\mathbf{r})}\,d\nu. \tag{5.7}$$

In equation (5.7), $\mathcal{E}$ is the energy density of the diffuse radiation field, and $\kappa_{\mathrm{P}}^{\mathrm{abs}}$ and $\kappa_{E}^{\mathrm{abs}}$ are the Planck and energy-weighted mean opacities defined by

$$\kappa_{\mathrm{P}}^{\mathrm{abs}} \equiv \frac{4\pi}{caT^4} \int_0^\infty \kappa_\nu^{\mathrm{abs}} B_\nu(T)\, d\nu, \tag{5.8}$$

$$\kappa_{E}^{\mathrm{abs}} \equiv \frac{1}{\mathcal{E}} \int_0^\infty \kappa_\nu^{\mathrm{abs}} \mathcal{E}_\nu\, d\nu. \tag{5.9}$$

Given the frequency dependence of the dust-absorption opacity $\kappa_\nu^{\mathrm{abs}}$, we can perform the integration indicated in equation (5.8) to obtain $\kappa_{\mathrm{P}}^{\mathrm{abs}}(T)$ as a tabulated function of T. Equation (5.7) then becomes an algebraic equation that determines $T(\mathbf{r})$ once we know $\mathcal{E}_\nu(\mathbf{r})$ and the input-source spectrum F_ν^*.

Notice that the zeroth-moment equation implies that the condition of radiative balance for the dust grains, equation (5.7), is exactly equivalent to the condition of radiative equilibrium, which requires that the divergence of the total radiative flux, stellar plus diffuse, equal zero; i.e.,

$$\nabla \cdot \mathcal{F}_{\mathrm{rad}} = -\nabla \cdot \left[\int_0^\infty \hat{\mathbf{r}} F_\nu^* e^{-\tau_\nu(\mathbf{r})}\, d\nu \right] = \rho \int_0^\infty \kappa_\nu F_\nu^* e^{-\tau_\nu^*}\, d\nu. \tag{5.10}$$

To obtain the last step, we have used the fact that the unattenuated light from the central source satisfies $\nabla \cdot [\hat{\mathbf{r}} F_\nu^*(\mathbf{r})] = 0$, whether or not the emission occurs spherically symmetrically. Thus the last part of equation (5.10) has a simple interpretation: the divergence of the diffuse radiation field $\nabla \cdot \mathcal{F}_{\mathrm{rad}}$ has a nonzero value only to the extent that residual rays from the central source remain for the dust to absorb. Once all the original directed photons from the source have been completely reprocessed (into the infrared) by the dust [effectively within a few (optical) mean-free paths relative to where the dust first appears at an inner radius $\mathbf{r}_d$ (see below)], the diffuse radiation field carries out a conserved total quantity of radiant energy, i.e., has zero local divergence. The net balance of absorption and emission does not imply that the diffuse radiation field remains unchanged thereafter. As the photons work their way out past the envelope of dust grains, the spectral distribution continues to be degraded to ever-longer wavelengths, until most of the luminosity ends up in the far infrared, where marginally optically thin conditions prevail.

SPHERICAL SYMMETRY

To proceed further, we adopt spherical symmetry. Let ϑ be the angle formed between the unit vectors $\hat{\mathbf{k}}$ and $\hat{\mathbf{r}}$ in the photon propagation and radial

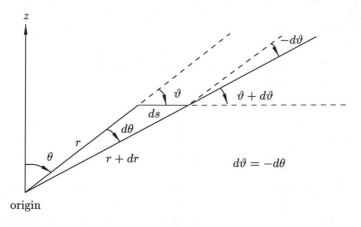

FIGURE 5.2
Geometry for radiative transfer with spherical symmetry.

directions. With the aid of Figure 5.2, we note that when a displacement ds occurs along a nonradial ray path, the angle ϑ also changes.

The differentials dr and $d\vartheta$ therefore have the following relationships to ds:

$$dr = (\cos\vartheta)ds, \qquad rd\vartheta = -(\sin\vartheta)ds. \tag{5.11}$$

The rate of change along a ray path can now be calculated as

$$\hat{\mathbf{k}} \cdot \boldsymbol{\nabla} \equiv \frac{d}{ds} = \frac{\partial r}{\partial s}\frac{\partial}{\partial r} + \frac{\partial\vartheta}{\partial s}\frac{\partial}{\partial\vartheta} = \cos\vartheta\frac{\partial}{\partial r} - \frac{1}{r}\sin\vartheta\frac{\partial}{\partial\vartheta}. \tag{5.12}$$

If we now introduce the variable

$$\mu \equiv \cos\vartheta = \hat{\mathbf{k}} \cdot \hat{\mathbf{r}}, \tag{5.13}$$

we may write equation (5.5) as

$$\mu\frac{\partial\mathcal{I}_\nu}{\partial r} + \frac{1}{r}(1-\mu^2)\frac{\partial\mathcal{I}_\nu}{\partial\mu} = \rho\kappa_\nu[B_\nu(T) - \mathcal{I}_\nu]. \tag{5.14}$$

With $d\Omega = 2\pi\sin\vartheta\,d\vartheta$, define the (scalar) angular moments of the diffuse radiation field:

$$\begin{pmatrix} c\mathcal{E}_\nu \\ \mathcal{F}_\nu \\ c\mathcal{P}_\nu \end{pmatrix} = 2\pi\int_{-1}^{+1}\begin{pmatrix} 1 \\ \mu \\ \mu^2 \end{pmatrix}\mathcal{I}_\nu\,d\mu. \tag{5.15}$$

If we multiply equation (5.14) by 1 and integrate over μ from -1 to $+1$ (ϑ from 0 to π), we obtain

$$\frac{\partial \mathcal{F}_\nu}{\partial r} + \frac{2}{r}\mathcal{F}_\nu = \rho\kappa_\nu[4\pi B_\nu(T) - c\mathcal{E}_\nu], \tag{5.16}$$

where we have integrated by parts to obtain the second term on the left-hand side. Similarly, if we multiply equation (5.14) by μ and integrate over $d\mu$, we obtain

$$c\frac{\partial \mathcal{P}_\nu}{\partial r} + \frac{c}{r}(3\mathcal{P}_\nu - \mathcal{E}_\nu) = -\rho\kappa_\nu\mathcal{F}_\nu. \tag{5.17}$$

Comparing equations (5.16) and (5.17) with (2.5) and (2.7), we can make the correspondence that in spherical symmetry

$$\nabla \cdot \mathbf{F}_\nu = \frac{1}{r^2}\frac{\partial}{\partial r}(r^2\mathcal{F}_\nu), \tag{5.18}$$

$$\nabla \cdot \mathbf{P}_\nu = \hat{\mathbf{r}}\left[\frac{\partial \mathcal{P}_\nu}{\partial r} + \frac{1}{r}(3\mathcal{P}_\nu - \mathcal{E}_\nu)\right]. \tag{5.19}$$

Equation (5.18) represents a familiar formula from vector calculus; equation (5.19) can be established by tensor calculus—or more revealingly, as we have done above—and represents a less-familiar formula for most people. In any case, equations (5.16) and (5.17) represent two ordinary differential equations (ODEs) that govern, frequency by frequency, the spatial dependences of $\mathcal{E}_\nu(r)$, $\mathcal{F}_\nu(r)$, and $\mathcal{P}_\nu(r)$. To obtain a solution of the set, we need a closure relation. One possible choice, whose form we motivate in the section following the next one, takes the expression

$$\mathcal{E}_\nu = 3\mathcal{P}_\nu - \frac{q_\nu}{c}\mathcal{F}_\nu, \tag{5.20}$$

with $q_\nu(r)$ being a (monochromatic) variable closure factor. If we wish to obtain the precise form of $q_\nu(r)$, we must proceed via iteration: guess $q_\nu(r)$; integrate the ODEs (5.16) and (5.17), with equation (5.20) providing closure; compute $T(r)$ from equation (5.7); plug into the formal solution for $\mathcal{I}_\nu$; perform angular integrations over $\mathcal{I}_\nu$ to obtain the angular moments $\mathcal{E}_\nu(r)$, $\mathcal{F}_\nu(r)$, $\mathcal{P}_\nu(r)$; solve equation (5.20) as an equation for a new $q_\nu(r)$; etc. As an approximate starting value, we could guess $q_\nu(r) = 2$ (see below).

FREQUENCY-INTEGRATED MOMENT EQUATIONS

Since we can meet the entire objective of the present radiative-transfer problem once we know $T(r)$, the iterative procedure outlined in the preceding paragraph contains an element of overkill. Surely the spatial dependence of the temperature $T(r)$ cannot depend sensitively on the details of

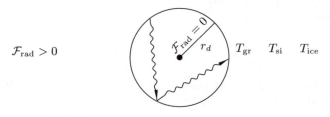

FIGURE 5.3
Configuration for envelope of gas and dust infalling onto a central protostar.

the frequency and angular dependences of the diffuse radiation field $\mathcal{I}_\nu$, since we ultimately integrate over these dependences. The value of an integral varies much less than the integrand itself as long as the integrand satisfies certain basic constraints—e.g., has fixed zeroth, first, and second moments. This philosophy suggests the following approximate procedure.

If we integrate over frequency ν, the first part of equation (5.10) yields the constancy of total luminosity:

$$4\pi r^2 \left[\mathcal{F}_{\mathrm{rad}}(r) + \int_0^\infty F_\nu^*(r) e^{-\tau_\nu(r)} \, d\nu \right] = L^*, \tag{5.21}$$

where L^* is the total luminosity to leave the central source,

$$L^* \equiv 4\pi r^2 \int_0^\infty F_\nu^*(r) \, d\nu = \int_0^\infty L_\nu^* \, d\nu = \text{constant}.$$

In equation (5.21), we identify the optical depth $\tau_\nu(r)$ to equal

$$\tau_\nu(r) = \int_{r_d}^r \rho(r') \kappa_\nu(r') \, dr', \tag{5.22}$$

if $r \geq r_d$ (the innermost radius to which graphite grains can survive), and $\tau_\nu(r) = 0$ otherwise ($r < r_d$). (Consult the schematic picture in Figure 5.3, based on the idea that the different components of the interstellar dust are successively destroyed as they flow inward toward the central source—say, a protostar.)

Notice that equation (5.21) requires the diffuse flux $\mathcal{F}_{\mathrm{rad}}$ to be zero inside the dust-free cavity ($r < r_d$). Naively, we might have thought that on the free (transparent) side of a hot wall (the dust destruction front for inflow—the dust formation front for outflow), a net flux of photons would occur away from the wall, leading to a $\mathcal{F}_{\mathrm{rad}} < 0$. This naive picture fails to account for the fact that the cavity is *spherical*; every photon leaving the wall will statistically be balanced by a photon coming in from the other side of the cavity. This elegant argument, first advanced by Milne in 1930 in a paper on planetary nebulae (shells of gas thrown off by a highly evolved star on the way to becoming a white dwarf), demonstrates that $\mathcal{F}_\nu$ and $\mathcal{F}_{\rm rad}$ are both identically zero everywhere inside the cavity (if the central source is approximated as being infinitesimally small). The integrated diffuse flux $\mathcal{F}_{\rm rad}$ rises outward from a zero value at $r = r_{\rm d}$ in exact pace with the exponential attenuation of the directed flux $F_{\rm rad}^*$, and so keeps the sum equal to $L^*/4\pi r^2$ (a generalization of the "$1/r^2$ law").

Similarly, if we integrate equation (5.17) over all frequencies, we obtain

$$\frac{\partial \mathcal{P}_{\rm rad}}{\partial r} + \frac{1}{r}(3\mathcal{P}_{\rm rad} - \mathcal{E}_{\rm rad}) = -\frac{\rho}{c}\kappa_F \mathcal{F}_{\rm rad}, \tag{5.23}$$

where κ_F is the flux-weighted mean opacity

$$\kappa_F \equiv \frac{1}{\mathcal{F}_{\rm rad}} \int_0^\infty \kappa_\nu \mathcal{F}_\nu \, d\nu. \tag{5.24}$$

Notice that if $\mathcal{F}_\nu$ were given by the radiation conduction approximation, equation (2.13), κ_F becomes the Rosseland mean, equation (2.15) [see equation (2.17)].

VARIABLE CLOSURE FACTORS

To solve equations (5.7) and (5.23) we need a closure relation, as well as expressions for κ_E and κ_F. For small r (but r still larger than $r_{\rm d}$), if the medium is optically thick, we may adopt Eddington's closure assumption:

$$\mathcal{P}_{\rm rad} = \mathcal{E}_{\rm rad}/3. \tag{5.25}$$

In such regions, we also anticipate

$$\kappa_E = \kappa_{\rm P}, \qquad \kappa_F = \kappa_{\rm R}, \tag{5.26}$$

where $\kappa_{\rm P}$ and $\kappa_{\rm R}$ are functions of T (and possibly also ρ) that can be discovered independently of a solution for $\mathcal{I}_\nu$ (or $\mathcal{E}_\nu$ and $\mathcal{F}_\nu$).

On the other hand, at large r, where the medium becomes optically thin, we expect photons to stream freely in the radial direction. When $\mathcal{I}_\nu$ becomes increasingly peaked like a delta function in the $\mu = 1$ direction, the three rows of equations (5.15) imply the monodirectional free-streaming approximation

$$\mathcal{P}_{\rm rad} = \mathcal{E}_{\rm rad} = \mathcal{F}_{\rm rad}/c. \tag{5.27}$$

Notice that when we use equation (5.27) in conjunction with the assumption $\rho\kappa_F r \ll 1$ (optically thin medium), equation (5.23) has the solution that $\mathcal{E}_{\mathrm{rad}}$ and $\mathcal{P}_{\mathrm{rad}}$, as well as $\mathcal{F}_{\mathrm{rad}}$, decline as r^{-2} when $r \to \infty$. The same conclusion holds in detail for the monochromatic quantities, so that all of the angular moments of $\mathcal{E}_\nu$, $\mathcal{F}_\nu$, $\mathcal{P}_\nu$, etc., have a common spectral content when $r \to \infty$.

An approximate closure relationship which spans both equations (5.25) and (5.27) is

$$\mathcal{E}_{\mathrm{rad}} = 3\mathcal{P}_{\mathrm{rad}} - \frac{2}{c}\mathcal{F}_{\mathrm{rad}}. \tag{5.28}$$

In optically thick regions, $\mathcal{F}_{\mathrm{rad}} \ll c\mathcal{E}_{\mathrm{rad}}$; so equation (5.28) tends to the Eddington approximation, equation (5.25). In optically thin regions, $\mathcal{F}_{\mathrm{rad}}$ approaches the free-streaming value $c\mathcal{E}_{\mathrm{rad}}$; so equation (5.28) tends to the monodirectional approximation, equation (5.27). Approximate formulae for κ_E and κ_F that could be used in conjunction with equation (5.28) might be

$$\kappa_E = \kappa_{\mathrm{P}}(T_{\mathrm{rad}}), \qquad \kappa_F = \kappa_{\mathrm{R}}(T_{\mathrm{rad}}), \tag{5.29}$$

where we characterize the spectral content of the ambient radiation field as a blackbody of color temperature,

$$T_{\mathrm{rad}} = \text{larger of } T \text{ and } T_{\mathrm{eff}}, \tag{5.30}$$

with T_{eff} being defined by the matter temperature $T(r)$ at a radius r which satisfies

$$\mathcal{F}_{\mathrm{rad}} = \sigma T^4. \tag{5.31}$$

With these approximations, we may integrate equation (5.23)—together with the condition of radiative equilibrium, equation (5.7)—subject to the outer boundary condition,

$$\mathcal{E}_{\mathrm{rad}} \to \mathcal{F}_{\mathrm{rad}}/c \qquad \text{as} \qquad r \to \infty. \tag{5.32}$$

With $T(r)$ obtained, iterative schemes can be invented to successively improve the solution. Such iterative schemes could be based, for example, on the introduction of a variable closure factor $q_F(r)$ defined so that [see equation (5.20)]

$$\mathcal{E}_{\mathrm{rad}} = 3\mathcal{P}_{\mathrm{rad}} - \frac{q_F}{c}\mathcal{F}_{\mathrm{rad}}. \tag{5.33}$$

In such a scheme, successive iterations between the exact moment equations [equations (5.16) and (5.17) in monochromatic form and equations (5.7) and (5.23) in integrated form] converge to a solution of arbitrarily high accuracy. An equivalent procedure (see Mihalas 1978) introduces a *variable Eddington factor* $f(r)$ defined as the ratio of $\mathcal{P}_{\mathrm{rad}}$ and $\mathcal{E}_{\mathrm{rad}}$. Figure 5.4 depicts a sample calculation for a model of a protostellar source. (See the listed references for further details.)

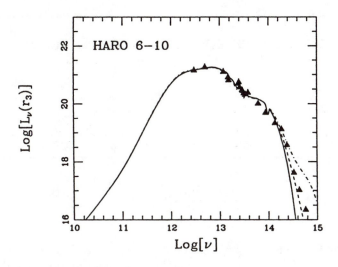

FIGURE 5.4
Emergent spectral energy distribution for a model of the protostellar source
Haro 6-10. The dashed curves indicate different amounts of correction for
scattering at near-infrared and optical frequencies. (Adapted from F.C. Adams
and F.H. Shu 1986, *Ap. J.*, **836**, 118.)

6

Thermodynamics and Statistical Mechanics

Reference: Huang, *Statistical Mechanics*, Chapters 7, 8, 9, 11.

Before we can appreciate any discussion of the rate equations that govern departures from LTE, we must first understand how matter and radiation behave under conditions of perfect thermodynamic equilibrium. The purpose of this chapter is to provide this basis from the point of view of thermodynamics and statistical mechanics.

REVIEW OF THERMODYNAMICS

The entropy-maximum principle, or, equivalently, the energy-minimum principle, underlies all of thermodynamics. The former constitutes the postulate that the entropy S of a closed system (fixed total energy E) reaches a maximum value when the system attains thermodynamic equilibrium.

Simple Example Isolated box of single-phase gas of volume V (see Figure 6.1). Partition it in your mind into two pieces: Maximize $S = S_1 + S_2$ by varying E_1 and V_1 subject to the constraints $E_1 + E_2 =$ constant and $V_1 + V_2 =$ constant. According to the first law of thermodynamics,

$$dE = TdS - PdV; \tag{6.1}$$

where P is the internal pressure. We have therefore,

$$dS = d(S_1 + S_2) = 0 = \frac{dE_1}{T_1} + \frac{P_1}{T_1}dV_1 + \frac{dE_2}{T_2} + \frac{P_2}{T_2}dV_2. \tag{6.2}$$

Since the constraints require $dE_2 = -dE_1$ and $dV_2 = -dV_1$, we obtain, upon collecting terms,

$$dS = \left(\frac{1}{T_1} - \frac{1}{T_2}\right)dE_1 + \left(\frac{P_1}{T_1} - \frac{P_2}{T_2}\right)dV_1 = 0. \tag{6.3}$$

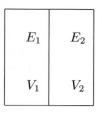

FIGURE 6.1
An isolated box with matter of uniform thermodynamic properties on each side
of an imaginary partition.

Because dE_1 and dV_1 can be varied independently, the system entropy is
maximized when $T_1 = T_2$ and $P_1 = P_2$, which are the conditions here for
thermal and mechanical equilibrium (thermodynamic equilibrium).

Conversely, if $T_1 \neq T_2$, then natural events must proceed so as to
increase S. With $T_1 > T_2$, for example, dS will be greater than zero if
$dE_1 < 0$; i.e., internal energy will flow from the hot region 1 to the cold
region 2. Similarly, if $T_1 = T_2$, but $P_1 > P_2$, the condition $dS > 0$ will lead
to $dV_1 > 0$; i.e., the high-pressure region 1 will expand into the low-pressure
region 2.

Clearly, we could have arrived at the same conclusions by considering
an (open) system constrained to have constant total entropy $S = S_1 + S_2$
and volume $V = V_1 + V_2$ by *minimizing* the energy $E = E_1 + E_2$. Why
energy minimum rather than maximum? Suppose E were not minimized
in thermodynamic equilibrium; then we could withdraw work, keeping S
constant, add back heat (integral of TdS) to get back original E, but now S
has increased. This implies that the original S was not maximized, yielding
a contradiction with the entropy-maximum principle unless the original E
had its minimum possible value. (Q.E.D.)

When we have the possibility of reactions (chemical or otherwise), we
generalize equation (6.1) of any species a to read

$$dE = TdS - PdV + \mu dN, \qquad (6.4)$$

where μ is the *chemical potential* per particle and N is the number of particles of species a in volume V. (We eventually choose V to be a small piece
of a large system.) Analysis of the simple example described previously now
leads to the expression

$$dS = \left(\frac{1}{T_1} - \frac{1}{T_2}\right) dE_1 + \left(\frac{P_1}{T_1} - \frac{P_2}{T_2}\right) dV_1 - \left(\frac{\mu_1}{T_1} - \frac{\mu_2}{T_2}\right) dN_1. \qquad (6.5)$$

If $T_1 = T_2$ and $P_1 = P_2$, but $\mu_1 > \mu_2$, the condition $dS > 0$ requires $dN_1 < 0$; i.e., particles will flow from a region of high chemical potential to one of low chemical potential. [This result explains why the sign convention in equation (6.4) is as written for the last term.]

Sometimes we will want to vary T rather than S, or P rather than V, or μ rather than N. These desires are accomplished by Legendre transformations:

enthalpy:
$$H = E + PV \quad \Rightarrow \quad dH = TdS + VdP + \mu dN, \qquad (6.6)$$

Helmholtz free energy:
$$F = E - TS \quad \Rightarrow \quad dF = -SdT - PdV + \mu dN, \qquad (6.7)$$

Gibbs free energy:
$$G = E - TS + PV \quad \Rightarrow \quad dG = -SdT + VdP + \mu dN, \quad (6.8)$$

grand potential:
$$\Omega = E - TS - \mu N \quad \Rightarrow \quad d\Omega = -SdT - PdV - Nd\mu. \quad (6.9)$$

For each of these "energies," there exists a corresponding energy-minimum principle that follows from the entropy-maximum principle.

Finally, for systems with only short-range forces, we have additive *extensive* properties and the Euler relation:

$$E = TS - PV + \mu N. \qquad (6.10)$$

We may regard the Euler relation (6.10) as resulting from the integration of equation (6.4) when we hold the intensive variables temperature T, pressure P, and chemical potential per particle μ constant. The setting of the constant of integration equal to zero (because we want the internal energy E to double if we double the entropy S, the volume V, and the number of particles N at fixed intensive properties) then establishes the physical content of the expression (6.10). The assumed linearity of extensive quantities holds only for systems with short-range interparticle forces; long-range forces (e.g., the self-gravitation of an astronomical system) usually introduce nonlinearities that prevent a scaled superposition of potential energies.

The existence of the general relationship (6.10) implies for equations (6.8) and (6.9)
$$G = \mu N \qquad \text{and} \qquad \Omega = -PV.$$

The last formulae represent *numerical* equalities only, and do not give the complete thermodynamic information that comes with knowing G and Ω as functions of their natural variables, $G(T, P, N)$ and $\Omega(T, V, \mu)$. However, the existence of a simple scaling law for thermodynamic systems (doubling the system doubles E, S, V, and N) implies that there are fewer independent variations among the *intensive* parameters T, P, and μ than we might have naively thought. In other words, when we held the set T, P, μ constant in the mental integration for E, we do not have arbitrary freedom to choose each of their separate values. For example, if we take the differential of the relation $G = \mu N$,

$$dG = \mu \, dN + N \, d\mu,$$

which is valid for a single-component system, and compare the result with equation (6.8), we find that μ cannot be varied independently of T and P, but must satisfy the Gibbs-Duhem relation:

$$N d\mu = -S dT + V dP. \tag{6.11}$$

Equation (6.11) represents a special example of the Gibbs phase rule that (apart from scaling) the number of thermodynamic degrees of freedom d of a c-component system containing p-phases is given by

$$d = c - p + 2. \tag{6.12}$$

A single-component system ($c = 1$) with a single phase ($p = 1$) has two degrees of freedom ($d = 2$) because (apart from scaling for N) we can vary, for example, P and T independently. If it has two phases (e.g., a liquid and a gaseous phase), $d = 1$, and the pressure P at equilibrium (e.g., the "vapor pressure") would be given if T is known. If it has three phases simultaneously in equilibrium (e.g., a solid, a liquid, and a gaseous phase), then there are no degrees of freedom left (the system can exist only at a unique "triple point" with specified P and T).

CLASSICAL STATISTICAL MECHANICS

Classical statistical mechanics, as developed by Gibbs to derive the properties of matter under conditions of thermodynamic equilibrium, considers a sequence of ensembles of systems ranging from:

> *Microcanonical ensemble:* composed of identical closed systems of N particles, whose evolution in $6N$-dimensional phase space (Γ space) satisfies Liouville's theorem in a fine-grained sense, but which is postulated to uniformly cover the energy hypersurface E = constant in a coarse-grained sense. The ergodic theorem hypothesizes that this will always occur asymptotically in time to any real ensemble of systems that contains "enough" interacting particles.

to:

> *Grand canonical ensemble:* composed of identical systems in contact with a much larger thermal and particle bath in which

number and energy exchange need not keep either N or E constant. Relaxing the constraints of the microcanonical ensemble (the most fundamental of all ensembles in that it is closest to ordinary mechanics) simplifies, as we shall see, many statistical-mechanical calculations.

DERIVATION OF PROPERTIES OF GRAND CANONICAL ENSEMBLE

Consider an ensemble of systems, each one composed of a small piece A and a much larger piece B, which together form a *closed* system of a large number of particles. Let D_{A+B} be the density of system points of the ensemble in Γ-space. Clearly, D_{A+B} is proportional to the probability of finding a system at a given location in Γ-space. If the ergodic hypothesis applies, D_{A+B} will (eventually) acquire a (coarse-grained) uniform (constant) value consistent with the constraints. In the meantime, according to Liouville's theorem, D_{A+B} depends only on the integrals of motion of the system. If subsystem $A \ll B$, and if the interaction between A and B is mild and short-range (leaving out the gravitational interaction as a proper topic for grand canonical ensembles), then we expect the law of independent probabilities to apply:

$$D_{A+B} = D_A D_B. \tag{6.13}$$

Figure 6.2 contains the geometric interpretation of what this means in terms of Γ-space volumes. (For the microcanonical ensemble, where $D =$ constant, the probability of finding a system in a given macrostate is simply proportional to the total Γ-space volume occupied by all microstates that correspond to the same macrostate.)

Taking logarithms of equation (6.13) yields

$$\ln D_{A+B} = \ln D_A + \ln D_B. \tag{6.14}$$

But since D_{A+B} approaches a constant in time, $\ln D_A$ and $\ln D_B$ must be additive combinations of the following quantities in the respective subsystems: number N, energy E, angular momentum $\mathbf{J}$, and linear momentum $\mathbf{P}$. In other words,

$$\ln D_A = \alpha_A + \beta(\mu N_A - E_A + \boldsymbol{\omega} \cdot \mathbf{J}_A + \mathbf{u} \cdot \mathbf{P}_A), \tag{6.15}$$

with an analogous expression for $\ln D_B$, where we require β, μ, $\boldsymbol{\omega}$, and $\mathbf{u}$ to be the *same* constants for both A and B (so that the sums will include the expressions $N_A + N_B$, $E_A + E_B$, $\mathbf{J}_A + \mathbf{J}_B$, and $\mathbf{P}_A + \mathbf{P}_B$ that are the conserved quantities of the total closed system). The properties of the microcanonical ensemble lead us to identify $\beta = 1/kT$. We shall shortly demonstrate that μ is the chemical potential.

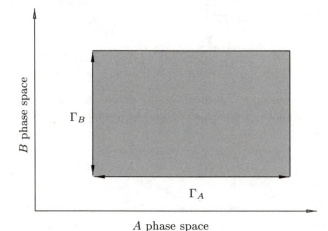

FIGURE 6.2
The nearly independent motions of two loosely coupled subsystems A and B
sweep out a hypervolume Γ_{A+B} in the combined Γ-space equal to the product
$\Gamma_A \Gamma_B$ of the individual Γ-spaces.

Equation (6.15) contains the standard result if we set $\boldsymbol{\omega} = 0$ and $\mathbf{u} = 0$.
(Bulk rotation and translation at a uniform angular rate $\boldsymbol{\omega}$ and uniform
velocity $\mathbf{u}$ can be transformed away without affecting any microscopic thermodynamic considerations. The energy E of the system is then the *internal
energy*.) The density of subsystems A in a grand canonical ensemble with
a given location in Γ_A-space is given by

$$D = \text{constant} \times \exp[(\mu\mathcal{N} - H_\mathcal{N})/kT], \qquad (6.16)$$

where we have suppressed the subscript notation A, and where $H_\mathcal{N}$ is the
Hamiltonian (energy) of the $\mathcal{N}$ particles in A. In equation (6.16), T and μ
refer to the reservoir B with which A is in porous thermal contact.

Suppose the small subsystem A to have a volume V (a small value),
and define the normalized probability density:

$$D_\mathcal{N} \equiv \frac{1}{Q} \exp[(\mu\mathcal{N} - H_\mathcal{N})/kT]. \qquad (6.17)$$

We calculate the normalization factor Q so that the total probability of all configurations equals unity:

$$1 = \sum_{\mathcal{N}=0}^{\infty} \int D_{\mathcal{N}} \, d\Gamma_{\mathcal{N}} = \frac{1}{Q} \sum_{\mathcal{N}=0}^{\infty} e^{\mu \mathcal{N}/kT} \int \exp[-H_{\mathcal{N}}(q,p)/kT] \, d^{3\mathcal{N}}q \, d^{3\mathcal{N}}p,$$
(6.18)

with q and p denoting a canonical set of particle coordinates and momenta. From equation (6.18), we wish to define the *grand canonical partition function*, as the expression:

$$Q(T,V,\mu) \equiv \sum_{\mathcal{N}=0}^{\infty} \int \exp\{[\mu \mathcal{N} - H_{\mathcal{N}}(q,p)]/kT\} \, d^{3\mathcal{N}}q \, d^{3\mathcal{N}}p, \qquad (6.19)$$

with the integral of q ranging over V. The integral over $6\mathcal{N}$-dimensional phase space, with $\mathcal{N}$ held fixed (but the energy $H_{\mathcal{N}}$ can vary), is called the *canonical partition function*.

To motivate the connection with thermodynamics, let us tentatively use information theory to identify the entropy S associated with the distribution $D_{\mathcal{N}}$:

$$S \equiv -k \sum_{\mathcal{N}=0}^{\infty} \int D_{\mathcal{N}} \ln D_{\mathcal{N}} d\Gamma_{\mathcal{N}}. \qquad (6.20)$$

To see that S corresponds to the *thermodynamic* entropy, note that the substitution of equation (6.17) into equation (6.20) gives

$$S = -k \sum_{\mathcal{N}=0}^{\infty} \int D_{\mathcal{N}} \left[\frac{1}{kT}(\mu \mathcal{N} - H_{\mathcal{N}}) - \ln Q \right] d\Gamma_{\mathcal{N}}, \qquad (6.21)$$

which we may write with an obvious notation as

$$S = -\frac{\mu}{T}\langle \mathcal{N} \rangle + \frac{1}{T}\langle H_{\mathcal{N}} \rangle + k \ln Q,$$

where the (grand canonical) ensemble average of any quantity X equals

$$\langle X \rangle \equiv \sum_{\mathcal{N}=0}^{\infty} \int X D_{\mathcal{N}} \, d\Gamma_{\mathcal{N}}. \qquad (6.22)$$

We identify the expectation values $\langle \mathcal{N} \rangle$ and $\langle H_{\mathcal{N}} \rangle$ with the thermodynamic variables N and E; thus equation (6.21) becomes

$$TS = -\mu N + E + kT \ln Q,$$

which we recognize as the thermodynamic relation (6.9) if we identify

$$\Omega \equiv -kT \ln[Q(T,V,\mu)]. \qquad (6.23)$$

To see that this identification is indeed correct, note that

$$\left(\frac{\partial\Omega}{\partial T}\right)_{V,\mu} = -k\ln Q - \frac{kT}{Q}\sum_{\mathcal{N}=0}^{\infty}\int\left(-\frac{1}{kT^2}\int[\mu\mathcal{N}-H_{\mathcal{N}}]\right.$$
$$\left.\times\exp\{[\mu\mathcal{N}-H_{\mathcal{N}}]/kT\}\right)d\Gamma_{\mathcal{N}}.$$

With the help of equations (6.17) and (6.22), we may write the above as

$$\left(\frac{\partial\Omega}{\partial T}\right)_{V,\mu} = -k\ln Q + \frac{\mu}{T}\langle\mathcal{N}\rangle - \frac{1}{T}\langle H_{\mathcal{N}}\rangle.$$

But the right-hand side equals $-S$ as calculated previously; therefore $(\partial\Omega/\partial T)_{V,\mu} = -S$, which is, indeed, the thermodynamic relation (6.9) between Ω and S.

RECIPE FOR USING THE GRAND CANONICAL PARTITION FUNCTION

(a) For a given $\mathcal{N}$-particle Hamiltonian, calculate the grand canonical partition function (6.19).

(b) Evaluate now the grand potential (6.23). The equation of state is derivable from $PV = -\Omega$, or from $P = -(\partial\Omega/\partial V)_{T,\mu}$; other thermodynamic variables are derivable from $S = -(\partial\Omega/\partial T)_{V,\mu}$, and $N = -(\partial\Omega/\partial\mu)_{T,V}$.

QUANTUM STATISTICAL MECHANICS

For a system of mildly interacting particles (e.g., a dilute gas), quantum mechanics introduces two fundamentally new considerations: (1) the concept of quantized single-particle states with energy level ϵ_j for $j = 1, 2, 3, \ldots$; and (2) the Pauli exclusion principle, which states that the occupation number $\mathcal{N}_j$ of the single-particle state j can take any value from 0 to ∞ for bosons, but can equal only 0 or 1 for fermions. The total energy of the system for either bosons or fermions can be written as

$$H_{\mathcal{N}} = \sum_j \mathcal{N}_j \epsilon_j$$

for any set $\mathcal{N}_j$ of given occupation numbers that add up to some total $\mathcal{N}$:

$$\sum_j \mathcal{N}_j = \mathcal{N}.$$

The recipe for going from classical statistical mechanics to quantum statistical mechanics then involves replacing

$$Q(T, V, \mu) = \sum_{\mathcal{N}=0}^{\infty} \int \exp[(\mu\mathcal{N} - H_{\mathcal{N}})/kT] \, d\Gamma_{\mathcal{N}}$$

by the expression

$$Q(T, V, \mu) = \sum_{\{\mathcal{N}_j\}} \exp\left[\sum_j \mathcal{N}_j(\mu - \epsilon_j)/kT\right], \qquad (6.24)$$

where the sum over $\{\mathcal{N}_j\}$ represents a sum over *all* possible sets of occupation numbers at particle energies $\{\epsilon_j\}$. The utility of the grand canonical ensemble arises because there is no constraint on the sum $\mathcal{N}_1 + \mathcal{N}_2 + \cdots \equiv \mathcal{N}$ being a *fixed* number.

Before we can use equation (6.24), we find it convenient to reverse the operation of the two sums, i.e., to perform the sum over statistics first and the sum over particle states last. We begin by noting that an exponential of sums equals a product of exponentials, i.e.,

$$\exp\left[\sum_j \mathcal{N}_j(\mu - \epsilon_j)/kT\right] = \prod_j \exp\left[\mathcal{N}_j(\mu - \epsilon_j)/kT\right].$$

We may now write equation (6.24) as

$$Q(T, V, \mu) = \sum_{\{\mathcal{N}_j\}} \left[\prod_j x_j^{\mathcal{N}_j}\right], \qquad (6.25)$$

where

$$x_j \equiv \exp[(\mu - \epsilon_j)/kT]. \qquad (6.26)$$

For any $\alpha_j(\mathcal{N}_j)$, we now note the identity that

$$\sum_{\{\mathcal{N}_j\}} \left[\prod_j \alpha_j(\mathcal{N}_j)\right] \equiv \prod_j \left[\sum_{\mathcal{N}_j} \alpha_j(\mathcal{N}_j)\right].$$

The proof consists of merely writing out what we mean by all the symbols in a system with occupation numbers $\{\mathcal{N}_j\}_{j=1}^{J}$ in J total (particle) energy states. In this case, the left-hand side of the identity means

$$\sum_{\mathcal{N}_1} \cdots \sum_{\mathcal{N}_J} [\alpha_1(\mathcal{N}_1) \ldots \alpha_J(\mathcal{N}_J)],$$

which, because the sums over $\mathcal{N}_1, \ldots, \mathcal{N}_J$ are independent of each other, equals

$$\left[\sum_{\mathcal{N}_1} \alpha_1(\mathcal{N}_1)\right] \cdots \left[\sum_{\mathcal{N}_J} \alpha_J(\mathcal{N}_J)\right],$$

what we mean by the right-hand side of the identity. Setting $\alpha_j(\mathcal{N}_j) = x_j^{\mathcal{N}_j}$, and applying our identity to equation (6.25), we obtain

$$Q(T, V, \mu) = \prod_j \sum_{\mathcal{N}_j} x_j^{\mathcal{N}_j}, \tag{6.27}$$

where x_j is defined by equation (6.26).

BOSE-EINSTEIN STATISTICS

If we have an ideal gas (which, however, may have internal degrees of free-dom), we may assume that individual particle energies ϵ_j are independent of $\mathcal{N}_j$. Bose-Einstein statistics applies to particles (bosons) for which there exists no restrictions whatsoever on $\mathcal{N}_j$; i.e., the sum over $\mathcal{N}_j$ runs from 0 to ∞. Since

$$\sum_{\mathcal{N}_j=0}^{\infty} x_j^{\mathcal{N}_j} = 1 + x_j + x_j^2 + \cdots = \frac{1}{1-x_j},$$

equation (6.27) acquires the simple expression:

$$Q = \prod_j \frac{1}{1-x_j}. \tag{6.28}$$

The grand potential now becomes

$$\Omega = -kT \ln Q = kT \sum_j \ln[1 - e^{(\mu-\epsilon_j)/kT}]. \tag{6.29}$$

With $N = -(\partial\Omega/\partial\mu)_{T,V}$, we obtain

$$N = \sum_j \frac{e^{(\mu-\epsilon_j)/kT}}{1 - e^{(\mu-\epsilon_j)/kT}} = \sum_j \frac{1}{e^{(\epsilon_j-\mu)/kT} - 1}. \tag{6.30}$$

But we expect the mean total number N of particles to equal $\sum_j \langle \mathcal{N}_j \rangle$, where $\langle \mathcal{N}_j \rangle$ is the mean occupation number in energy state j; thus we can identify

$$\langle \mathcal{N}_j \rangle = \frac{1}{e^{(\epsilon_j-\mu)/kT} - 1}.$$

In the continuum limit, if we are speaking of elementary particles,

$$\sum_j \ldots \rightarrow (2s+1)\frac{V}{h^3} \int \ldots d^3p, \tag{6.31}$$

where $2s+1$ represents the number of independent spin states. For each spin state, therefore, the expected occupation number under thermodynamic equilibrium is given by equation (1.15), with the particle kinetic energy in general given by the relativistic formula

$$\epsilon \equiv (p^2c^2 + m^2c^4)^{1/2} - mc^2. \tag{6.32}$$

We use the convention of subtracting out the rest energy mc^2 in ϵ so that it does not appear in formulae for the "chemical potential" μ. [This convention needs re-examination for situations where particles and antiparticles of finite rest mass can be created and destroyed; see Problem Set 2. Moreover, for compound particles, e.g., atoms or molecules that have additional internal degrees of freedom, we need to replace the factor $(2s+1)$ in equation (6.31) by a sum over internal energy levels (see next chapter).]

FERMI-DIRAC STATISTICS

For an ideal gas consisting of fermions, Pauli's exclusion principle restricts $\mathcal{N}_j$ to be either 0 or 1. Thus equation (6.27) yields

$$Q = \prod_j \left[1 + e^{(\mu-\epsilon_j)/kT}\right].$$

The grand potential now becomes

$$\Omega = -kT \ln Q = -kT \sum_j \ln \left[1 + e^{(\mu-\epsilon_j)/kT}\right].$$

With $N = -(\partial\Omega/\partial\mu)_{T,V} \equiv \sum_j \langle \mathcal{N}_j \rangle$, we obtain the identification,

$$\langle \mathcal{N}_j \rangle = \frac{1}{e^{(\epsilon_j-\mu)/kT} + 1}.$$

In the continuum limit (6.31), we obtain for each spin state the expected occupation number of equation (1.17), with ϵ given by the relativistic formula (6.32).

The derivation of these familiar results completes the present chapter. In the following chapter, we will use the results to calculate the conditions of ionization equilibrium (Saha's equation) and nuclear statistical equilibrium.

7

Reaction Equilibria

Reference: M. Saha 1920, *Phil. Mag.*, **40**, 472, 809; E.M. Burbidge, G.R. Burbidge, W.A. Fowler, and F. Hoyle 1957, *Rev. Mod. Phys.*, **29**, 547.

At the end of Chapter 6 we treated the thermodynamics of the external (translational) degrees of freedom of particles (bosons or fermions possibly moving relativistically). In this chapter, we wish explicitly to include the internal degrees of freedom of these particles, and, in particular, to consider the coupling between the internal and external states under conditions of thermodynamic equilibrium. To fix ideas, we could think of the internal states as the electronic energy levels of atoms, and of the coupling with the continuum states as recombination and ionization. But our formulation holds also for the formation and dissociation of molecules (chemical reactions) with or without excited rotational and vibrational levels, as well as for the fusion and breakup of atomic nuclei (nuclear reactions) with or without excited nuclear states.

EXAMPLES OF REACTIONS

A common "reaction" in astrophysics involves the excitation (collisional or radiative) of a hydrogen atom H(1) in the ground electronic state to a higher level, say, H(2):

$$H(1) \rightleftharpoons H(2). \tag{7.1}$$

The energy required to drive the forward reaction (or released by the backward one) may come from collisions, or from photons, or from some catalyst that appears on both sides of the equation. The nature of the mediating agent can affect the rate at which the system approaches thermodynamic equilibrium, but the properties at equilibrium are independent of such details, which is why we can write the reaction in the abbreviated form of equation (7.1).

Another important reaction involves a hydrogen atom H(1) (in the ground electronic state) recombining with an electron e^- to give the H^-

ion (having a single bound state), plus the release of an amount of energy equal to $\chi = 0.754\,\mathrm{eV}$:

$$H(1) + e^- \rightleftharpoons H^-. \tag{7.2}$$

A third (the first step of the "triple alpha process") involves the nuclear fusion of two helium-4 nuclei (alpha particles) to give one beryllium-8 nucleus:

$$2\,\mathrm{He}^4 \rightleftharpoons \mathrm{Be}^8. \tag{7.3}$$

In each case, we can write the reaction in a generic form, by bringing all A species $\{X^{(a)}\}_{a=1}^A$ to the left-hand side:

$$\sum_{a=1}^A \nu^{(a)} X^{(a)} = 0, \tag{7.4}$$

where $\nu^{(a)}$ is a positive or negative integer depending on whether $X^{(a)}$ is a reactant or a product. Which species we label as "reactants" and which as "products" does not matter, since only the relative signs count when "forward" and "backward" reactions are balanced in thermodynamic equilibrium. To fix ideas, if we let $X^{(1)}$ denote He^4 and $X^{(2)}$ denote Be^8 in reaction (7.3), we have $\nu^{(1)} = 2$ and $\nu^{(2)} = -1$, but the content of equation (7.4) remains unchanged if we make $\nu^{(1)} = -2$ and $\nu^{(2)} = 1$.

GENERALIZED LAW OF MASS ACTION

When the numbers in different species may change, we obtain the condition of thermodynamic equilibrium by minimizing the Gibbs free energy G. Since

$$dG = -S\,dT + V\,dP + \sum_{a=1}^A \mu^{(a)} dN^{(a)}, \tag{7.5}$$

we have $dG = 0$ at constant T and total P when

$$\sum_{a=1}^A \mu^{(a)} dN^{(a)} = 0, \tag{7.6}$$

where the set of A numbers $\{N^{(a)}\}_{a=1}^A$ cannot vary independently but must follow Dalton's rule concerning integer proportions:

$$dN^{(1)} : dN^{(2)} : \ldots : dN^{(A)} = \nu^{(1)} : \nu^{(2)} : \ldots : \nu^{(A)}.$$

If we introduce a common scale,

$$dN^{(a)} = \nu^{(a)} dN \qquad \text{for} \qquad a = 1 \text{ to } A, \tag{7.7}$$

we have the freedom to vary only dN, and equation (7.6) gives the result,

$$\sum_{a=1}^{A} \nu^{(a)} \mu^{(a)} = 0. \tag{7.8}$$

Notice that the chemical potentials of material catalysts (agents not changed by the reaction), which enter equally on both sides of equation (7.4), cancel out in equation (7.8). The wall of a thermodynamic enclosure can be considered a catalyst that converts, say, one high-energy photon $h\nu$ into two lower-energy ones $h\nu'$ and $h\nu''$:

$$h\nu + \text{wall} \rightleftharpoons h\nu' + h\nu'' + \text{wall}. \tag{7.9}$$

If μ_{ph} represents the chemical potential of the photon in (7.9), equation (7.8) requires

$$\mu_{\text{ph}} - 2\mu_{\text{ph}} = 0; \quad \text{i.e. } \mu_{\text{ph}} = 0, \tag{7.10}$$

which is the standard conclusion. Thus we do not need to include photons (or neutrinos) in the generic equation (7.4) because they have zero chemical potential and contribute nothing to equation (7.8) when they can be freely created and destroyed in thermodynamic equilibrium. Finally, notice that the additivity of chemical potentials of different species implies that a reaction chain of the type,

$$A + B \rightleftharpoons D + E,$$

$$C + D \rightleftharpoons F + G,$$

has a net result,

$$A + B + C \rightleftharpoons E + F + G,$$

with a net law in equation (7.8) identical to that obtained by summing the individual laws in the (arbitrarily long) chain. In doing equilibrium calculations, we may choose either the latter route, or we may jump directly to the net law. The *principle of detailed balance* underlies this fundamental freedom. As we will discuss more deeply in a later chapter, when thermodynamic equilibrium prevails, invoking detailed balance allows us to analyze any reaction equation as an isolated basic process.

The preceding comment also has a flip side, namely, that thermodynamic equilibrium need not apply to all species of a system in order for us to use the arguments of statistical mechanics for some of them. Thus, in discussing chemical equilibrium in a laboratory flask, chemists can assume that the nuclear states of matter are frozen. In discussing nuclear statistical equilibrium inside a presupernova star, astrophysicists need not worry about the statistical states occupied by the constituent quarks. In assembling her hierarchy of compound particles, nature has been kind to keep very distinct the energy levels accessible to different temperature and pressure regimes.

CHEMICAL POTENTIAL OF IDEAL GASES WITH INTERNAL DEGREES OF FREEDOM

In order to make use of equation (7.8), we need to have expressions for the chemical potential $\mu^{(a)}$, i.e., the individual terms in Euler's relation for G:

$$G = \sum_{a=1}^{A} \mu^{(a)} N^{(a)}. \tag{7.11}$$

This task is easy only for ideal gases, which we now assume, with the generalization, however, that we allow internal degrees of freedom (or there would be no reactions). Since the A species are distinguishable from one another, we may write the grand canonical partition function of the system as

$$Q = \prod_{a=1}^{A} Q^{(a)}, \quad \text{with} \quad Q^{(a)} = \prod_{j} \sum_{\mathcal{N}_j^{(a)}} \exp\left[\mathcal{N}_j^{(a)}(\mu^{(a)} - \epsilon_j^{(a)})/kT\right],$$

$$\tag{7.12}$$

where (for full generality) we have allowed each species a to have a set of combined external and internal energy levels $\{\epsilon_j^{(a)}\}$. Alternatively, we may treat each internal energy level as a different species, as in "reaction" (7.1). Both procedures lead to the same results. In any case, the grand potential of the system now becomes

$$\Omega = -kT \ln Q = \sum_{a=1}^{A} \Omega^{(a)} \quad \text{where} \quad \Omega^{(a)} = -kT \ln Q^{(a)}. \tag{7.13}$$

The mean number of species a in volume V can be extracted from the thermodynamic relation,

$$N^{(a)} = -\left(\frac{\partial \Omega^{(a)}}{\partial \mu^{(a)}}\right)_{T,V}. \tag{7.14}$$

According to the calculations of the last chapter, for an ideal gas, we may write the mean number of species a in volume V as

$$N^{(a)} = \sum_{j} \frac{1}{e^{(\epsilon_j^{(a)} - \mu^{(a)})/kT} \pm 1}, \tag{7.15}$$

where the upper sign applies to fermions and the lower sign to bosons. For the present application, we need to include in $\epsilon_j^{(a)}$ not only the energy of the translational degrees of freedom [see equation (6.32)], but also that of the internal degrees of freedom,

$$\epsilon_j^{(a)} = \epsilon_{\text{trans}}^{(a)} - \chi_j^{(a)}, \tag{7.16}$$

where we use χ generically to denote the binding energy of the (possibly excited) state (electronic, chemical, nuclear, or whatever).

In most circumstances where we are interested in applying equation (7.8), but not all (e.g., electron-positron pair creation in the early universe or in some supernova models), we can take the limit of low occupation numbers and non-relativistic thermal motions:

$$N^{(a)} = \sum_j e^{(\mu^{(a)} - \epsilon_j^{(a)})/kT}, \tag{7.17}$$

where

$$\epsilon_j^{(a)} = \frac{|\mathbf{p}^{(a)}|^2}{2m^{(a)}} - \chi_j^{(a)}. \tag{7.18}$$

Notice that we have set the zero point of the energy for a particle with zero momentum in the free state rather than at the ground level of the bound state. We interpret the sum over states $\sum_j$ now to be a sum over the (quantized) internal levels ℓ times an integral over the translational degrees of freedom:

$$N^{(a)} = e^{\mu^{(a)}/kT} \left[\sum_\ell g_\ell^{(a)} e^{\chi_\ell^{(a)}/kT} \right] \frac{V}{h^3} \left[\int_0^\infty e^{-p^2/2m^{(a)}kT} 4\pi p^2 \, dp \right], \tag{7.19}$$

where $g_\ell^{(a)}$ is the statistical weight associated with the internal degree of freedom (number of degenerate states that have the same binding energy $\chi_\ell^{(a)}$). The integral over the Maxwellian divided by h^3 can be performed in the usual manner to yield $(\lambda_T^{(a)})^{-3}$, where $\lambda_T^{(a)}$ is the thermal de Broglie wavelength of species a:

$$\lambda_T^{(a)} \equiv h/(2\pi m^{(a)} kT)^{1/2}. \tag{7.20}$$

We define the number density $n^{(a)} \equiv N^{(a)}/V$, and solve equation (7.19) to obtain

$$\mu^{(a)} = kT \ln \left[\frac{n^{(a)}}{\mathcal{Z}^{(a)}(T)} \right], \tag{7.21}$$

where we have defined the partition function associated with the internal and external states:

$$\mathcal{Z}^{(a)}(T) \equiv (\lambda_T^{(a)})^{-3} \sum_\ell g_\ell^{(a)} e^{\chi_\ell^{(a)}/kT}. \tag{7.22}$$

With equation (7.21), equation (7.8) becomes, upon taking antilogarithms,

$$\prod_{a=1}^A \left[\frac{n^{(a)}}{\mathcal{Z}^{(a)}(T)} \right]^{\nu^{(a)}} = 1, \quad \text{i.e.,} \quad \prod_{a=1}^A [n^{(a)}]^{\nu^{(a)}} = \prod_{a=1}^A [\mathcal{Z}^{(a)}(T)]^{\nu^{(a)}}. \tag{7.23}$$

Chemists call equation (7.23) the *law of mass action*; we have generalized its meaning to incorporate reactions other than chemical ones.

BOLTZMANN'S LAW

As our first example, we apply equation (7.23) to reaction (7.1). Here the internal part of the partition function (7.22), when summed over the degenerate sublevels, produces $g^{(a)}e^{\chi^{(a)}/kT}$ for $a = 1,\ 2$ ($= n$, the principal quantum number), with $g^{(n)} = 2n^2$. Since the thermal de Broglie wavelengths are the same for the two species H(1) and H(2), upon moving superscripts to the more usual place as subscripts, we have Boltzmann's law:

$$\frac{n_2}{n_1} = \frac{g_2}{g_1}e^{-(E_2-E_1)/kT}, \tag{7.24}$$

where $E_2 - E_1 = \chi_1 - \chi_2$ is the energy difference between the *excitation above ground* of the $n = 2$ and $n = 1$ levels of atomic hydrogen. We have not set the ground state energy $E_1 = 0$ in equation (7.24), because we wish to emphasize that it applies to the relative populations of *any* two states 1 and 2 of a given atom or molecule, provided we correctly specify the statistical weight factors g. Standard elementary derivations of equation (7.24) usually gloss over the assumption that atoms and molecules typically exist only in environments where the thermal bath of free states with which the bound states interact can be approximated to be *Maxwellian*, i.e., the classical limit of Fermi-Dirac and Bose-Einstein statistics.

Sometimes we want to express the level population n_ℓ of some excited level ℓ relative to the total n_{tot} summed over all levels of excitation. For this purpose, it is convenient to introduce the level partition function:

$$Z \equiv \sum_n g_n e^{-E_n/kT}. \tag{7.25}$$

With this notation, we may write

$$\frac{n_\ell}{n_{\text{tot}}} = \frac{g_\ell}{Z}e^{-E_\ell/kT}. \tag{7.26}$$

If we apply this equation to atomic hydrogen, we encounter an immediate problem when we try to sum the partition function in equation (7.25) from $n = 1$ to ∞. For atomic hydrogen, $g_n = 2n^2$; whereas $E_n = (1 - n^{-2})I$, where $I = 13.6\,\text{eV}$ is the ionization potential from the ground state. Thus $e^{-E_n/kT}$ is no smaller than $e^{-I/kT}$, a finite number, whereas the sum of $2n^2$ diverges rapidly.

The divergence at large n is not peculiar to atomic hydrogen, but merely reflects the large number of microstates (i.e., phase space) available to atoms in highly excited states, which overwhelms the disadvantage of a small but finite exponential Boltzmann factor. States with large n correspond, however, to very large atoms. The approximation of atoms as isolated entities must fail for large n. For example, Debye and Huckel showed (see Chapter 1 of Volume II) that free electrons with a number density n_e

and temperature T tend to shield ions of mass m_i and charge $Z_i e$ [not to be confused with equation (7.25)], and to give each ion an effective potential that departs from the Coulomb value at large r:

$$\phi = \frac{Z_i e}{r} e^{-r/L_D}. \tag{7.27}$$

Here L_D is the Debye length

$$L_D = \left[\left(\frac{\omega_{pe}}{v_{Te}} \right)^2 + \sum_i \left(\frac{\omega_{pi}}{v_{Ti}} \right)^2 \right]^{-1/2}, \tag{7.28}$$

and v_{Te}, ω_{pe} and v_{Ti}, ω_{pi} are the thermal speed and plasma frequency of the electrons and ions, respectively:

$$v_{Te} \equiv \left(\frac{kT}{m_e} \right)^{1/2}, \quad \omega_{pe} \equiv \left(4\pi n_e \frac{e^2}{m_e} \right)^{1/2};$$

$$v_{Ti} \equiv \left(\frac{kT}{m_i} \right)^{1/2}, \quad \omega_{pi} \equiv \left[4\pi n_i \frac{(Z_i e)^2}{m_i} \right]^{1/2}. \tag{7.29}$$

Notice that the dependences on m_e and m_i cancel out in the formula for L_D, which is why the repelling of ions can contribute as much as the attraction of electrons.

If we expand the exponential in equation (7.27) for small r/L_D, we obtain the physical picture that the effective potential surrounding an ion has the approximation $\phi \approx Z_i e(r^{-1} - L_D^{-1})$; consequently, the continuum energy level of an electron behaves as if it had been lowered by a constant amount $Z_i e^2 / L_D$. This removes the high-lying states of hydrogenic energies $Z_i e^2 / 2n^2 a_Z$, where $a_Z \equiv \hbar^2 / Z_i m_e e^2$ represents the analog of the Bohr radius for a nuclear charge of Z_i, and limits $n \leq n_{max}$, where

$$n_{max} \equiv (L_D / 2a_Z)^{1/2}. \tag{7.30}$$

If $n_{max} < 1$, i.e., if the Debye length becomes less than the equivalent Bohr diameter, bound states become impossible (see Problem Set 2). When we apply equation (7.30) to the atmospheres of stars, we generally find that terms with low n dominate the partition function (7.25) at low T because of the smallness of the Boltzmann exponential for excited states. As T increases, large values of n begin to become important, but not before the hydrogen becomes almost completely ionized (see below). Thus truncating the partition function very precisely almost never becomes a practical problem.

SAHA'S EQUATION

As a second example, we apply equation (7.23) to reaction (7.2). Here we have only one bound state for H^-; so $Z_{H^-} = g_{H^-} e^{\chi/kT}$, and $Z_{H(1)} = g_{H(1)}$, with $g_{H^-} = 1$ and $g_{H(1)} = 2$. The thermal de Broglie wavelengths are nearly the same for H^- and $H(1)$, but $Z_e = 2$ for a free electron because of its two internal-spin degrees of freedom. Saha's equation for the ionization of H^- therefore gives

$$\frac{n_{H^-}}{n_e n_{H(1)}} = \frac{g_{H^-}}{g_{H(1)} \mathcal{Z}_e} e^{\chi/kT}, \tag{7.31}$$

where $\mathcal{Z}_e$ is the total partition function for free electrons (including both spin and translational degrees of freedom):

$$\mathcal{Z}_e \equiv 2(\lambda_{Te})^{-3}, \tag{7.32}$$

with λ_{Te} being the electron thermal de Broglie wavelength:

$$\lambda_{Te} \equiv h/(2\pi m_e kT)^{1/2}. \tag{7.33}$$

Notice that we have deliberately written equation (7.31) in a form reminiscent of the Boltzmann law, equation (7.24). The similarity derives from their common origins in the generalized law of mass action. Notice also that the Boltzmann exponential factor, $e^{\chi/kT}$, favors the bound state (H^-), but the large amount of phase space available for exploration by free electrons favors the "ionized" state $H(1)$ [remember $H(1)$ is H^- with the *removal* of an electron], because the dimensionless factor $n_e \lambda_{Te}^3 \ll 1$ if the electron gas is not degenerate (consult Problem Set 2).

For species with more than one internal bound state (e.g., the ionization of atomic hydrogen or some more complex atom), the generalized law of mass action can be applied, level by level in their relation to the continuum, or all at once (by summing over the internal degrees of freedom). If we do the latter, we recover Saha's equation in the form usually found in astronomical textbooks:

$$\frac{n_{i+1}}{n_i} = \frac{2Z_{i+1}}{(n_e \lambda_{Te}^3)Z_i} e^{-I_i/kT}, \tag{7.34}$$

where n_{i+1} and n_i are the total number of atoms in ionization stage $i + 1$ and i, with I_i being the ionization potential from the ground state of i to the ground state of $i + 1$. In equation (7.34), Z_{i+1} and Z_i are the partition functions associated with summing over all possible excited levels of ionization stage $i + 1$ and i [see equation (7.25)].

We present yet another, and more revealing, way to write equation (7.34). For a nondegenerate background of free electrons, $-\mu_e = kT\xi$, where $\xi \equiv \ln(2/n_e \lambda_{Te}^3)$ is a large number, of order 10–30 in many applications.

The quantity ξ will substantially exceed unity if the mean number of electrons in a thermal de Broglie cube is very small, $n_e \lambda_{Te}^3 \ll 1$; i.e., if the electron gas is indeed nondegenerate [as was assumed in order to derive equation (7.34)]. On the other hand, ξ does not have huge variations, because it depends only logarithmically on n_e and T. In any case,

$$\frac{n_{i+1}}{n_i} = \frac{Z_{i+1}}{Z_i} e^{-(\mu_e + I_i)/kT}. \tag{7.35}$$

Since the ratio of level partition functions will generally be a number of order unity, significant fractional ionization of any species occurs only when $-\mu_e \approx I_i$, i.e., at a temperature that satisfies $T \approx I_i/k\xi$, which is more than one order of magnitude smaller than the naive expectation $T \approx I_i/k$. In other words, the increase in phase space accessible to a freed electron largely offsets the penalty in binding energy paid to produce the ionization. Because of this advantage, it is, in a sense, almost easier to ionize an atom than to excite it. For example, the ionization potential $I = 13.6\,\mathrm{eV}$ of atomic hydrogen corresponds to $I/k = 158,000\,\mathrm{K}$, but because $\xi \approx 15$ or so in an A star's atmosphere (where $T_{\mathrm{eff}} \approx 10,000\,\mathrm{K}$), we find that the stars exhibiting the strongest (visible) hydrogen lines (an alphabetical sequence) actually lie intermediate in photospheric temperatures, ones high enough to excite atomic hydrogen partially to $n = 2$ (from where resonant absorption of photons can produce the optically visible lines of the Balmer series), but not enough to completely ionize it. This fundamental insight by Saha proved crucial in the historical unraveling of the surface temperature sequence of stellar spectroscopic types: OBAFGKM.

NUCLEAR STATISTICAL EQUILIBRIUM

As a final application, consider the schematic nuclear reaction wherein we assemble Z protons [not to be confused with equation (7.25)] and $A - Z$ neutrons to get a nucleus (Z, A):

$$Z\mathrm{p} + (A - Z)\mathrm{n} - (Z, A) = 0. \tag{7.36}$$

The generalized law of mass action (7.23) applied to this reaction yields the equation of *nuclear statistical equilibrium*:

$$\frac{n(Z, A)}{n_{\mathrm{p}}^Z n_{\mathrm{n}}^{A-Z}} = [\lambda_{T(Z,A)}^{-3} Z_{(Z,A)}] e^{Q(Z,A)/kT} (\lambda_{T\mathrm{p}}^3/2)^A, \tag{7.37}$$

where $Q(Z, A)$ is the binding energy of the ground nuclear state of (Z, A), and $Z_{(Z,A)}$ is the partition function associated with its excited nuclear states, where we have approximated the thermal de Broglie wavelength of the neutron to equal that of the proton, and where we have assigned each proton and neutron two independent spin states. Although equation (7.37) would seem to require an improbable A-body reaction, we can in fact use this equation to relate to each other relative nuclear abundances that form by other, faster routes. (Recall our comments about reaction chains.) In particular, just before a supernova explosion takes place, the matter near the center of the star is close to nuclear statistical equilibrium (with a heavy dominance of iron-peak nuclei), provided we exclude from the calculation those species which are made by weak interactions.

To apply equation (7.37) in the early days of nuclear physics, people often used von Weizsacker's semi-empirical binding-energy formula:

$$Q(Z, A) = aA - bA^{2/3} - cZ(Z - 1)/A^{1/3} - eA(1 - 2Z/A)^2 + f\delta/A, \quad (7.38)$$

where $a = 15.7\,\text{MeV}$, $b = 17.8\,\text{MeV}$, $c = 0.712\,\text{MeV}$, $e = 23.6\,\text{MeV}$, $f = 132$ MeV are constants, and $\delta = +1$, 0, or -1 depending on whether (Z, A) are even-even, odd-even, or odd-odd. The term proportional to a arises from attractive (volume) binding; b from surface deficits of neighboring nucleons; c from repulsive Coulomb effects of the protons; e from exchange (isotope) effects; and f from paired spins (or lack of). (For a qualitative explanation of the effects, see Shu 1982, Chapter 6.) Comparison of the predictions of equation (7.37) for various postulated environments (see the discussion of Clayton 1968, pp. 533–545) with the observed cosmic abundances of the heavy elements (near and beyond the iron peak) provided early researchers of the supernova phenomenon with invaluable guidance. Nowadays, for accurate work, we use tabular data compiled by nuclear physicists in place of von Weizsacker's semi-empirical formula (7.38).

8

Detailed Balance and Einstein-Milne Relations

Reference: Mihalas, *Stellar Atmospheres*, pp. 77–80, 94–96.

The methods of statistical mechanics can tell us how systems behave in conditions of near thermodynamic equilibrium, and such information constitutes much of the input physics for theories of the structure and evolution of the early universe and the interiors of stars. They are less helpful in studies of the interstellar medium and active galactic nuclei, where extreme conditions of irradiation and matter rarefaction may lead to large departures from local thermodynamic equilibrium. For realistic investigations of these subjects, as well as for more accurate descriptions of stellar atmospheres, we often need to use *rate equations* to obtain detailed information about the local state of the system.

At first sight, elementary collisional processes would seem to require the experimental measurement or theoretical calculation of two *cross sections* (one for the forward rate and one for the backward), whereas photoprocesses seem to need three *rate coefficients* (one for spontaneous emission, one for stimulated emission, and one for true absorption). In fact, as we will show, they require only one independent specification each—general principles specify the other quantities in terms of any known one. This fortuitous circumstance saves us a lot of work, or at worst, lets us choose the best-determined value for a set of related rate coefficients. The resulting relations hold as long as the individual nuclear, atomic, or molecular species can be considered isolated except at the moment of interaction; in particular, they hold under conditions of thermodynamic equilibrium, where the character of the general relations are most easily established. The guiding principle in each case is the *principle of detailed balance*, which states: *In thermodynamic equilibrium, every elementary process is statistically balanced by its exact reverse.*

ELASTIC COLLISIONS

Historically, Boltzmann was the first to discover a specific example of this principle in his discussion of how elastic collisions would establish a Maxwellian distribution in a classical gas of molecules. Consider the distribution function $f(\mathbf{x}, \mathbf{p}_1, t)$ in 6-D phase space defined so that

$$f(\mathbf{x}, \mathbf{p}_1, t)\, d^3x\, d^3p_1 \qquad (8.1)$$

equals the number of (material) particles of a certain type at time t with position and momentum centered within d^3x and d^3p_1 about $\mathbf{x}$ and $\mathbf{p}_1$. As derived in books on kinetic theory (e.g., Chapter 3 of Huang 1963), Boltzmann's transport equation reads

$$\frac{\partial f}{\partial t} + \mathbf{v}_1 \cdot \frac{\partial f}{\partial \mathbf{x}} + \mathbf{F} \cdot \frac{\partial f}{\partial \mathbf{p}_1} = \left(\frac{\delta f}{\delta t}\right)_c, \qquad (8.2)$$

where $\mathbf{F}$ gives the *smooth* body forces that can change the momentum $\mathbf{p}$ of a particle, and the collision term $(\delta f/\delta t)_c$ contains the effects of *irregular and short-range* intermolecular forces.

Boltzmann's great insight allowed him to write down an explicit expression for the collision integral for a rarefied (classical) gas where we need be concerned with only two-body interactions:

$$\left(\frac{\delta f}{\delta t}\right)_c = \int d^3p_2 \int d\Omega\, \sigma(\Omega)|\mathbf{v}_1 - \mathbf{v}_2|[f(\mathbf{p}_1')f(\mathbf{p}_2') - f(\mathbf{p}_1)f(\mathbf{p}_2)], \quad (8.3)$$

where $\sigma(\Omega)$ is the differential cross section for the scattering process $(\mathbf{p}_1, \mathbf{p}_2) \rightarrow (\mathbf{p}_1', \mathbf{p}_2')$ and involves two angles that are integrated in the integral over solid angle $d\Omega$. The differential cross section for the reverse process $\sigma(\Omega')$ has been assumed to have the same value. This assumption can be justified by quantum-mechanical arguments as long as certain general symmetry principles hold. Conversely, we can argue that the requirement that detailed balance produce the known thermodynamics distribution (see below) demonstrates that all elastic collision processes must have this property.

Two angles—or an impact parameter and one angle (about which scattering by featureless molecules would have axial symmetry)—define the process because energy and momentum conservation

$$\epsilon(\mathbf{p}_1') + \epsilon(\mathbf{p}_2') = \epsilon(\mathbf{p}_1) + \epsilon(\mathbf{p}_2), \qquad (8.4)$$

$$\mathbf{p}_1' + \mathbf{p}_2' = \mathbf{p}_1 + \mathbf{p}_2, \qquad (8.5)$$

give four constraints on the two possible primed vectors $\mathbf{p}_1'$ and $\mathbf{p}_2'$ when $\mathbf{p}_1$ and $\mathbf{p}_2$ are known. In equation (8.4), $\epsilon(\mathbf{p})$ is the function of $\mathbf{p}$ given by equation (6.32) if we allow for relativistic effects. In equation (8.2),

$|\mathbf{v}_1 - \mathbf{v}_2| = |\mathbf{v}_1' - \mathbf{v}_2'|$ (a collision invariant even at relativistic velocities) is the relative speed of the two colliding particles, with the velocity $\mathbf{v}$, momentum $\mathbf{p}$, and kinetic energy ϵ of a particle of mass m being related through the famous formulae:

$$\mathbf{p} = \gamma m \mathbf{v}, \qquad \epsilon = (\gamma - 1)mc^2, \qquad \text{where} \qquad \gamma \equiv \left(1 - \frac{v^2}{c^2}\right)^{-1/2}. \quad (8.6)$$

The assumption of short-range interactions is implicit in the derivation of equation (8.3), since we do not distinguish between the positions $\mathbf{x}$ of the two particles. We shall soon uncover another hidden assumption that concerns quantum statistics. For now, however, we will discuss how the principle of detailed balance arises for thermodynamic equilibrium in equation (8.2). When thermodynamic equilibrium holds, the left-hand and right-hand sides of equation (8.2) individually equal zero. For this to be true for all $\mathbf{p}$ requires not only that the collisional integral equal zero in net, but that the integrand vanish for all possible values of the relevant variables. To achieve this result requires

$$f(\mathbf{p}_1')f(\mathbf{p}_2') = f(\mathbf{p}_1)f(\mathbf{p}_2) \qquad (8.7)$$

for all possible collision pairs. If we take the logarithm of equation (8.7), we deduce that $\ln f(\mathbf{p}_1) + \ln f(\mathbf{p}_2)$ must be a conserved quantity in all possible collisions; i.e., $\ln f(\mathbf{p})$ must be a linear sum of the independent collisional invariants:

$$\ln f(\mathbf{p}) = \alpha - \beta(\epsilon - \mathbf{u} \cdot \mathbf{p}), \qquad (8.8)$$

where α, β, and $\mathbf{u}$ are constants that characterize the system as a whole, and not individual molecules.

In a frame for which the bulk velocity $\mathbf{u}$ of the fluid is zero, we obtain from equation (8.8) the result that detailed balance requires for the equilibrium distribution function

$$f \propto e^{-\epsilon/kT}, \qquad (8.9)$$

where we have made the identification $\beta = 1/kT$ on the usual basis (i.e., that $3kT/2$ defines the mean kinetic energy of a particle for a nonrelativistic perfect gas). From this mechanistic perspective, the Maxwell-Boltzmann formula (8.9) gives the equilibrium outcome because it constitutes the only distribution function that can be statistically preserved through countless numbers of random collisions.

CORRECTION FOR STIMULATED AND SUPPRESSED SCATTERING

The fact that we recovered the Maxwell-Boltzmann formula (8.9) as the equilibrium distribution function, rather than the Bose-Einstein or Fermi-Dirac distributions, suggests that the collision integral (8.3) contains a classical assumption that can stand generalization. By defining the occupation number

$$\mathcal{N}(\mathbf{p}) = h^3 f(\mathbf{p})/(2s+1), \tag{8.10}$$

where s is the spin of the particle (assumed elementary for simplicity), we easily verify that the replacement of the term

$$f(\mathbf{p}_1')f(\mathbf{p}_2') - f(\mathbf{p}_1)f(\mathbf{p}_2)$$

by the expression

$$f(\mathbf{p}_1')f(\mathbf{p}_2')[1 \pm \mathcal{N}(\mathbf{p}_1)][1 \pm \mathcal{N}(\mathbf{p}_2)] - f(\mathbf{p}_1)f(\mathbf{p}_2)[1 \pm \mathcal{N}(\mathbf{p}_1')][1 \pm \mathcal{N}(\mathbf{p}_2')] \tag{8.11}$$

in $(\delta f/\delta t)_c$ will recover the correct distribution under conditions of thermodynamic equilibrium (Bose-Einstein for upper sign, Fermi-Dirac for lower sign). The proof follows by noting that at thermodynamic equilibrium, we require expression (8.11) to equal zero, i.e.,

$$\frac{\mathcal{N}(\mathbf{p}_1')\mathcal{N}(\mathbf{p}_2')}{[1 \pm \mathcal{N}(\mathbf{p}_1')][1 \pm \mathcal{N}(\mathbf{p}_2')]} = \frac{\mathcal{N}(\mathbf{p}_1)\mathcal{N}(\mathbf{p}_2)}{[1 \pm \mathcal{N}(\mathbf{p}_1)][1 \pm \mathcal{N}(\mathbf{p}_2)]},$$

for all collision pairs. This condition requires $\ln[\mathcal{N}/(1 \pm \mathcal{N})]$ to be a sum of the conserved quantities in an elastic collision:

$$\ln\left[\frac{\mathcal{N}(\mathbf{p})}{1 \pm \mathcal{N}(\mathbf{p})}\right] = \beta(\mu - \epsilon + \mathbf{u} \cdot \mathbf{p}),$$

where β, μ, and $\mathbf{u}$ again relate to constants of the system as a whole. In a frame where $\mathbf{u} = 0$, with the identification $\beta = 1/kT$, a little algebra now recovers the Bose-Einstein and Fermi-Dirac distributions:

$$\mathcal{N}(\mathbf{p}) = \frac{1}{e^{(\epsilon-\mu)/kT} \mp 1}.$$

Stimulated (or suppressed) scattering underlies the physics of the replacement recommended by equation (8.11); the presence of a boson (or a fermion) already in the final state stimulates (or suppresses) the scattering there of an identical incident particle by a factor $1 + \mathcal{N}$ (or by $1 - \mathcal{N}$). In other words, the classical assumption that the scattering probability depends only on the products of the two f's of the *initial* states ignores the quantum statistics of the actual particles (bosons or fermions) present in the universe. In particular, if almost all the allowable states are locally filled by a Fermi-Dirac gas (as can happen to the degenerate electrons in a white dwarf), so that the factor $1 - \mathcal{N}$ is nearly zero, the fermion may have to travel a long way before it can scatter to an available unfilled state. The corresponding enhancement of the collisional mean-free-path length can make heat conduction by degenerate electrons competitive with radiative diffusion in the interiors of highly evolved stars (that basically contain white dwarfs at their cores). In Volume II we will see that the "moment equations" associated with Boltzmann's transport equation can form a basis for compressible fluid dynamics. For now, we content ourselves with the demonstration that stimulated effects cancel in the problem of coherent scattering of radiation.

THE ABSENCE OF STIMULATED EFFECTS
IN COHERENT SCATTERING

With the comments of the previous section, we see that we should, in principle, have written the coherent scattering terms in equation (1.19) as

$$\rho \kappa_\nu^{\text{sca}} \oint \phi_\nu(\hat{\mathbf{k}}, \hat{\mathbf{k}}') \left\{ I_\nu(\hat{\mathbf{k}}') \left[1 + \mathcal{N}(\hat{\mathbf{k}}) \right] - I_\nu(\hat{\mathbf{k}}) \left[1 + \mathcal{N}(\hat{\mathbf{k}}') \right] \right\} d\Omega, \quad (8.12)$$

where we define the photon occupation number in each state of polarization by the symbol

$$\mathcal{N}(\hat{\mathbf{k}}) = \left(\frac{c^2}{2h\nu^3} \right) I_\nu(\hat{\mathbf{k}}), \quad (8.13)$$

with a similar expression when the propagation direction is $\hat{\mathbf{k}}'$ instead of $\hat{\mathbf{k}}$. In equation (8.12) the term $I_\nu(\hat{\mathbf{k}}')[1 + \mathcal{N}(\hat{\mathbf{k}})]$ represents the stimulated scattering of photons from any direction $\hat{\mathbf{k}}'$ into the beam direction $\hat{\mathbf{k}}$; whereas the term $I_\nu(\hat{\mathbf{k}})[1 + \mathcal{N}(\hat{\mathbf{k}}')]$ represents the stimulated enhancement of scattering of photons out of the beam direction $\hat{\mathbf{k}}$ into any other direction $\hat{\mathbf{k}}'$. We have also assumed time-reversal symmetry:

$$\phi_\nu(\hat{\mathbf{k}}', \hat{\mathbf{k}}) = \phi_\nu(\hat{\mathbf{k}}, \hat{\mathbf{k}}').$$

If we substitute equation (8.13) into (8.12), we obtain an exact cancelation of the terms proportional to $I_\nu(\hat{\mathbf{k}})I_\nu(\hat{\mathbf{k}}')$. If we further use equation (6.18), we recover the coherent scattering terms as they were naively written down in equation (1.19). Because as many photons are stimulated to scatter into any beam as out of it, stimulated effects yield no net consequence for *coherent scattering*. The same result does not hold if the scattering (say, by energetic electrons) removes or adds energy to the photons; the inclusion of changes in the frequency of the scattered photons (important for some considerations in x-ray astronomy) leads to the so-called *Kompaneet's equation*. (See Rybicki and Lightman 1979 for further discussion of this point.) For the development followed in this chapter, we move on now to extend the principle of detailed balance to inelastic collisions and photoprocesses.

INELASTIC COLLISIONS

Consider the level populations n_1 and n_2 of atoms (or molecules) in two energy levels $E_1 < E_2$. In thermodynamic equilibrium, the relative populations can be maintained, for example, by detailed balance between inelastic collisions and superelastic collisions with thermal electrons:

$$n_1\sigma_{12}[n_e v f_e(v)\, 4\pi v^2 dv] = n_2\sigma_{21}[n_e v' f_e(v')\, 4\pi v'^2 dv'], \qquad (8.14)$$

where σ_{12} and σ_{21} are the excitation and de-excitation cross sections and where we have ignored the motion of the atoms in comparison with those of the much lighter electrons. (With two Maxwellians, we get formulae involving reduced masses and center-of-mass energies.) We assume nonrelativistic conditions and adopt $f_e(v)$ to be the *velocity distribution* normalized to unity:

$$f_e(v) = (m_e/2\pi kT)^{3/2}\exp(-m_e v^2/2kT). \qquad (8.15)$$

Consistent with this assumption, thermodynamic equilibrium requires

$$n_2/n_1 = (g_2/g_1)e^{-E_{21}/kT}. \qquad (8.16)$$

Energy conservation relates v and v':

$$\frac{1}{2}m_e v^2 = \frac{1}{2}m_e v'^2 + E_{21}, \qquad (8.17)$$

where

$$E_{21} \equiv E_2 - E_1.$$

If we substitute these expressions into equation (8.14), we obtain the desired relationship between excitation and de-excitation cross sections,

$$g_1\epsilon\sigma_{12}(\epsilon) = g_2\epsilon'\sigma_{21}(\epsilon'), \qquad (8.18)$$

where we have allowed the cross sections to be dependent on the kinetic energy ϵ of the colliding electron, with the relation (8.17) translating into

$$\epsilon = \epsilon' + E_{21}. \qquad (8.19)$$

Notice that equation (8.18) implies that the excitation cross section σ_{12} by electron impact is zero until a threshold energy E_{21} is reached, but no such threshold holds, of course, for the de-excitation cross section σ_{21}.

Notice also that although we have used thermodynamic equilibrium populations and the argument of detailed balance to derive equation (8.18), the final relationship does not depend on the temperature T. Indeed, since σ_{12} and σ_{21} are *atomic constants*, they *cannot* depend on any of the statistical properties of the colliding particles, and equation (8.18) must hold *a priori*, independently of whether or not the impacting electrons even have a thermal distribution (Maxwellian or Fermi-Dirac).

EINSTEIN COEFFICIENTS FOR BOUND-BOUND PHOTOPROCESSES

In 1916 Einstein provided a brilliant new derivation of Planck's radiation formula for a blackbody that anticipated later developments in quantum electrodynamics by many years. Einstein's idea was that detailed balance involving radiative processes required *three* terms, not two. In addition to spontaneous emission and true absorption, there must also exist *stimulated emission*. (Recall that the distinction between bosons and fermions did not become clear until Dirac's work in 1926.) Consider *bound-bound transitions* involving atoms with level populations n_1 and n_2 per unit volume in the bound states 1 and 2. In the presence of a radiation field I_ν, with mean intensity J_ν, Einstein proposed that, averaged over all angles, the rate per unit volume of upward transitions from 1 to 2 by *photoabsorption* equals

$$n_1 B_{12} J_\nu, \tag{8.20}$$

where energy conservation for the transition requires (from Einstein's analysis of the photoelectric effect and Bohr's analysis of the atom):

$$h\nu = E_2 - E_1 \equiv h\nu_{12}.$$

On the other hand, downward transitions occur by two processes, *spontaneous* emission at a rate (per unit volume) given by

$$n_2 A_{21}, \tag{8.21}$$

and *stimulated* (or *induced*) emission at a rate (per unit volume) given by

$$n_2 B_{21} J_\nu. \tag{8.22}$$

The quantities A_{21}, B_{21}, and B_{12} are called Einstein's coefficients for spontaneous emission, stimulated emission, and photoabsorption. We treat the line absorption cross section as a delta function in frequency: $\alpha_\nu = a_{12}\delta(\nu - \nu_{12})$ for the atom, with

$$a_{12} \equiv \frac{\pi e^2}{m_e c} f_{12}, \tag{8.23}$$

where f_{12} equals a dimensionless number called the *oscillator strength* of the line for reasons that will become apparent in later chapters. (Notice the units: $[\alpha_\nu] = \text{cm}^2$ per atom, and $[a_{12}] = \text{cm}^2\,\text{Hz}$.) We now have

$$B_{12} = 4\pi a_{12}/h\nu, \qquad (8.24)$$

since $4\pi \int \alpha_\nu (J_\nu/h\nu)\, d\nu$ equals the total rate of absorption of photons by an atom. In practice, the uncertainty principle (if nothing else) gives actual absorption processes finite linewidth, but we shall consider this refinement at a later juncture. In any case, like B_{12}, A_{21} and B_{21} must also be properties solely of the atom and cannot depend on the statistical properties of the material system or of the ambient radiation field.

In thermodynamic equilibrium, $J_\nu = B_\nu(T)$, with

$$B_\nu(T) = \frac{2h\nu^3/c^2}{e^{h\nu/kT} - 1}, \qquad (8.25)$$

while n_2/n_1 is given by Boltzmann's law:

$$n_2/n_1 = (g_2/g_1)e^{-h\nu/kT}. \qquad (8.26)$$

For detailed balance, we set equation (8.20) equal to the sum of equations (8.21) and (8.22). After a little manipulation, we obtain

$$(g_2/g_1)A_{21}(e^{h\nu/kT} - 1) = (2h\nu^3/c^2)[e^{h\nu/kT}B_{12} - (g_2/g_1)B_{21}]. \qquad (8.27)$$

Since A_{21}, B_{21}, and B_{12} are atomic constants independent of T, we require

$$A_{21} = (g_1/g_2)(2h\nu^3/c^2)B_{12}, \qquad (8.28)$$

$$B_{21} = (2h\nu^3/c^2)^{-1}A_{21} = (g_1/g_2)B_{12}, \qquad (8.29)$$

which, with equations (8.23) and (8.24), gives all three Einstein coefficients in terms of the f-value, f_{12}.

Although we have used thermodynamic conditions to make our arguments, the relations (8.28), (8.29), and (8.24) have general validity [independent of, e.g., equation (8.26)] and must result (to all orders of perturbation theory) from a direct quantum-mechanical calculation of the microprocesses. In particular, when we do consider the quantum theory of radiation processes later in this text, we will see that the ability to create photons spontaneously where none existed before requires, at critical junctures, a replacement of the photon occupation number $\mathcal{N}$ that results from a semiclassical treatment of electromagnetic fields by the term $(1 + \mathcal{N})$. From a field-theoretic point of view, then, the new term is *spontaneous emission*, not stimulated emission. It is only everyday intuition that finds the latter term more surprising than the former. Moreover, for a fully quantized formulation of radiation processes (not covered in this text), spontaneous and stimulated emission enter automatically without being fed in by hand; so detailed balance really does involve equating only one forward rate and one backward rate.

CORRECTION FOR STIMULATED EMISSION

If we pretend that stimulated emission behaves like negative absorption, we subtract expression (8.22) from expression (8.20) (actually, their non-angle-averaged versions) and obtain the net term as

$$B_{12}[1 - (g_1 n_2 / g_2 n_1)] I_\nu,$$

where we have used equation (8.29). Examining equation (8.24), we see therefore that "correction for stimulated emission" involves replacing the contribution of the true absorption α_ν by the recipe

$$\alpha_\nu^c = \alpha_\nu (1 - g_1 n_2 / g_2 n_1). \tag{8.30}$$

If in addition, we make the LTE assumption, equation (8.26), we see that the recipe boils down to multiplying the absorption opacity by the factor $(1 - e^{-h\nu/kT})$ for all transitions and frequencies where the LTE assumption has validity. On the other hand, if non-LTE conditions produce a "population inversion," with $n_2/g_2 > n_1/g_1$ and $\alpha_\nu^c < 0$, then masing can take place with exponential amplification $e^{|\tau_\nu|}$ of radiation rather than exponential attenuation $e^{-|\tau_\nu|}$.

EINSTEIN-MILNE RELATIONS

As an example of photoprocesses involving transitions to and from the continuum, we consider the radiative destruction and formation of H^-:

$$H^- + h\nu \rightleftharpoons H(1) + e^-. \tag{8.31}$$

The discussion of this reaction is particularly simple because we need to deal with only one bound state on each side and the continuum. We wish to derive the relationships between the Einstein-Milne coefficients $\mathcal{A}$ and $\mathcal{B}$ associated with spontaneous and induced emission to the continuum (with the recombination of H and e to H^-) and the bound-free absorption coefficient p_ν. In thermodynamic equilibrium at temperature T, we have detailed balance:

$$n(H^-) p_\nu B_\nu(T) \, d\nu = [v f_e(v) \, 4\pi v^2 \, dv] n_e n(H[1]) [\mathcal{A} + \mathcal{B} B_\nu(T)], \tag{8.32}$$

where p_ν is related to the photoabsorption cross section α_ν (per H^- ion):

$$p_\nu = 4\pi\alpha_\nu / h\nu, \tag{8.33}$$

and where energy conservation requires

$$h\nu = \chi + m_e v^2 / 2, \tag{8.34}$$

with $\chi = 0.754$ eV. Note that $h\,d\nu = m_{\mathrm{e}}v\,dv$. If we substitute in equation (8.32) the expressions for $f_{\mathrm{e}}(v)$, $B_\nu(T)$, and $n(\mathrm{H}[1])/n(\mathrm{H}^-)$ given by the Maxwell-Boltzmann, Planck, and Saha formulae, equations (8.15), (1.14), and (7.31), we obtain, upon cancelation of common factors,

$$p_\nu \left(\frac{2h\nu^3}{c^2}\right) = \frac{g_1}{g_{\mathrm{H}^-}} \left(\frac{m_{\mathrm{e}}}{h}\right)^2 8\pi v^2 \left[\mathcal{A}\left(1 - e^{-h\nu/kT}\right) + \mathcal{B}\left(\frac{2h\nu^3}{c^2}e^{-h\nu/kT}\right)\right].$$

(8.35)

But $\mathcal{A}$, $\mathcal{B}$, and p_ν are *atomic* constants independent of T; therefore,

$$\mathcal{A} = \frac{2h\nu^3}{c^2}\mathcal{B},$$

(8.36)

$$\mathcal{B} = \left[\left(\frac{h^2}{16\pi m_{\mathrm{e}}}\right)\frac{g_{\mathrm{H}^-}}{g_1}\right]\left(\frac{p_\nu}{\epsilon}\right).$$

(8.37)

In equation (8.37) ϵ is the electron kinetic energy and is related to the photon energy $h\nu$ via

$$\epsilon = h\nu - \chi.$$

(8.38)

Equations (8.36) and (8.37) constitute the desired Einstein-Milne relations that allow us to express any of the rates in terms of one cross section α_ν [see equation (8.33)]. They apply to the photoionization of any bound state (replacing H^-) to an ionized state (replacing $\mathrm{H}[1]$) if we simply use the appropriate statistical weights and photoabsorption cross section α_ν.

The correction for stimulated emission can now be written as

$$\alpha_\nu^{\mathrm{c}} = \alpha_\nu[1 - (b_{\mathrm{H}^-})^{-1}e^{-h\nu/kT}],$$

(8.39)

where the *departure coefficient* for H^- is defined by

$$b_{\mathrm{H}^-} \equiv [n_{\mathrm{H}^-}g_{\mathrm{H}(1)}\mathcal{Z}_{\mathrm{e}}/n_{\mathrm{e}}n_{\mathrm{H}(1)}g_{\mathrm{H}^-}]e^{-\chi/kT},$$

(8.40)

and equals the ratio of the actual abundance of H^- to the LTE value that it would have if it satisfied Saha's equation (7.31). In LTE, $b_{\mathrm{H}^-} = 1$, and the correction for stimulated emission amounts to the usual multiplication by $1 - e^{-h\nu/kT}$. If non-LTE effects led to a small departure coefficient at reasonable T, a continuum laser could result in principle, but the effect might be unnoticeable if $\alpha_\nu^{\mathrm{c}}n_{\mathrm{H}^-}$ is small (although negative).

9

Fluid Description, Rate Equations, and Sobolev Approximation

Reference: Mihalas and Mihalas, *Foundations of Radiation Hydrodynamics*; G. Rybicki and D. Hummer 1983, *Ap. J.*, **274**, 380.

In this chapter, we derive the general form of the governing equations for competing rate processes. Concurrently, we obtain an outline of what it takes to complete a specification of the state of the combined matter and radiation fields of gaseous astrophysical systems. Finally, we consider the problem of the formation of spectral lines in a moving medium. We begin with a discussion of those aspects of the problem where we expect LTE still to hold.

LOCAL MAXWELLIAN VELOCITY DISTRIBUTIONS

Under almost all circumstances, mutual collisions will establish thermal distributions for the translational degrees of freedom of a gas. Because photons carry relatively little momentum compared to matter (until we get to gamma rays), radiation does not couple well with the external degrees of freedom in a direct sense; soft photons interact primarily with the internal states. The configuration of internal states may affect the external ones via inelastic and superelastic collisions (see Chapter 8)—providing cooling or heating agents for the gas—but on a more basic level, inelastic collisions (or, more accurately, the corresponding rates) usually cannot compete with elastic scatterings in determining the functional form of the distribution function. (As a homespun example, consider the fact that each air molecule in this room suffers $\sim 10^9$ collisions per second. If a significant fraction of these collisions were inelastic—leading to an eventual radiative loss of heat—the air would cool down almost instantaneously. Since it does not do so, elastic collisions must vastly dominate inelastic ones.) The claim holds especially strongly in astrophysics for free electrons; their frequent Coulomb interactions with other charged species (but primarily with themselves) generally allow us safely to assume that they have Maxwellian distributions.

For neutral atoms or molecules, Boltzmann's transport equation (8.2) yields the same conclusion. Suppose the distribution function f departed significantly from a thermal distribution. Then the time-rate of change of f because of elastic collisions would have the order of magnitude

$$\left(\frac{\delta f}{\delta t}\right)_c \sim n\sigma\langle w\rangle f,$$

where σ is a typical elastic scattering cross section and $\langle w\rangle$ is a typical random speed of encounter. In kinetic theory, we refer to the combination

$$\nu_c \equiv n\sigma\langle w\rangle \qquad (9.1)$$

as the *collision frequency*. For neutral atoms and molecules in even as rarefied an environment as an interstellar cloud, n might be $\sim 10^3\,\mathrm{cm}^{-3}$, $\sigma \sim 10^{-15}\,\mathrm{cm}^2$, and $\langle w\rangle \sim 10^5\,\mathrm{cm\ s}^{-1}$; thus the collision frequency $\nu_c \sim 10^{-7}$ s^{-1}, or one collision every four months. Four months sounds very long to go without an encounter, but it is in fact much shorter than the typical evolutionary lifetimes of gas clouds (of the order of 10^7 yr), or the intervals between processes (e.g., the passage of a shockwave) which can introduce strong departures from Maxwellian distributions for the thermal gas:

$$f\,d^3p = n(2\pi kT/m)^{-3/2}\exp(-m|\mathbf{v}-\mathbf{u}|^2/2kT)\,d^3v. \qquad (9.2)$$

In equation (9.2) we have specialized to nonrelativistic and nondegenerate conditions with $\mathbf{p} = m\mathbf{v}$. We have also allowed the possibility that the gas has a mean velocity $\mathbf{u}$ relative to the lab frame. In this notation, $\mathbf{v}$ is the total particle speed; $\mathbf{u}$, the bulk (or fluid) velocity; and $\mathbf{w} \equiv \mathbf{v} - \mathbf{u}$, the random velocity.

The Maxwellians so established will generally hold only locally, in the sense that the density n, bulk (or fluid) velocity $\mathbf{u}$, and temperature T can vary with position $\mathbf{x}$ and time t. To smooth out unbalanced pressure differences, for example, over a length scale L will take a typical hydrodynamical time

$$t_{\mathrm{dyn}} \equiv L/\langle w\rangle. \qquad (9.3)$$

(We do not bother to distinguish between such niceties as $\langle w\rangle$ and $\langle w^2\rangle^{1/2}$ at the present crude level of discussion.) With a typical length scale $L \sim 10^{18}$ cm and $\langle w\rangle \sim 10^5\,\mathrm{cm\ s}^{-1}$, $t_{\mathrm{dyn}} \sim 10^{13}$ s, about 3×10^5 yr, much longer than the aforementioned 4 months. To smooth out temperature gradients by heat conduction (by random particle motions) will take even longer, typically a diffusive time (see Chapter 2)

$$t_{\mathrm{diff}} \equiv L^2/\mathcal{D}, \qquad (9.4)$$

where $\mathcal{D} \sim \ell\langle w\rangle/3$, with ℓ being the mean-free-path length for particle collisions, $\ell \equiv 1/n\sigma$. Thus t_{diff} is longer than even the hydrodynamical time t_{dyn}, roughly by an additional factor L/ℓ ($\sim 10^6$ in this example).

FLUID DESCRIPTION AND RATE EQUATIONS

The fact that $\ell \ll L$ allows the possibility of a *fluid description*, a great simplification over a kinetic approach based on a direct attack on the Boltzmann equation. In Volume II we will show how a systematic expansion in the small parameter ℓ/L (the Chapman-Enskog procedure) allows a contraction of equation (8.2) to a manageable set of fluid equations. For example, we will derive explicit expressions for terms like $\mathbf{f}_{\rm vis}$, Ψ, and $\mathbf{F}_{\rm cond}$ that appear below. At present, we merely assume the intuitive validity of the resulting expressions, and comment that the existence of a fluid velocity $\mathbf{u}$ common to all species allows us to write a generic rate equation for each species a:

$$\frac{\partial n^{(a)}}{\partial t} + \boldsymbol{\nabla} \cdot (n^{(a)}\mathbf{u}) = \mathcal{P}^{(a)} - \mathcal{D}^{(a)}, \tag{9.5}$$

where the rates per unit volume $\mathcal{P}^{(a)}$ equals the sum of all the ways in which species a can be produced, and $\mathcal{D}^{(a)}$, of all the ways in which it can be destroyed. [In the presence of large-scale electromagnetic fields, we may need to relax the assumption that all species move alike. For such fields, we may have to introduce the concept of different fluid velocities for electrons, ions, and neutrals, and write down separate equations of motion (see below) for each. We defer consideration of this problem (*magnetohydrodynamics*) until Volume II. Finally, we may also need to consider that dust may move differently from gas; this constitutes a specialized topic best left to specialized discussions.]

When LTE for the species a (which might simply represent an excited state of a certain element) cannot be assumed, equation (9.5) replaces the appropriate generalized law of mass action for that species. Conversely, LTE conditions typically apply when that species couples sufficiently strongly to the translational degrees of freedom (assumed to exist at a temperature T) so that the corresponding production and destruction terms on the right-hand side outweigh any other contributions on either side of the equation and must therefore nearly balance each other.

We can obtain the mass density ρ of the entire fluid from a sum over all species,

$$\rho = \sum_a m^{(a)} n^{(a)}. \tag{9.6}$$

We often find it convenient to introduce the concept of a mean molecular mass m defined by

$$m \equiv \frac{\sum_a m^{(a)} n^{(a)}}{\sum_a n^{(a)}}, \tag{9.7}$$

but we need to remember that m may not stay sensibly constant if appreciable changes in $n^{(a)}$ occur.

EQUATIONS THAT GOVERN THE BULK FLUID PROPERTIES

We may schematically write down the equations that govern the motion of a fluid by applying the generic equation (2.6). In particular, (non-relativistic) mass conservation for the system requires

$$\frac{\partial \rho}{\partial t} + \nabla \cdot (\rho \mathbf{u}) = 0. \tag{9.8}$$

Occasionally equation (9.8) is written in the form

$$\rho^{-1} \frac{D\rho}{Dt} = -\nabla \cdot \mathbf{u}, \tag{9.9}$$

where D/Dt equals the *substantial* (or Lagrangian) derivative that follows the motion of a fluid element:

$$\frac{D}{Dt} \equiv \frac{\partial}{\partial t} + \mathbf{u} \cdot \nabla. \tag{9.10}$$

Equation (9.9) demonstrates that the divergence of the velocity field, $\nabla \cdot \mathbf{u}$, yields the fractional rate of change of the volume per unit mass ρ^{-1}.

The fluid velocity satisfies a momentum conservation relation,

$$\frac{\partial}{\partial t}(\rho \mathbf{u}) + \nabla \cdot (\rho \mathbf{u}\mathbf{u}) = -\nabla P + \mathbf{f}_{\text{vis}} + \mathbf{f}_{\text{rad}} + \mathbf{f}_{\text{ext}}, \tag{9.11}$$

where P is gas pressure, $\mathbf{f}_{\text{vis}}$ represents the possible presence of internal viscous forces per unit volume, $\mathbf{f}_{\text{rad}}$, the force per unit volume due to radiation, and $\mathbf{f}_{\text{ext}}$, the force per unit volume due to all other agents (e.g., gravity). If the translational degrees of freedom for each species a possess a local Maxwellian distribution at a common temperature T, the total gas pressure P satisfies

$$P = \sum_a P^{(a)}, \tag{9.12}$$

where $P^{(a)}$ gives the partial pressure associated with species a:

$$P^{(a)} = n^{(a)} kT. \tag{9.13}$$

Notice that we may also write equation (9.12) as

$$P = \rho \mathcal{R} T \quad \text{where} \quad \mathcal{R} \equiv k/m, \tag{9.14}$$

but the "gas constant" $\mathcal{R}$ would not be a true constant if m varied.

Consistent with the assumption that each species has a Maxwellian velocity distribution, we may take the part of the internal energy density of the gas associated with the translational degrees of freedom *only* to equal

$$\mathcal{E}_{\text{trans}} = \frac{3}{2} P. \tag{9.15}$$

This internal energy satisfies a restricted version of the first law of thermodynamics:

$$\frac{\partial \mathcal{E}_{\text{trans}}}{\partial t} + \boldsymbol{\nabla} \cdot (\mathcal{E}_{\text{trans}} \mathbf{u}) = -P\boldsymbol{\nabla} \cdot \mathbf{u} - \boldsymbol{\nabla} \cdot \mathbf{F}_{\text{cond}} + \Psi + \Gamma_{\text{trans}} - \Lambda_{\text{trans}}, \quad (9.16)$$

where $\mathbf{F}_{\text{cond}}$ is the heat flux carried by matter conduction, Ψ is the volumetric rate of heat production by the viscous dissipation of ordered motions (if $\mathbf{u}$ has spatial gradients), Γ_{trans} is the volumetric rate of gain of heat into the system by all other agents, and Λ_{trans} is the corresponding volumetric loss of heat. The terms Γ_{trans} and Λ_{trans} include only the contributions (e.g., due to interactions with radiation) that add or subtract energy from the translational degrees of freedom. In equation (9.16), $P\boldsymbol{\nabla} \cdot \mathbf{u}$ represents the rate of doing work by the gas.

In principle, a similar equation could also be written down for the internal energy per unit volume $\mathcal{E}_{\text{int}}$ associated with the internal degrees of freedom:

$$\frac{\partial \mathcal{E}_{\text{int}}}{\partial t} + \boldsymbol{\nabla} \cdot (\mathcal{E}_{\text{int}} \mathbf{u}) = \Gamma_{\text{int}} - \Lambda_{\text{int}}. \quad (9.17)$$

If the internal degrees of freedom existed also in LTE, we would gain some savings (implicitly adopted as the convention in the theory of stellar interiors and, sometimes, stellar atmospheres) by adding equations (9.16) and (9.17):

$$\frac{\partial \mathcal{E}}{\partial t} + \boldsymbol{\nabla} \cdot (\mathcal{E} \mathbf{u}) = -P\boldsymbol{\nabla} \cdot \mathbf{u} - \boldsymbol{\nabla} \cdot \mathbf{F}_{\text{cond}} + \Psi + \Gamma - \Lambda, \quad (9.18)$$

where $\mathcal{E} \equiv \mathcal{E}_{\text{trans}} + \mathcal{E}_{\text{int}}$, with similar expressions for Γ and Λ.

The convenience of the last procedure rests with cases when the internal degrees of freedom also exist in LTE. For such systems, the fundamental law of thermodynamics holds:

$$de = Tds - Pd(\rho^{-1}) + \rho^{-1} \sum_a \mu^{(a)} \, dn^{(a)}, \quad (9.19)$$

where ρ^{-1} is the specific volume (volume per unit mass), $e = \rho^{-1}\mathcal{E}$ is the specific internal energy, and s is the specific entropy (including translational and internal degrees of freedom). If the relative abundances of the different species satisfy the generalized law of mass action, then equation (9.19) becomes

$$de = Tds - Pd(\rho^{-1}). \quad (9.20)$$

We may now use the equation of continuity (9.8) to rewrite (9.18) as the matter heat equation:

$$\rho T \frac{Ds}{Dt} = -\boldsymbol{\nabla} \cdot \mathbf{F}_{\text{cond}} + \Psi + \Gamma - \Lambda. \quad (9.21)$$

If, in addition, radiation constitutes the only source and sink of energy exchange with the matter (i.e., if $\Gamma = \Gamma_{\text{rad}}$ and $\Lambda = \Lambda_{\text{rad}}$), and if the radiation field exists in equilibrium at the same temperature T, then we may write the zeroth moment equation for it (allowing for the presence of matter motion) as

$$\rho T \frac{Ds_{\text{rad}}}{Dt} = -\mathbf{\nabla} \cdot \mathbf{F}_{\text{rad}} - \Gamma_{\text{rad}} + \Lambda_{\text{rad}}, \qquad (9.22)$$

where $s_{\text{rad}} = 4aT^3/3\rho$ is the entropy of blackbody radiation per unit mass of fluid, and $\Lambda_{\text{rad}} = \rho j$ and $\Gamma_{\text{rad}} = \rho \kappa_P caT^4$ (with κ_P equal to the Planck-mean opacity corrected for stimulated emission) are the radiative emissivity and absorptivity of the matter per unit volume. Equation (9.22) is the generalization of equations (3.11) and (3.12) when we wish to follow the system over such long time scales (e.g., stellar evolutionary ages) that radiative equilibrium no longer holds exactly and the distribution of matter does not remain perfectly static. Nevertheless, in Chapter 2 we pointed out the near equality between the very large terms Λ_{rad} and Γ_{rad} in the interiors of stars; consequently, it is never a good idea to try to subtract these terms directly to deduce the slow evolution of any very optically thick medium (where matter and radiation have good thermal coupling). On the other hand, if we add equations (9.21) and (9.22), we obtain exact cancelation of the troublesome terms and the total heat equation,

$$\rho T \frac{Ds_{\text{tot}}}{Dt} = -\mathbf{\nabla} \cdot (\mathbf{F}_{\text{cond}} + \mathbf{F}_{\text{rad}}), \qquad (9.23)$$

where $s_{\text{tot}} \equiv s + s_{\text{rad}}$. Equation (9.23) contains no terms that are especially difficult to treat, and its generalization in what follows constitutes another of the standard equations used in stellar evolution theory [where we use Lagrangian mass coordinates to handle the derivative D/Dt, and where we want to eliminate all other references to the Eulerian velocity $\mathbf{u}$, such as the term $-P\mathbf{\nabla} \cdot \mathbf{u}$ in the internal energy equation (9.18)].

To allow for the possibility of species abundances and nuclear reactions that do not satisfy the conditions of LTE, we generalize equation (9.23) to the form

$$\rho T \frac{Ds_{\text{tot}}}{Dt} = -\mathbf{\nabla} \cdot (\mathbf{F}_{\text{cond}} + \mathbf{F}_{\text{rad}}) + \rho \varepsilon - \sum_a \mu^{(a)} \frac{Dn^{(a)}}{Dt}, \qquad (9.24)$$

with ε being the net rate of nuclear release of energy per unit mass (i.e., $\rho \varepsilon = \Gamma_{\text{nuc}} - \Lambda_{\text{nuc}}$) available to heat *both* the gas and radiation field. In equation (9.24), the change of the "chemical energy" of species a, $\mu^{(a)} Dn^{(a)}/Dt$, does not take into account the nuclear degrees of freedom, which have been incorporated into $\rho \varepsilon$. We normally do not see the remaining equation of stellar-interiors theory written down in the form of equation (9.24) because usually when nuclear reactions are important (e.g., stars in stable phases of nuclear burning), the terms containing time derivatives are small; whereas, when the release of stored heat $\rho T Ds_{tot}/Dt$ is all-important (e.g, stars in the pre-main-sequence phase of gravitational contraction), few nuclear reactions occur either to contribute appreciably to $\rho\varepsilon$ or to make $\sum_a \mu^{(a)} Dn^{(a)}/Dt$ significantly different from zero. All the terms in equation (9.24) come fully into play only during the last stages of the evolution of a presupernova star, when everything happens extremely quickly.

Obvious advantages accrue to *not* carrying out the addition of the separate equations for matter and radiation for applications to other situations— e.g., the interstellar medium. As long as LTE still holds for the translational degrees of freedom, however, we may still apply partial thermodynamic arguments. (Other fields also follow this practice; for example, in treating chemical equilibria, we apply thermodynamics to the electronic configurations of the participating atoms, but not to their nuclear states.) Thus, if we define e_{trans} as the internal energy per unit mass associated with the translational degrees of freedom alone,

$$e_{trans} \equiv \rho^{-1}\mathcal{E}_{trans} = \frac{3P}{2\rho}, \tag{9.25}$$

we may use the equation of continuity (9.8) to rewrite equation (9.16) as

$$\frac{3}{2}P\frac{D}{Dt}\left[\ln\left(\frac{P}{\rho^{5/3}}\right)\right] = -\boldsymbol{\nabla}\cdot F_{cond} + \Psi + \Gamma_{trans} - \Lambda_{trans}. \tag{9.26}$$

If, in addition, the mean molecular weight m suffered no variations (i.e., if the changes in $n^{(a)}$ were negligible), the left-hand side of equation (9.26) would also equal $\rho T Ds_{trans}/Dt$, where s_{trans} is the sum of the translational specific entropies of all species; however, the validity of equation (9.26) does not depend on this further assumption.

SUMMARY

Given expressions for $\mathbf{f}_{vis}$, $\mathbf{F}_{cond}$, and Ψ (Volume II) and for $\mathbf{f}_{rad}$, $\mathcal{P}^{(a)}$, and $\mathcal{D}^{(a)}$ (previous chapters), as well as Γ_{trans}, Λ_{trans}, and possibly Γ_{int}, Λ_{int} (next chapter), we have a complete set of equations for the solution of the physical and dynamical state of the gas. Since the matter exchanges both energy and momentum with the radiation, we will generally need to give a simultaneous treatment of the equation of radiative transfer in order to compute the terms involving photoprocesses in Γ_{trans}, etc. In its full glory, therefore, the combined set of equations for matter *and* radiation forms a closed but formidable system of nonlinearly coupled relations. Rarely is this system attacked in complete generality. As a rule of thumb, people who wish to understand the radiative transport in some detail make simplifying assumptions for the matter field—e.g., a static medium, or one in LTE, or one having appreciable spatial symmetry—whereas people who have a greater interest in the dynamics often adopt simplifying assumptions concerning the thermal consequences of radiative coupling (or lack of it)—e.g., isothermal or adiabatic variations, or optically thin conditions, or LTE abundances. The combined solution of both aspects of the problem, treated with relatively complete realism, constitutes one of the forefront research areas of computational astrophysics—a subject known as *radiative hydrodynamics*. In our formal development, we have now finished deriving the equations that govern the complete problem (except for generalizations to allow near-relativistic velocities). We will give some examples, but defer discussion of specific examples to specialized texts.

RADIATIVE TRANSFER FOR SPECTRAL LINES
IN RAPIDLY MOVING MEDIA

One subject for which fluid motion clearly represents a very important effect concerns the formation of spectral lines. In a static rarefied gas, the local source contribution to the width of an observed spectral line involves a convolution of the natural Lorentz profile (see Chapter 23) and the Gaussian function associated with the thermal motions of individual atoms or molecules. In higher-pressure environments, other line-broadening mechanisms—such as the Stark effect—may come into play. Also important under many circumstances may be chaotic macroscopic motions generically labeled "turbulence" in the astrophysical literature.

On the face of it, including the possibility of nonzero bulk motion only adds to the complications. In fact, when the macroscopic fluid motions are systematic and dominate the random microscopic motions, the problem of radiative transfer in spectral lines can be greatly simplified. The properties of the radiation field at any point in space usually depend on the conditions applicable at distant regions with which the former couple by direct photon propagation. If, however, the distant regions move at substantially different velocities, the Doppler effect acting on line photons emitted there will shift the frequency outside the resonant range where they can effectively interact with the local atoms or molecules. Under certain monotonicity conditions for the flow field (to be stated below), the *Sobolev approximation* applies, and it becomes possible to analyze the radiative transport as if all the contributions to the formation of a spectral line at any given frequency occurred *locally*. The resultant spatial decoupling introduces considerable analytical advantages that we shall now proceed to exploit.

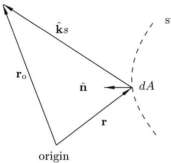

FIGURE 9.1
Geometry for line transfer problem in a moving medium: $\mathbf{r}_o$ represents the observer's position; $\mathbf{r}$, the source position where emitting atoms or molecules move at velocity $\mathbf{u}(\mathbf{r})$; $\hat{\mathbf{k}}$ the direction of photon propagation along a ray path s that connects $\mathbf{r}$ and $\mathbf{r}_o$. Contributions to a given observed frequency ν come from a surface of constant line-of-sight velocity $\hat{\mathbf{k}} \cdot \mathbf{u}$.

MONOTONIC FLOW FIELDS

Consider line radiation from gas at position $\mathbf{r}$ moving at speed $\mathbf{u}$ (see Figure 9.1). Let $\hat{\mathbf{k}}$ denote the propagation direction for photons from $\mathbf{r}$ to reach the position of the observer, $\mathbf{r}_o$, with s being the distance between these two points:

$$\mathbf{r}_o = \mathbf{r} + \hat{\mathbf{k}}s. \tag{9.27}$$

Let ℓ be the length scale over which the differential velocity along a characteristic line of sight varies by an amount equal to the thermal velocity v_T: $\ell = v_T/\hat{\mathbf{k}} \cdot \boldsymbol{\nabla}(\hat{\mathbf{k}} \cdot \mathbf{u})$; L, the length scale over which a fluid variable like the density or the velocity varies by an amount comparable to itself. If $\ell \ll L$ and $|\mathbf{u}| \ll c$, we may suppose the moving gas to emit line radiation at a single frequency given by the nonrelativistic formula

$$\nu = \nu_0 \left(1 + \frac{\hat{\mathbf{k}} \cdot \mathbf{u}}{c}\right), \tag{9.28}$$

where ν_0 equals the rest frequency. If we regard ν as the independent variable rather than $\mathbf{r}$, we denote the position $\mathbf{r}$ where equation (9.28) holds by $\mathbf{r}_\nu$. For a purely contractive or purely expansive flow (as might occur in a stellar wind or a supernova explosion), Rybicki and Hummer showed in a 1978 paper generalizing earlier work by Sobolev, Castor, and others, that there exists at most only one value of the separation distance s along any given line of sight $\hat{\mathbf{k}}$, such that the Doppler-shifted frequency given by equation (9.28) will have a specified value ν. In other words, the root $\mathbf{r}_\nu$ is either unique or nonexistent.

Proof: Consider the change in line frequency ν as we move along the line of sight:

$$\frac{d\nu}{ds} = \frac{\nu_0}{c}\hat{\mathbf{k}} \cdot \boldsymbol{\nabla}(\hat{\mathbf{k}} \cdot \mathbf{u}). \tag{9.29}$$

We wish to investigate the flow conditions under which $d\nu/ds$ stays systematically either positive or negative for all s. For such a case, the curve $\nu(s)$ plotted against s can cross a given value ν only once. We concern ourselves only with the limit when the observer's position lies essentially at an infinite distance from all sources (see below). In this limit, $\hat{\mathbf{k}}$ equals a *constant* vector, and we may use Cartesian tensors to write

$$\hat{\mathbf{k}} \cdot \boldsymbol{\nabla}(\hat{\mathbf{k}} \cdot \mathbf{u}) = \hat{k}_i \hat{k}_j \frac{\partial u_i}{\partial x_j}.$$

Split the rate of strain tensor $\partial u_i / x_j$ into its symmetric and antisymmetric parts:

$$\frac{\partial u_i}{\partial x_j} = S_{ij} + A_{ij},$$

where

$$S_{ij} \equiv \frac{1}{2}\left(\frac{\partial u_i}{\partial x_j} + \frac{\partial u_j}{\partial x_i}\right), \qquad A_{ij} \equiv \frac{1}{2}\left(\frac{\partial u_i}{\partial x_j} - \frac{\partial u_j}{\partial x_i}\right).$$

Except for the further removal of the trace, one refers to S_{ij} in fluid dynamics as half of the *deformation rate tensor*; to A_{ij}, as half of the *vorticity* (see Chapter 4 of Volume II). The contraction of an antisymmetric tensor A_{ij} and a symmetric one $\hat{k}_i \hat{k}_j$ yields nothing. We can understand this result in the context of our immediate problem as follows. The tensor A_{ij} would constitute all there is to $\partial u_i/\partial x_j$ if the fluid possessed only uniform rotation. The component of velocity along any line of sight does not vary with distance s for a uniformly rotating medium; thus the term A_{ij} contributes nothing to $d\nu/ds$.

In consequence of the last result, we need only examine the contraction of S_{ij} and $\hat{k}_i \hat{k}_j$:

$$\frac{d\nu}{ds} = \frac{\nu_0}{c}\hat{k}_i S_{ij} \hat{k}_j.$$

It is well known that a local rotation of coordinate axis can transform any symmetric tensor S_{ij} to a diagonal form S'_{mp}. Under such a change of coordinates, the vector $\hat{k}_i$ transforms to $\hat{k}'_m$, but the scalar quadratic form $\hat{k}_i S_{ij} \hat{k}_j$ remains invariant:

$$\hat{k}_i S_{ij} \hat{k}_j = \hat{k}'_m S'_{mp} \hat{k}'_p = S'_{11}(k'_1)^2 + S'_{22}(k'_2)^2 + S_{33}'(k'_3)^2,$$

where the last expression obtains because the tensor S'_{mp} is diagonal. The individual elements of this tensor represent *extensional strains*:

$$S'_{11} = \frac{\partial u'_1}{\partial x'_1}, \qquad S'_{22} = \frac{\partial u'_2}{\partial x'_2}, \qquad S'_{33} = \frac{\partial u'_3}{\partial x'_3}.$$

By definition, for a purely expansive flow, the principal axes components S'_{11}, S'_{22}, S'_{33} are all *positive*; for a purely contractive flow, all *negative*. Since $|\hat{\mathbf{k}}|^2 = (\hat{k}'_1)^2 + (\hat{k}'_2)^2 + (\hat{k}'_3)^2 = 1$ cannot vanish, we see that $d\nu/ds$ has only one sign for purely expansive or purely contractive flows. In other words, ν increases or decreases monotonically with increasing s for these two special types of flows (the only ones we will consider henceforth in this chapter), and there can exist at best one point along any given line of sight contributing to a given value of the line frequency. (Q.E.D.)

LTE ASSUMPTION AND SOBOLEV APPROXIMATION

To simplify the considerations further, we assume here that the relative level populations for the line radiation under consideration exist in LTE. The Sobolev approximation then yields the monochromatic specific intensity reaching the observer as (see below):

$$I_\nu(\hat{\mathbf{k}}, \mathbf{r}_o) = B_\nu(T) \left(1 - e^{-\tau_\nu}\right). \tag{9.30}$$

In the above, the optical depth in the line, τ_ν, has the value:

$$\tau_\nu = n_\ell \left(\frac{a_{\ell u}}{\nu_0}\right) \left[\frac{c}{|\hat{\mathbf{k}} \cdot \boldsymbol{\nabla}(\hat{\mathbf{k}} \cdot \mathbf{u})|}\right] \left(1 - e^{-h\nu/kT}\right), \tag{9.31}$$

where $a_{\ell u}$ equals the atomic or molecular cross section of the line integrated over all frequencies [see equation (8.23) and Chapter 23]:

$$a_{\ell u} \equiv \frac{\pi e^2}{m_e c} f_{\ell u}. \tag{9.32}$$

In equations (9.30) and (9.31), we evaluate the number density n_ℓ of atoms or molecules in the lower level, the temperature T, and the Doppler coherence length $\propto c/\hat{\mathbf{k}} \cdot \boldsymbol{\nabla}(\hat{\mathbf{k}} \cdot \mathbf{u})$, at the position $\mathbf{r}_\nu$ that produces the given frequency ν. The factor $1 - \exp(-h\nu/kT_\nu)$ represents the LTE correction for stimulated emission, and the term $\hat{\mathbf{k}} \cdot \boldsymbol{\nabla}(\hat{\mathbf{k}} \cdot \mathbf{u}) > 0$ for expansive flows and < 0 for contractive flows.

If no position $\mathbf{r}_\nu$ exists along the line of sight that contributes to the frequency ν of the line-formation process, we take $\tau_\nu = 0$. Equation (9.30) then lacks any contribution from that line of sight to I_ν. To derive equation (9.30), we have ignored the possible presence of a background source of continuous light that might cause the line to form as an absorption feature. Inclusion of such a continuous background does not pose any great difficulties of principle.

Derivation of Equation (9.30)

In the absence of a background source of continuum light, the formal solution of the equation of transfer under conditions of LTE reads (see Chapter 3):

$$I_\nu(\hat{\mathbf{k}}, \mathbf{r}_o) = \int_0^\infty B_\nu[T(\mathbf{r})] \exp[-\tau_\nu(\hat{\mathbf{k}}, \mathbf{r}_o, s)] \rho(\mathbf{r}) \kappa_\nu(\hat{\mathbf{k}}, \mathbf{r}) \, ds, \qquad (9.33)$$

where

$$\tau_\nu(\hat{\mathbf{k}}, \mathbf{r}_o, s) \equiv \int_0^s \rho(\mathbf{r}') \kappa_\nu(\hat{\mathbf{k}}, \mathbf{r}') \, ds' \qquad (9.34)$$

represents the flow optical depth along a ray path from $\mathbf{r}$ to $\mathbf{r}_o$, with [see equation (9.27)]

$$\mathbf{r} = \mathbf{r}_o - s\hat{\mathbf{k}}, \qquad \mathbf{r}' = \mathbf{r}_o - s'\hat{\mathbf{k}}.$$

In the Sobolev approximation, which holds to a high degree of approximation for any relatively *cool* flow, we may take the line opacity $\kappa_\nu(\hat{\mathbf{k}}, \mathbf{r})$ for photons at position $\mathbf{r}$ propagating in the direction $\hat{\mathbf{k}}$ to be a Dirac delta function in frequency [see equation (9.28)]:

$$\rho(\mathbf{r}) \kappa_\nu(\hat{\mathbf{k}}, \mathbf{r}) = n_\ell(\mathbf{r}) a_{\ell u} \left[1 - e^{-h\nu/kT(\mathbf{r})}\right] \delta[\nu - \nu_0(1 + \hat{\mathbf{k}} \cdot \mathbf{u}/c)]. \quad (9.35)$$

If we substitute equation (9.35) into equation (9.33), we obtain zero for the integral unless the (unique) point $\mathbf{r}_\nu$ lies along the path of integration from $s = 0$ to ∞. Assuming the latter situation to hold, we may use the delta function to evaluate $B_\nu[T(\mathbf{r})]$ at $\mathbf{r} = \mathbf{r}_\nu$ and pull this term outside the integral. The remaining integral is of the form,

$$B_\nu[T(\mathbf{r}_\nu)] \int_0^{\tau_\nu(\hat{\mathbf{k}}, \mathbf{r}_o, \infty)} e^{-\tau_\nu} \, d\tau_\nu,$$

where we have used equation (9.34) to write $\rho \kappa_\nu \, ds$ as $d\tau_\nu$. We easily carry out the above integration, obtaining for equation (9.33), the solution (9.30). In that expression, $\tau_\nu \equiv \tau_\nu(\hat{\mathbf{k}}, \mathbf{r}_o, \infty)$:

$$\tau_\nu = \int_0^\infty \rho(\mathbf{r}) \kappa_\nu(\hat{\mathbf{k}}, \mathbf{r}) \, ds,$$

and we have removed the prime in the dummy integration variable as being superfluous. Upon further substituting the expression (9.35) and introducing the transformation of integration variables

$$ds = \frac{d\nu}{d\nu/ds},$$

where $d\nu/ds$ is given by equation (9.29), we obtain the identification (9.31). (Q.E.D.)

The integral in equation (9.33) should give zero if there exists no such point $\mathbf{r}_\nu$ along the given line of sight. We can arrange this result for the solution (9.30) simply by defining τ_ν equal to 0 for this situation.

OBSERVED SPECTRAL ENERGY DISTRIBUTION

The observer sees a radiative flux given by

$$\mathbf{F}_\nu = \int \hat{\mathbf{k}} I_\nu \, d\Omega_o,$$

where the integration is taken over all solid angles of photon propagation direction $\hat{\mathbf{k}}$ at the position of the observer ($\hat{\mathbf{k}}$ will vary during the integration if we take the observer to be located at a finite distance to start). For calculational convenience, we transform to an integration over propagation directions at the source by first noting that an element of radiating area dA with unit normal $\hat{\mathbf{n}}$ will contribute to the observed frequency ν if dA corresponds to a piece of the surface of constant $\hat{\mathbf{k}} \cdot \mathbf{u}$. The normal to such a surface is given by

$$\hat{\mathbf{n}} = \frac{\mathbf{\nabla}(\hat{\mathbf{k}} \cdot \mathbf{u})}{|\mathbf{\nabla}(\hat{\mathbf{k}} \cdot \mathbf{u})|}; \tag{9.36}$$

and the solid angle subtended by dA at the position of the observer will equal

$$d\Omega_o = \frac{|\hat{\mathbf{k}} \cdot \hat{\mathbf{n}}| \, dA}{s^2}.$$

From the origin (center of the source), the same element of area dA subtends a solid angle

$$d\Omega = \frac{|\hat{\mathbf{r}} \cdot \hat{\mathbf{n}}| \, dA}{r^2};$$

thus we have the transformation

$$d\Omega_o = \left(\frac{|\hat{\mathbf{k}} \cdot \hat{\mathbf{n}}|}{|\hat{\mathbf{r}} \cdot \hat{\mathbf{n}}|}\right) \left(\frac{r^2}{s^2}\right) d\Omega.$$

We may now compute the radiative flux at the position of the observer as

$$\mathbf{F}_\nu = \int \hat{\mathbf{k}} \left[\frac{|\hat{\mathbf{k}} \cdot \mathbf{\nabla}(\hat{\mathbf{k}} \cdot \mathbf{u})|}{|\hat{\mathbf{r}} \cdot \mathbf{\nabla}(\hat{\mathbf{k}} \cdot \mathbf{u})|}\right] \left(\frac{r^2}{s^2}\right) I_\nu \, d\Omega,$$

where we have made use of equation (9.36) for $\hat{\mathbf{n}}$. In the limit $r_o \to \infty$, s becomes r_o and $\hat{\mathbf{k}}$ equals a constant vector. If the observer places a detector perpendicular to the incident beam, he or she will deduce the (unresolved) source to have an effective monochromatic "luminosity"

$$L_\nu(\hat{\mathbf{k}}) = 4\pi r_0^2 \hat{\mathbf{k}} \cdot \mathbf{F}_\nu = 4\pi \int \left[\frac{|\hat{\mathbf{k}} \cdot \boldsymbol{\nabla}(\hat{\mathbf{k}} \cdot \mathbf{u})|}{|\hat{\mathbf{r}} \cdot \boldsymbol{\nabla}(\hat{\mathbf{k}} \cdot \mathbf{u})|} \right] r^2 I_\nu \, d\Omega, \qquad (9.37)$$

where the integration occurs over all solid angles subtended at the center of the source. Notice that the above definition for L_ν depends on the direction of photon propagation to the distant observer $\hat{\mathbf{k}}$. Except for a source possessing spherical symmetry, $L_\nu(\hat{\mathbf{k}})$, when integrated over all ν, does *not* represent the total energy radiated by the source per unit time into *all* photon propagation directions.

In the integration over source direction in equation (9.37), $\mathbf{r}$ must be evaluated at the position $\mathbf{r}_\nu$. The evaluation of equation (9.37) for a series of different values of ν to obtain the theoretical line profile then occurs in a straightforward, if somewhat inconvenient, manner.

LINE STRENGTH

We may obtain a more computationally efficient procedure by transforming independent variables from (ν, θ, ϕ) to (r, θ, ϕ), where θ and ϕ are the usual spherical polar angles characterizing the direction of $\mathbf{r}$. We motivate the discussion by considering the problem of computing line strengths. For line strengths, we wish to evaluate the total "luminosity" in the line (assumed isolated in frequency space for sake of simplicity):

$$L(\hat{\mathbf{k}}) = \int_0^\infty L_\nu(\hat{\mathbf{k}}) \, d\nu, \qquad (9.38)$$

where $L_\nu(\hat{\mathbf{k}})$ is given by equation (9.37). To perform the integration indicated in equation (9.38) in (r, θ, ϕ)-space, we note the transformation

$$d\nu \, d\theta \, d\phi = J \, dr \, d\theta \, d\phi,$$

where J equals the Jacobian

$$J \equiv \left| \frac{\partial(\nu, \theta, \phi)}{\partial(r, \theta, \phi)} \right|.$$

With $\nu = \nu(\hat{\mathbf{k}}; r, \theta, \phi)$ given by equation (9.28), we easily evaluate, for fixed $\hat{\mathbf{k}}$,

$$J = \left| \left(\frac{\partial \nu}{\partial r} \right)_{\theta, \phi} \right| = \frac{\nu_0}{c} |\hat{\mathbf{r}} \cdot \boldsymbol{\nabla}(\hat{\mathbf{k}} \cdot \mathbf{u})|. \qquad (9.39)$$

If we now substitute the above considerations into equation (9.38), we obtain

$$L = \int_0^\infty r^2 \, dr \int_0^\pi \sin\theta \, d\theta \int_0^{2\pi} 4\pi \left(\frac{\nu_0}{c}\right) I_\nu |\hat{\mathbf{k}} \cdot \mathbf{\nabla}(\hat{\mathbf{k}} \cdot \mathbf{u})| \, d\phi. \qquad (9.40)$$

OPTICALLY THIN EMISSION

Equation (9.40) achieves a particularly simple form when line formation occurs under optically thin conditions, $\tau_\nu \ll 1$. Under these conditions we may write

$$4\pi \left(\frac{\nu_0}{c}\right) I_\nu = 4\pi \left(\frac{\nu_0}{c}\right) B_\nu(T) \tau_\nu$$

$$= 4\pi \left(\frac{2h\nu^3}{c^2}\right) a_{\ell u} n_\ell e^{-h\nu/kT} \left[\frac{1}{|\hat{\mathbf{k}} \cdot \mathbf{\nabla}(\hat{\mathbf{k}} \cdot \mathbf{u})|}\right]. \qquad (9.41)$$

If we identify $4\pi(2h\nu^3/c^2)a_{\ell u}$ as $(g_u/g_\ell)A_{u\ell}h\nu_0$, where $A_{u\ell}$ equals the Einstein coefficient for spontaneous emission (see Chapter 8), and if we use Boltzmann's law to rewrite $n_\ell \exp(-h\nu/kT)$ as $(g_\ell/g_u)n_u$, we may express equation (9.41) as

$$4\pi \left(\frac{\nu_0}{c}\right) I_\nu = \frac{n_u A_{u\ell} h\nu_0}{|\hat{\mathbf{k}} \cdot \mathbf{\nabla}(\hat{\mathbf{k}} \cdot \mathbf{u})|}.$$

If we now substitute the above expression into equation (9.40), we obtain

$$L^{\text{thin}} = \int_0^\infty r^2 dr \int_0^\pi \sin\theta \, d\theta \int_0^{2\pi} d\phi \, n_u A_{u\ell} h\nu_0 \equiv N_u A_{u\ell} h\nu_0, \qquad (9.42)$$

where we have defined the integral of n_u over the entire source volume V to equal

$$N_u \equiv \int_V n_u \, dV. \qquad (9.43)$$

Equation (9.42) states the obviously correct result that the line luminosity for optically thin emission can be obtained as the rate of the total number of atoms or molecules in the upper state N_u undergoing spontaneous emission. The validity of this statement does not depend on the LTE assumption.

ESCAPE PROBABILITY FORMALISM

If we cannot assume that the flow is optically thin, Boltzmann's law and the Einstein relation for $A_{u\ell}$ still allow us to express equation (9.40) as

$$L = A_{u\ell} h\nu_0 \int_V \frac{n_u}{\tau_\nu} \left(1 - e^{-\tau_\nu}\right) \, dV.$$

The mean-value theorem lets us write the above in the suggestive form:

$$L = N_u \beta A_{u\ell} h \nu_0, \tag{9.44}$$

with the *escape probability* β defined as the density-weighted average of the optical depth factors:

$$\beta \equiv \left\langle \frac{1}{\tau_\nu} \left(1 - e^{-\tau_\nu} \right) \right\rangle. \tag{9.45}$$

We refer to the quantity β as the escape probability, because it multiplies the Einstein coefficient $A_{u\ell}$ in equation (9.44), making the effective rate of downward transitions, in the presence of photon trapping due to finite optical depth, equal to $\beta A_{u\ell}$ rather than the value $A_{u\ell}$ that applies under transparent conditions. If we were to obtain β by an actual integration over the source volume weighted by n_u/N_u, equation (9.44) would represent an exact result—within the limitations set by the LTE correction for stimulated emission and by the Sobolev approximation. Its real utility, however, lies in approximate estimates that guess a "typical" point to evaluate the optical depth factors (e.g., in applications to molecular clouds). For optically thin regions, $\beta \approx 1$; for optically thick regions, $\beta \approx \langle \tau_\nu^{-1} \rangle$.

TRANSFORMATION TO INTEGRATION OVER SOURCE VOLUME

Apart from the approximate treatment allowed by the escape probability formalism, equation (9.39) represents the most important result derived from the above discussion. If we divide up the source volume into spatial elements, $dV = r^2 \, d\Omega \, dr$, each volume element contributes to L_ν, an amount equal to

$$dL_\nu(\hat{\mathbf{k}}) = 4\pi \left(\frac{\nu_0}{c} \right) I_\nu \hat{\mathbf{k}} \cdot \boldsymbol{\nabla}(\hat{\mathbf{k}} \cdot \mathbf{u}) \left(\frac{dV}{d\nu} \right), \tag{9.46}$$

where

$$|d\nu| = J \, |dr|,$$

and J is given by equation (9.39), whereas ν is given by equation (9.28), *when we regard (r, θ, ϕ) as the independent variables.* This observation allows an obvious finite-element technique for computing L_ν by summing up contributions over the source volume. In this form, we notice that we have not made any use of the LTE assumption of our hypothetical problem; thus the last part of our discussion applies with perfect generality to three-dimensional steady flows as long as I_ν is computed without restrictions.

FLOWS WITH SPHERICAL SYMMETRY

Suppose the flow field has spherical symmetry:

$$\mathbf{u} = \hat{\mathbf{r}}u(r).$$

Without loss of generality, we may then orient our coordinate system so that the line of sight occurs parallel to the z-axis:

$$\hat{\mathbf{k}} \cdot \mathbf{u} = u(r)\cos\theta \equiv u(r)\mu,$$

where we define in the usual way,

$$\mu \equiv \cos\theta.$$

We now have

$$\hat{\mathbf{r}} \cdot \boldsymbol{\nabla}(\hat{\mathbf{k}} \cdot \mathbf{u}) = \frac{\partial}{\partial r}[u(r)\cos\theta] = u'(r)\mu,$$

and the Doppler condition becomes

$$\nu = \nu_0 \left[1 + \frac{1}{c}u(r)\mu\right]. \tag{9.47}$$

For fixed ν, we have therefore the relation,

$$u'(r)\mu\, dr + u(r)\, d\mu = 0, \qquad \Rightarrow \qquad d\mu = -\frac{u'(r)\mu}{u(r)}\, dr.$$

The differential spectral energy distribution for an optically thin flow,

$$dL_\nu^{\text{thin}} = \frac{cn_u A_{u\ell}h\nu_0 r^2\, d\mu d\phi}{\nu_0|\hat{\mathbf{r}} \cdot \boldsymbol{\nabla}(\hat{\mathbf{k}} \cdot \mathbf{u})|},$$

now becomes

$$dL_\nu^{\text{thin}} = \frac{cr^2 n_u(r)A_{u\ell}h\nu_0}{\nu_0|u(r)|}\, dr d\phi. \tag{9.48}$$

The corresponding expression for an optically thick flow,

$$dL_\nu^{\text{thick}} = 4\pi B_\nu(T)\frac{|\hat{\mathbf{k}} \cdot \boldsymbol{\nabla}(\hat{\mathbf{k}} \cdot \mathbf{u})|}{|\hat{\mathbf{r}} \cdot \boldsymbol{\nabla}(\hat{\mathbf{k}} \cdot \mathbf{u})|}\, r^2\, d\mu d\phi,$$

reads

$$dL_\nu^{\text{thick}} = 4\pi B_\nu[T(r)] \left|\mu^2 u'(r) + (1-\mu^2)\frac{u(r)}{r}\right| r^2 \frac{dr}{|u(r)|}\, d\phi. \tag{9.49}$$

At each r, equation (9.47) states that pieces of a spherical shell at different μ, as μ ranges from -1 to $+1$, contribute to the frequency range,

$$\nu_0 - \nu_r \le \nu \le \nu_0 + \nu_r, \tag{9.50}$$

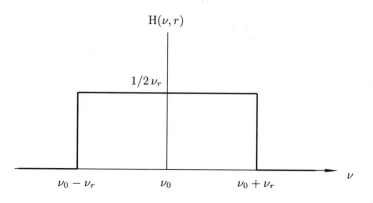

FIGURE 9.2
At each r, the "top-hat" function $H(\nu, r) = 0$ for ν outside of the range $\nu_0 \mp \nu_r$, and it has a constant value inside the range that makes the area between H and the ν-axis equal to unity.

where

$$\nu_r \equiv \frac{1}{c}\nu_0|u(r)|. \tag{9.51}$$

Integrating equation (9.48) over ϕ from 0 to 2π and over r from 0 to ∞ now yields

$$L_\nu^{\text{thin}} = N_u A_{u\ell} h\nu_0 \mathcal{H}(\nu), \tag{9.52}$$

where $\mathcal{H}(\nu)$ equals the profile function:

$$\mathcal{H}(\nu) \equiv \frac{1}{N_u} \int_0^\infty n_u(r) H(\nu, r) \, 4\pi r^2 \, dr, \tag{9.53}$$

with $H(\nu, r)$ equal to a "top-hat" function:

$$H(\nu, r) = \frac{1}{2\nu_r} \quad \text{for} \quad \nu_0 - \nu_r \le \nu \le \nu_0 + \nu_r, \tag{9.54}$$

and $H(r, \nu) = 0$ for ν outside the stated range (see Figure 9.2). Note that $\leftarrow$ $H(\nu, r)$ satisfies the normalization condition for any r,

$$\int_0^\infty H(\nu, r) \, d\nu = 1;$$

therefore, since N_u is given by equation (9.43),

$$N_u = \int_0^\infty n_u(r)\, 4\pi r^2 dr,$$

$\mathcal{H}(\nu)$ also satisfies a normalization condition,

$$\int_0^\infty \mathcal{H}(\nu)\, d\nu = 1.$$

Similarly, with equation (9.47) implying that $\mu^2 = [(\nu - \nu_0)/\nu_r]^2$, the spectral energy distribution for the optically thick case becomes

$$L_\nu^{\text{thick}} = 4\pi \int_0^\infty B_\nu[T(r)] \left| \left(\frac{\nu - \nu_0}{\nu_r}\right)^2 u'(r) \right. \tag{9.55}$$

$$\left. + \left[1 - \left(\frac{\nu - \nu_0}{\nu_r}\right)^2\right] \frac{u}{r} \right| \frac{\nu_r}{|u(r)|} H(\nu, r)\, 4\pi r^2 dr. \tag{9.56}$$

FREE EXPANSION OF A SUPERNOVA REMNANT

As an example, consider the case when the matter consists of "shrapnel" flying from a point explosion (e.g., the ejecta from Supernova 1987A):

$$u(r) = r/t \qquad \text{for} \qquad r \leq r_{\text{m}} \equiv v_{\text{m}} t,$$

with t equal to the time since the explosion and v_{m} equal to the maximum observed ejecta velocity. For simplicity, we follow McCray and assume uniform conditions inside the supernova remnant (an allowed mode of evolution for a "Hubble flow"): $n_u = 3N_u(t)/4\pi r_{\text{m}}^3$ independent of r for $r \leq r_{\text{m}}$ with $n_u = 0$ for $r \geq r_{\text{m}}$, and $T(r) = \text{constant} = T_{\text{m}}$ independent of r and t. The total number of atoms or molecules in the upper state $N_u(t)$ declines with time t because the underlying source of excitation—energy input by radioactivity—decays exponentially. For this case,

$$H(\nu, r) = \frac{ct}{2\nu_0 r} \qquad \text{for} \qquad \nu_0\left(1 - \frac{r}{ct}\right) \leq \nu \leq \nu_0\left(1 + \frac{r}{ct}\right),$$

and it equals zero otherwise. For any ν, therefore, nonzero contributions come in the r integration only from the interval between $r_1 = |\nu - \nu_0|ct$ and $r_2 = v_{\text{m}}t \equiv r_{\text{m}}$, with the integral giving nothing if the lower limit r_1 exceeds the upper limit r_2.

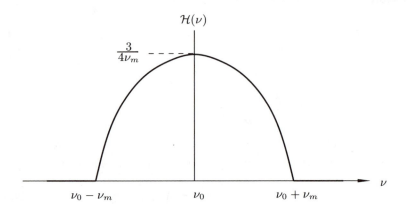

FIGURE 9.3
The profile-function $\mathcal{H}(\nu) = (3/4\nu_{\mathrm{m}})[1 - (\nu - \nu_0)^2/\nu_{\mathrm{m}}^2]$ has a parabolic shape for ν in the range $\nu_0 \mp \nu_{\mathrm{m}}$, equals zero outside this range, and has unit area.

The case of optically thin emission is now given by equation (9.42), with the line profile function having a parabolic form (see Figure 9.3):

$$\mathcal{H}(\nu) = \frac{3}{4\nu_{\mathrm{m}}}\left[1 - \left(\frac{\nu - \nu_0}{\nu_{\mathrm{m}}}\right)^2\right] \qquad \text{for} \qquad \nu_0 - \nu_{\mathrm{m}} \le \nu \le \nu_0 + \nu_{\mathrm{m}},$$

where we have defined

$$\nu_{\mathrm{m}} \equiv \nu_0 \frac{v_{\mathrm{m}}}{c}.$$

Interestingly enough, the optically thick case computed from equation (9.55) has the same parabolic shape:

$$L_\nu^{\text{thick}} = \frac{16\pi^2}{3}\nu_{\mathrm{m}}B_\nu(T_{\mathrm{m}})r_{\mathrm{m}}^2\mathcal{H}(\nu),$$

but the line strength increases with time t as $r_{\mathrm{m}}^2 \propto t^2$ because of the expanding emitting area, rather than decreasing as $N_u(t)$ because of the declining excitation by radioactivity. Observationally, the strengths of certain lines from Supernova 1987 actually did change from increasing in time to decreasing in time as the expansion of the remnant caused the conditions to go from optically thick to optically thin.

The identical line shapes for optically thick and thin emission (which requires for verification higher spectral resolution than available for most observations of Supernova 1987) arises for the simple reason that the surfaces of constant line-of-sight velocity corresponds in this case to planar sections through the remnant, and our assumptions of constant n_u and T have made these planar surfaces uniformly bright for both cases. Since $\mathcal{H}(\nu_0) = 3/4\nu_m$, notice that the monochromatic luminosity at line center for the optically thick case is given by

$$L_{\nu_0}^{\text{thick}} = 4\pi r_m^2 [\pi B_{\nu_0}(T_m)],$$

which equals the value for a blackbody emitter with outward flux $\pi B_{\nu_0}(T_m)$ and surface area $4\pi r_m^2$. The explanation for this result lies in the zero-velocity surface corresponding to a planar cut through the origin, the cut occupying the entire projected area of the remnant.

STEADY WIND COASTING AT CONSTANT SPEED

A steady stellar wind coasting at constant speed, $u(r) = v_m$, provides another easily computed example. If we see only the part of the flow moving at its terminal velocity v_m, then the optically thin emission is given by equation (9.42) with a rectangular line-profile function:

$$\mathcal{H}(\nu) = \frac{1}{2\nu_m} \qquad \nu_0 - \nu_m \leq \nu \leq \nu_0 + \nu_m,$$

where again

$$\nu_m \equiv \nu_0 \frac{v_m}{c}.$$

We get a single "top-hat" function rather than a superposition of them because $H(\nu, r)$ is the same at all r for $u(r) = $ constant. Notice that this result holds independent of the functional form of $n_u(r)$; we require only that excitation to the upper state falls off rapidly enough with increasing r so that the integral

$$N_u = \int_0^\infty n_u(r)\, 4\pi r^2\, dr$$

converges to a finite value. If a source of continuum light exists in the system (e.g., a star), the part of the spectrum blueshifted with respect to the stellar velocity (i.e., corresponding to gas headed toward the observer and therefore lying in front of the star) may go into absorption. Line spectra that show emission in the red wing and absorption in the blue are said to possess "P Cygni profiles," a signature for stellar winds (Figure 9.4). (The red wing will not have a rectangular shape if line formation occurs, as is likely in many cases, in the acceleration zone of the wind.)

For a constant-velocity stellar wind, we obtain from equation (9.55) the optically thick line-emission as

$$L_\nu^{\text{thick}} = \left\{ \frac{32\pi^2}{3} \nu_m \int_0^\infty B_\nu[T(r)]\, r\, dr \right\} \mathcal{H}(\nu),$$

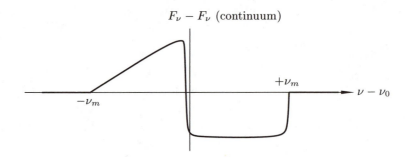

FIGURE 9.4
P Cygni line-profiles show emission in the red wing and absorption in the blue, and form the typical signature for a stellar wind.

where the line-profile function again has a parabolic shape:

$$\mathcal{H}(\nu) = \frac{3}{4\nu_{\mathrm{m}}} \left[1 - \left(\frac{\nu - \nu_0}{\nu_{\mathrm{m}}} \right)^2 \right] \qquad \nu_0 - \nu_{\mathrm{m}} \le \nu \le \nu_0 + \nu_{\mathrm{m}}.$$

Notice that this shape holds independently of the radial dependence of the flow temperature $T(r)$ because every spherical shell contributes the same parabolic line profile. The clean separation of the line-strength and line-profile problems in both optically thin and thick cases is probably unique to the constant-velocity spherical flow. In any case, by examining a variety of spectral lines formed under a range of conditions, from optically thin to optically thick, one can, in principle, deduce valuable information about the density, temperature, and flow kinematics of the emitting source.

10

Ionization Balance of H ɪɪ Regions

Reference: Osterbrock, *Astrophysics of Gaseous Nebulae and Active Galactic Nuclei*, Chapter 2.

In this chapter we will illustrate how rate equations of the kind introduced in the first part of Chapter 9 can be used to treat H ɪɪ regions. Our treatment here will emphasize the aspects dealing with ionization balance. We will ignore the equally interesting question of their thermal structure and mechanical evolution.

THE MACROSCOPIC PROBLEM

In 1939 Bengt Stromgren pointed out that the diffuse matter in interstellar space should exist in two sharply divided forms: (1) H ɪ regions, where the hydrogen is almost 100 percent atomic; (2) H ɪɪ regions, where the hydrogen is almost 100 percent ionized. He also briefly wondered whether interstellar hydrogen might exist in a molecular form, H_2, but concluded that this would be most unlikely if the mechanisms for the formation and destruction of H_2 were similar to the radiative processes that he had considered for H ɪ. In fact, the processes are completely different; so Stromgren failed to predict the ubiquitous existence of molecular clouds. However, in all other respects, his paper proved seminal in modern astronomical thinking about the physical state of interstellar gas.

Stromgren's basic point can be made in a very straightforward manner. Imagine a hot star, say, of spectral type O or B, situated in a region of constant density n_0 (see Figure 10.1.) For simplicity, suppose the gas to be made purely of hydrogen. (In fact, the present cosmic abundance of hydrogen is more like 70 percent by mass.) Let N_u be the total output of ultraviolet photons (beyond the Lyman limit) pouring from the star, with N_u having the units of number of photons per second. What would be the radius—now called the Stromgren radius R_S—of the sphere inside of which the star could keep the hydrogen nearly completely ionized in a

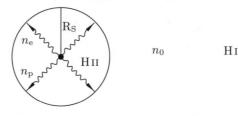

FIGURE 10.1
Photoionization of atomic hydrogen gas of uniform density surrounding a hot young star will produce a spherical H II region, whose radius equals the Stromgren value R_S.

steady state? (A fundamental point that we will gloss over to start with, but which Stromgren proved, concerns the assumption that the transition between being nearly 100 percent ionized and nearly 100 percent neutral would be made very sharply.)

For overall charge neutrality, the electron density n_e inside the H II region must equal the proton density n_p, which itself (nearly) equals n_0 by assumption:

$$n_e = n_p. \tag{10.1}$$

Electrons and protons meet in two-body encounters and recombine at a rate per unit volume proportional to the product $n_e n_p$; call the proportionality constant α (not to be confused with a photoabsorption cross section), now known as the *recombination* coefficient. Thus

$$\text{volumetric rate of recombinations} = \alpha n_e n_p. \tag{10.2}$$

The total rate of recombinations in a volume $V = 4\pi R_S^3/3$ must equal $\alpha n_e n_p V$, and in a steady state, this must equal the rate N_u of the out-pouring of ultraviolet photons from the central star (assuming that each photon is capable of producing only a single ionization). Setting the two rates equal and solving for R_S, we obtain

$$R_S = (3N_u/4\pi\alpha n_0^2)^{1/3}, \tag{10.3}$$

where we have made use of equation (10.1) and the approximation $n_p \approx n_0$.

An O5 star releases Lyman continuum photons at rate of about $N_u = 5 \times 10^{49}$ s^{-1}. If we adopt a value for the recombination coefficient $\alpha = 3 \times 10^{-13}$ cm^3 s^{-1} and a density $n_0 = 10^2$ cm^{-3}, we obtain $R_S \sim 1.6 \times 10^{19}$ cm ~ 5 pc, which would constitute a healthy piece of most interstellar clouds. Thus the formation of O stars gives rise to important agents for ionizing and eventually dispersing interstellar clouds.

In carrying out the above calculation, we have made a number of implicit assumptions that we would now like to justify. For example, why do we only need to consider photons beyond the Lyman limit, i.e., why can we assume that the hydrogen atoms are virtually all in the ground electronic state? Second, where does the value $\alpha = 3 \times 10^{-13}\,\mathrm{cm^3\,s^{-1}}$ come from?

BOUND-BOUND PHOTOABSORPTION CROSS SECTIONS

To show that hydrogen atoms in interstellar space quickly fall to the ground electronic level, we need to know the spontaneous transition probabilities for excited states. The formalism developed in Chapter 8 allows us to compute these rates if we know the corresponding bound-bound photoabsorption cross sections; so we will begin our discussion with the latter quantities. Later in this text we will develop the techniques for calculating photoabsorption cross sections, but for now we content ourselves with merely quoting well-documented results.

The cross section α_ν for the bound-bound transition of any atom from an initial level 1 to some final level 2 with the absorption of a photon of energy $h\nu_{21} = E_2 - E_1$ equals

$$\alpha_\nu = \frac{\pi e^2}{m_e c} f_{12}\phi_{12}(\nu), \tag{10.4}$$

where f_{12} is the *oscillator strength* of the transition. Upon integration over ν, this formula reproduces equation (8.23), since the function $\phi_{12}(\nu)$ describes the frequency profile of the line radiation, with

$$\int_{\Delta\nu} \phi_{12}(\nu)\,d\nu = 1, \tag{10.5}$$

when we integrate over over a narrow frequency interval $\Delta\nu$ centered on ν_{12}. In the development of Chapter 8 we assumed that $\phi_{12}(\nu)$ can be taken (in the rest frame of the atom) essentially as a delta function, $\phi_{12}(\nu) = \delta(\nu - \nu_{12})$. In fact, quantum mechanics requires natural lines to have finite widths. As we will demonstrate in a future chapter, in the rest frame of an isolated atom, the profile function is Lorentzian:

$$\phi_{12}(\nu) = \frac{1}{\pi}\left[\frac{\Gamma_{21}/4\pi}{(\nu - \nu_{21})^2 + (\Gamma_{21}/4\pi)^2}\right] \equiv \mathcal{L}_{21}(\nu), \tag{10.6}$$

where $\Gamma_{21}/4\pi$ is the frequency width associated with the transition. A finite width arises because the transition takes a finite time $\Delta t \sim \Gamma_{21}^{-1}$ to complete, and by the uncertainty principle, the energy difference between the two states involved must be uncertain by a characteristic amount that satisfies $(h\Delta\nu)(\Delta t) \sim \hbar/2$, i.e., $\Delta\nu \sim \Gamma_{21}/4\pi$.

We emphasize that equation (10.6) holds only for a single atom; a collection of atoms will generally have a (much) broader linewidth, governed by such things as thermal Doppler shifts. We also leave as an exercise for the reader the verification that the Lorentz profile does satisfy the normalization condition:

$$\int_0^\infty \mathcal{L}_{21}(\nu)\, d\nu = 1, \tag{10.7}$$

when $\Gamma_{21}/4\pi \ll \nu_{21}$. For a line with a pure Lorentz profile, the photoabsorption cross section at the center of the line equals

$$\alpha_\nu(\nu = \nu_{21}) = 4\pi f_{12} r_e(c/\Gamma_{21}), \tag{10.8}$$

where $r_e \equiv e^2/m_e c^2 = 2.82 \times 10^{-13}$ cm is the classical radius of the electron and c/Γ_{21} is the distance a photon travels in a mean "lifetime" Γ_{21}^{-1}. The first term arises because the photon interacts with an electron; the second because the electron is attached to an atom. If the electron were free, we would get another factor $\sim r_e$, i.e., the Thomson cross section, $\sigma_T = 8\pi r_e^2/3$.

SPONTANEOUS EMISSION PROBABILITY

Intuitively, we expect the approximate identification

$$\Gamma_{21} \approx A_{21}; \tag{10.9}$$

in a later chapter we will formalize and improve upon this identification for the natural linewidth. For now, however, we are motivated simply to look into the problem of how to calculate A_{21}. We begin by noting that equation (8.24) can be taken to read

$$B_{12} = \left(\frac{4\pi}{h\nu_{12}}\right) \int_{\Delta\nu} \alpha_\nu\, d\nu = \left(\frac{4\pi}{h\nu_{21}}\right)\left(\frac{\pi e^2}{m_e c}\right) f_{12}. \tag{10.10}$$

The corresponding rate for spontaneous emission now has the expression

$$A_{21} = \left(\frac{8\pi^2 e^2}{m_e c^3}\right) \nu_{21}^2 \left(\frac{g_1}{g_2}\right) f_{12}. \tag{10.11}$$

Notice that A_{21} can also be written as

$$A_{21} = [8\pi^2(g_1/g_2)f_{12}](r_e/c)\nu_{21}^2, \tag{10.12}$$

which makes explicit that it has the units of a total rate, s^{-1}.

Since the light-travel time across the classical radius of the electron r_e/c is $\sim 10^{-23}$ s, and since optical transitions are characterized by frequencies $\nu_{21} \sim 10^{15}\ \text{s}^{-1}$, we see that allowed (or permitted) transitions at visible wavelengths, which have oscillator strengths of order unity (or only somewhat smaller), have Einstein A-values $\sim 10^8$ s^{-1}. Such transitions are considered "permitted" because an excited atom (or molecule) *under terrestrial conditions* will de-excite radiatively for a significant fraction of the time, despite being bombarded by collisions with atmospheric molecules at a mean collisional rate $\sim 10^9$ s^{-1}. We will see from a later discussion that if the atom (or molecule) retains the same intrinsic size, $f_{12} \propto \nu_{21}$. Notice then that the dependence of equation (10.12) on the cube of the frequency ν_{21}^3 (or on the square if f_{12} does not diminish with decreasing ν_{21}) implies that the corresponding radio-frequency transitions have much smaller Einstein A's.

Before we leave this section, we comment that we have followed the standard discussion to derive Einstein A's from measured or calculated f-values, a natural strategy from the perspective of a laboratory spectroscopist. From the point of view of an atomic physicist, however, the calculation of a theoretical value of the Einstein A may seem more fundamental, in which case we could work backwards to derive an f. Problem Set 6 provides a calculation of the Einstein A for the Lyman-alpha transition, where we achieve an equivalent formula to equation (10.11), but which possesses greater elegance of appearance simply because the derivation lies closer to the actual radiation mechanism. Clearly, many routes exist, in principle, to get the basic atomic data for species more complex than hydrogen. Within measurement and approximation errors, they should all ideally give consistent results. When this does not happen, usually someone has misjudged the accuracy of the experiment or of the theoretical calculation.

IMPORTANCE OF PERMITTED ELECTRONIC TRANSITIONS TO THE ISM

Consider the application of equation (10.12) to the probability of a hydrogen atom in an upper level n radiatively de-exciting to a lower level n'. If we assume statistical equilibrium in the ℓ sublevels (usually okay, but not always),

$$A_{nn'} = 8\pi^2 \left(\frac{n'^2}{n^2}\right) \left(\frac{r_e}{c}\right) \nu_1^2 \left(\frac{1}{n'^2} - \frac{1}{n^2}\right)^2 f_{n'n}, \qquad (10.13)$$

where $h\nu_1 = e^2/2a_0 = 13.6\,\text{eV}$ is the energy of the Lyman limit, with $a_0 \equiv \hbar^2/m_e e^2 = 0.529 \times 10^{-8}$ cm being the Bohr radius. It is conventional to express oscillator strengths in terms of a semiclassical value calculated by Kramers:

$$f_{n'n} = \left(\frac{2^5}{3\pi\sqrt{3}}\right)\left[\frac{n'^3 n}{(n^2 - n'^2)^3}\right] g_{nn'}, \qquad (10.14)$$

where $g_{nn'}$ is a quantum-mechanical correction called the *Gaunt factor*. Carrying out the multiplication, we obtain $A_{nn'} = 1.57 \times 10^{10}[n^3 n'(n^2 - n'^2)]^{-1} g_{nn'}$ s^{-1}. For hydrogen, the Gaunt factors ≈ 1 at large n and n' (because quantum mechanics becomes like classical mechanics at high quantum numbers); even for the transition $2 \rightarrow 1$, $g_{21} = 0.717$. Numerical values for some common transitions (see Problem Set 5) are: Lyman alpha, $A_{21} = 4.68 \times 10^8$ s^{-1}; Lyman beta, $A_{31} = 5.54 \times 10^7$ s^{-1}; Hα (Balmer alpha), $A_{32} = 4.39 \times 10^7$ s^{-1}; Hβ (Balmer beta), $A_{42} = 8.37 \times 10^6$ s^{-1}. The most probable radio-frequency transition involves the so-called "alpha transition" with $n = n' + 1$. For example, the 109α recombination line of hydrogen (with $n' = 109$), in which H II regions were mapped in the early days of radio astronomy, occurs at a wavelength ≈ 6 cm and with a transition probability ≈ 0.38 s^{-1}.

Compared to interstellar collision rates, the above radiative transition probabilities are very large. Any atomic hydrogen accidentally finding itself in the $n = 2$ (or higher) level (through a collision with an energetic electron or through recombination from the free state) has therefore a strong tendency to de-excite radiatively to the ground state (or to cascade there). *As a consequence, the overwhelming majority of the hydrogen atoms that exist in interstellar space must reside in the ground electronic state.* This statement holds as well for atoms other than hydrogen. It does not hold for the rotational levels of molecules, which are much easier to excite than the higher electronic states. The most common route for hydrogen atoms to leave their ground electronic configuration (apart from combining to form hydrogen molecules H$_2$) involves photoionization by ultraviolet photons. Near a hot bright star (of spectral type O or B), this process can nearly ionize the ambient hydrogen completely, producing an H II region. The same process heats up the gas to $\sim 10^4$ K. We describe below the situation in more detail.

HYDROGEN BOUND-FREE PHOTOABSORPTION CROSS SECTIONS

In the preceding discussion, we have answered the question why for the process of photoionization, we effectively only need to consider hydrogen atoms in their ground electronic state. In contrast, to consider the recombination of electrons and protons to bound states, we need to allow for the possibility that they can recombine to any excited level n, from where they will then quickly cascade to $n = 1$. For the latter calculation, we will need

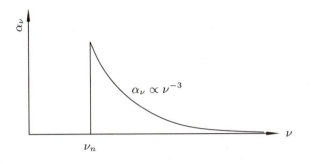

FIGURE 10.2
Frequency dependence of the bound-free cross section of atomic hydrogen for transitions from level n to the continuum.

the bound-free cross section for the n-th excited state of atomic hydrogen $H(n)$, which has a ν^{-3} dropoff from a threshold value for frequencies ν above the ionization limit $\nu_n = \nu_1/n^2$:

$$\alpha_\nu(n) = n\alpha_1(\nu_n/\nu)^3 g_n(\nu), \qquad (10.15)$$

while $\alpha_\nu(n) = 0$ for frequencies below the limit $\nu < \nu_n$ (see Figure 10.2). In equation (10.15), $h\nu_1 = 13.6\,\text{eV}$ again represents the Lyman limit, $g_n(\nu)$ is a bound-free Gaunt factor, and α_1 is the Kramers absorption cross section at the Lyman edge:

$$\alpha_1 \equiv \frac{64}{3\sqrt{3}} \left(\frac{e^2}{\hbar c}\right) \pi a_0^2, \qquad (10.16)$$

where we recognize the combination $e^2/\hbar c$ as the fine-structure constant $\approx 1/137$. The mnemonic for remembering this result is that the photoionization cross section at the Lyman edge is of the order ten times the geometric cross section of the atom multiplied by the electromagnetic coupling constant $e^2/\hbar c$. Numerically, $\alpha_1 = 7.91 \times 10^{-18}\,\text{cm}^2$.

Detailed calculations yield hydrogen bound-free Gaunt factors that are very close to being unity. This statement holds with high precision for large n, but even g_1 (averaged over appropriate frequencies) has typical values like 0.9. We now apply these results to the problem of H II regions.

APPLICATION TO H II REGIONS

At some distance r from a star of radius R_* and monochromatic luminosity L_ν^*, we can introduce the concept of a vacuum radiation field that results

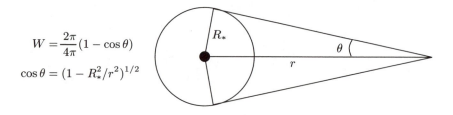

$$W = \frac{2\pi}{4\pi}(1 - \cos\theta)$$

$$\cos\theta = (1 - R_*^2/r^2)^{1/2}$$

FIGURE 10.3
Geometry for computing the dilution factor W.

from the dilution of the mean specific intensity J_ν^* leaving the star in the *outward* direction:

$$J_\nu^{\mathrm{vac}} = W J_\nu^* \qquad \text{where} \qquad J_\nu^* = \frac{1}{\pi} F_\nu^* = \frac{L_\nu^*}{4\pi^2 R_*^2}, \tag{10.17}$$

with F_ν^* being the radiative flux at the stellar surface. Note the relation $F_\nu^* = \pi J_\nu^*$, which holds at the bounding surface of an opaque body. The dilution factor W can be calculated as the fraction of the total solid angle that is subtended at r by the star (see Figure 10.3).

$$W = \frac{1}{2}\left[1 - \left(1 - \frac{R_*^2}{r^2}\right)^{1/2}\right]. \tag{10.18}$$

Notice that when $r \gg R_*$, $W = R_*^2/4r^2$, i.e., πR_*^2 divided by $4\pi r^2$.

In general, there will also be some attenuation of the ultraviolet photons produced by absorption by the intervening atomic hydrogen, $J_\nu = J_\nu^{\mathrm{vac}} e^{-\tau_\nu}$. We include this factor in a rough way by computing the attenuation at the Lyman edge; thus the total reduction over the stellar value equals

$$J_\nu = W_1 J_\nu^* \qquad \text{with} \qquad W_1 = W e^{-\tau_1} \quad \text{where} \quad \tau_1 \equiv \alpha_1 \int_{R_*}^r n(\mathrm{H}[1])\,dr, \tag{10.19}$$

and we have set the Gaunt factor $g_1(\nu)$ equal to unity. The absorption of Lyman continuum photons proceeds until the photons get almost completely depleted at $r = R_{\mathrm{S}}$, at which point W_1 is essentially zero (see below). Moreover, as long as the stellar temperature T_* is smaller than $13.6\,\mathrm{eV}/k = 158{,}000\,\mathrm{K}$, Lyman continuum photons come from the Wien tail of a blackbody curve (only a very rough approximation for an actual hot stellar photosphere; see Osterbrock 1989, Figure 5.8). Consequently, the bulk of the photoionizations occurs near the Lyman limit, explaining why we may approximate each ultraviolet photon as able to perform only one ionization. In practice, the frequency dependence of the absorption coefficient will lead to a progressive depletion of the softer photons and a hardening of the spectrum of the radiation as it propagates farther and farther from the star, but we ignore this complication here.

The photoabsorption of Lyman continuum photons yields a destruction rate of hydrogen per unit volume given by

$$\mathcal{D}_{bf}(H[1]) = n(H[1])W_1 p_A(1), \qquad (10.20)$$

where $p_A(1)$ is the normalized absorption photorate from level 1 to free states:

$$p_A(1) = 4\pi\alpha_1 \int_{\nu_1}^{\infty} (h\nu)^{-1} J_\nu^* \, d\nu = \frac{\alpha_1}{\pi R_*^2} N_u. \qquad (10.21)$$

In equation (10.21), we have again replaced $\alpha_\nu(1)$ by α_1, and we have used $J_\nu^* = L_\nu^*/4\pi^2 R_*^2$ to write the total rate of release of ultraviolet photons beyond the Lyman limit as

$$N_u = \int_{\nu_1}^{\infty} (h\nu)^{-1} L_\nu^* \, d\nu. \qquad (10.22)$$

Offsetting the photoionizations are recombinations of protons and electrons. Since $W_1 \ll 1$ at large distances from the star (where we are interested in making detailed applications), we may ignore the stimulated process. The spontaneous recombination to level n gives a rate per unit volume for the bound-free production of $H(n)$:

$$\mathcal{P}_{bf}(H[n]) = n_e n_p \int_{0}^{\infty} \mathcal{A}(n) \, v f_e(v) 4\pi v^2 dv, \qquad (10.23)$$

where $\mathcal{A}$ is the Einstein-Milne coefficient [see equation (8.36)]:

$$\mathcal{A}(n) = \left(\frac{h^2 \nu^2}{2\pi m_e c^2}\right)\left(\frac{g_n}{g_p}\right)\left(\frac{\alpha_\nu(n)}{h\nu - h\nu_n}\right), \qquad (10.24)$$

with $g_n/g_p = 2n^2$. (There are $2n^2$ bound sublevels for each spin orientation of the proton.)

With the relation,

$$\frac{1}{2}m_e v^2 = h\nu - h\nu_n, \qquad (10.25)$$

we perform the integration in equation (10.23) to obtain

$$\mathcal{P}_{bf}(H[n]) = n_e n_p k_S(n), \qquad (10.26)$$

where $k_S(n)$ is the rate constant

$$k_S(n) = n^2 \lambda_{Te}^3 e^{h\nu_n/kT} p_S(n), \qquad (10.27)$$

with

$$p_S(n) \equiv 4\pi \int_{\nu_n}^{\infty} \frac{\alpha_\nu(n)}{h\nu} \frac{2h\nu^3}{c^2} e^{-h\nu/kT} \, d\nu.$$

(10.28)

For large n, $h\nu_n$ does not remain large in comparison with kT, and it becomes important to include the ν dependence of $\alpha_\nu(n)$, which explains why we cannot simply evaluate this cross section at the ionization edge.

Each recombination to level n is quickly followed by a cascade down to the ground state because the Einstein A values are so large. Thus we need to sum the recombination rate constants over all states from $n = 1$ to $n = n_{max}$, with the latter being a function of local temperature and electron density (see the discussion in Chapter 8 about Debye cutoffs):

$$\mathcal{P}_{bf}(H[1]) = n_e n_p \sum_{n=1}^{n_{max}} k_S(n).$$

(10.29)

Unlike the example with hydrogen-level partition functions, no difficulty emerges here when the upper limit is extended to infinity. The resulting sum is known as the total recombination coefficient

$$\alpha_{tot} \equiv \sum_{n=1}^{\infty} k_S(n).$$

(10.30)

For practical conditions ("Case B"), recombination from the continuum to the level $n = 1$ can be ignored. Such a recombination produces a Lyman continuum photon to which the hydrogen gas is typically very optically thick. Therefore the gas continues to reprocess this photon locally (the *on-the-spot* approximation) until it finally breaks down to a Lyman-alpha photon plus a Balmer photon plus lower-frequency photons (see Shu 1982, Figure 11.9). Thus the effective recombination coefficient that normally enters in equation (10.2) involves a sum only from $n = 2$:

$$\alpha = \sum_{n=2}^{\infty} k_S(n).$$

(10.31)

IONIZATION STRUCTURE OF AN H II REGION

In a steady state, we require $\mathcal{D} = \mathcal{P}$. Since the bound-free contributions dominate all other terms (e.g., collisional ionization of H by fast electrons), we may equate $\mathcal{D}_{bf}(H[1]) = \mathcal{P}_{bf}(H[1])$. This procedure yields the local-ionization balance equation:

$$n(H[1])W_1 p_A(1) = n_e n_p \alpha.$$

(10.32)

Charge neutrality and proton conservation require

$$n_e = n_p = n_0 - n(H[1]).$$ (10.33)

Taken together with equation (10.19), equations (10.32) and (10.33) yield the following equation to solve for the fractional abundance of atomic hydrogen, $f(r) \equiv n(H[1])/n_0$, as a function of spatial position:

$$\exp\left(-\alpha_1 n_0 \int_{R_*}^{r} f \, dr\right) = [\alpha n_0 / W p_A(1)][(1-f)^2/f].$$ (10.34)

For $r \gg R_*$, $W = R_*^2/4r^2$, and we may use equations (10.21) and (10.3) to write

$$\frac{\alpha n_0}{W p_A(1)} = \frac{4\pi \alpha n_0 r^2}{\alpha_1 N_u} = \frac{3r^2}{\alpha_1 n_0 R_S^3} = \frac{3}{\tau_S}\left(\frac{r}{R_S}\right)^2,$$

where

$$\tau_S \equiv R_S n_0 \alpha_1$$ (10.35)

is a large dimensionless parameter equal to the optical depth at the Lyman edge through the entire H II region *as if the nebula were composed completely of neutral atomic hydrogen*. In equation (10.35) and in what follows, we consider R_S to be *defined* by equation (10.3).

We assume that τ_S can be taken to be a constant independent of r, i.e., that R_S defined through equation (10.3) via the recombination coefficient α, which depends on the gas temperature T, does not vary with spatial position. This turns out to be a good approximation because thermal-balance calculations (as well as observations) indicate that H II regions are effectively isothermal. The result arises for the following reason. The photoabsorption heating rate is proportional to $n(H[1])W_1 p_A(1)$ times the mean excess kinetic energy ϵ_A of the ejected photoelectron. (For $kT_* \ll h\nu_1$, one can show $\epsilon_A \approx kT_*$.) In accordance with equation (10.32), this heating rate is proportional, in equilibrium, to the product $n_e n_p$. But under optically thin conditions, the cooling rate due to inelastic two-body collisions is also proportional to $n_e n_p$ (times a function of the gas temperature T). Consequently, when heating balances cooling, the density dependence drops out, and a definite temperature T results. Typically, T works out to be $\sim 10^4$ K.

If we define the dimensionless radial variable

$$z \equiv r/R_S,$$ (10.36)

we may now write equation (10.34) as

$$\exp\left(-\tau_S \int_0^z f \, dz\right) = \frac{3z^2}{\tau_S}(1-f)^2 f^{-1}.$$ (10.37)

Problem Set 3 asks you to solve this equation to obtain f as a function of z, in particular, to derive Stromgren's conclusion that $f \approx 0$ for $z < 1$, and $f \approx 1$ for $z > 1$ when $\tau_S \gg 1$.

ANALYTIC EVALUATION OF SUMS AND INTEGRALS

For numerical estimates, we need to evaluate the integrals involved in the definitions of the rate coefficients. If we set the Gaunt factor in equation (10.15) equal to 1, we readily obtain

$$p_S(n) = \frac{8\pi}{c^2} n\alpha_1 \nu_n^3 E_1(x_n), \qquad \text{where} \qquad x_n \equiv \frac{h\nu_n}{kT}, \qquad (10.38)$$

with E_1 being the first exponential integral. The corresponding expression for $k_S(n)$ reads

$$k_S(n) = k_0 x_n^{3/2} e^{x_n} E_1(x_n), \qquad (10.39)$$

with

$$k_0 \equiv \pi^{-1/2}(e^2/\hbar c)^3 c\alpha_1 = 5.2 \times 10^{-14} \text{ cm}^3 \text{ s}^{-1}. \qquad (10.40)$$

We follow M. Seaton (1959, *M.N.R.A.S.*, **119**, 81) in replacing sums of any function F by integrals through the application of the approximate formula (see Abramowitz and Stegun 1965),

$$\sum_{n=\mathcal{N}}^{N} F(n) \approx \frac{1}{2}[F(\mathcal{N}) + F(N)] + \int_{\mathcal{N}}^{N} F(n)dn. \qquad (10.41)$$

By transforming the integral to one over $x \equiv x_n$, we obtain for $\mathcal{N} = 2$ and $N \to \infty$,

$$\alpha = k_0 x_2^{1/2} \left[x_2 e^{x_2} E_1(x_2) \left(\frac{1}{2} + \frac{1}{x_2} \right) + \gamma + \ln x_2 \right], \qquad (10.42)$$

where $\gamma = 0.5772\ldots$ is Euler's constant. Equation (10.42) represents the lowest order approximation to the standard "Case B" rate constants of Seaton. For x_2 not very different from 1, α has the order of magnitude of k_0. To evaluate it more accurately, we need to be able to calculate $x_2 = h\nu_2/kT = (39{,}500 \text{ K}/T)$, i.e., the temperature T of the H II region. This task is considered in Osterbrock (1989, Chapter 3); please see his discussion.

CLASSICAL THEORY
OF RADIATION PROCESSES

11

Classical Electrodynamics

With this chapter, we begin Part II of this volume: the classical theory of radiation processes. This topic provides a natural bridge between the macroscopic description of radiation transport that we discussed in Part I of the book and the microscopic description at a quantum level that we shall develop during Part III.

MAXWELL'S EQUATIONS

The classical theory of electromagnetic fields has its basis in Maxwell's equations. In Maxwell's original formulation, four vector fields characterize the subject: an electric field $\mathbf{E}$ and a magnetic field $\mathbf{B}$, plus their associated counterparts, the displacement vector $\mathbf{D}$ and the magnetic intensity $\mathbf{H}$. The latter are required to take into account the secondary influences (induced polarization charge and magnetic moment) that electric and magnetic fields have on a background material medium that is otherwise electrically neutral and nonmagnetic. If one formulates the theory with the vacuum state as the background medium (the great contribution of Lorentz), only the microscopic fields $\mathbf{E}$ and $\mathbf{B}$ appear, and Maxwell's equations take the simplified form (in cgs units) of: Coulomb's law,

$$\boldsymbol{\nabla} \cdot \mathbf{E} = 4\pi\rho_{\mathrm{e}}; \tag{11.1}$$

Faraday's law of induction,

$$\boldsymbol{\nabla} \times \mathbf{E} = -\frac{1}{c}\frac{\partial \mathbf{B}}{\partial t}; \tag{11.2}$$

Gilbert's discovery of the nonexistence (we believe) of magnetic monopoles,

$$\boldsymbol{\nabla} \cdot \mathbf{B} = 0, \tag{11.3}$$

113

and Maxwell's generalization of Ampere's law,

$$\mathbf{\nabla} \times \mathbf{B} = \frac{4\pi}{c}\mathbf{j}_e + \frac{1}{c}\frac{\partial \mathbf{E}}{\partial t}. \tag{11.4}$$

In equations (11.1) and (11.4), ρ_e and $\mathbf{j}_e$ represent *all* the electric charge and flowing current in the system, including the contributions, if any, from the polarization and magnetization of neutral atoms and molecules that are usually handled by introducing the additional vectors $\mathbf{D}$ and $\mathbf{H}$.

We have followed the standard procedure of writing equations (11.1)–(11.4) in the forms that best remind us that the complete characterization of a vector field requires the joint specification of its divergence and curl. From this perspective, the distributions of electric charge ρ_e and electric current $\mathbf{j}_e$ constitute the ultimate material sources for the generation of electric and magnetic fields. These two quantities are not independent of each other, since the divergence of equation (11.4), together with equation (11.1), recovers Benjamin Franklin's finding of the conservation of total electric charge:

$$\frac{\partial \rho_e}{\partial t} + \mathbf{\nabla} \cdot \mathbf{j}_e = 0. \tag{11.5}$$

Indeed, it was Maxwell's mathematical formulation using the field concepts envisaged empirically by Faraday which led Maxwell to realize that Ampere's law in its original form needed an additional term, the so-called "displacement current," $(1/c)\partial \mathbf{E}/\partial t$, if the laws of electrodynamics were to be consistent with the conservation of charge, equation (11.5).

From the point of view that field equations (like the equations of hydrodynamics) should allow us to calculate the future time-evolution of the system given its present spatial configuration, we more naturally write Maxwell's equations (11.1)–(11.4) as consisting of the fundamental set:

$$\frac{\partial \mathbf{E}}{\partial t} = c\mathbf{\nabla} \times \mathbf{B} - 4\pi\mathbf{j}_e, \tag{11.6}$$

$$\frac{\partial \mathbf{B}}{\partial t} = -c\mathbf{\nabla} \times \mathbf{E}, \tag{11.7}$$

$$\frac{\partial \rho_e}{\partial t} = -\mathbf{\nabla} \cdot \mathbf{j}_e. \tag{11.8}$$

From this viewpoint, equations (11.1) and (11.3) may be regarded as *initial conditions* to be imposed on the system, because taking the divergence of equations (11.6) and (11.7), together with the use of (11.8), gives

$$\frac{\partial}{\partial t}(\mathbf{\nabla} \cdot \mathbf{E} - 4\pi\rho_e) = 0, \tag{11.9}$$

$$\frac{\partial}{\partial t}(\mathbf{\nabla} \cdot \mathbf{B}) = 0, \tag{11.10}$$

which demonstrates that if equations (11.1) and (11.3) held initially, they would hold for all time.

POYNTING'S THEOREM

In principle, Maxwell's equations (with the imposition of appropriate boundary and initial conditions) could be solved to obtain $\mathbf{E}$ and $\mathbf{B}$ if only we knew the detailed distributions of electric charge and current: ρ_e and $\mathbf{j}_e$. In astrophysics, we may generally (but not always) concern ourselves with the contributions from only free electrons and ions. Moreover, in the classical approximation, we can specify, in principle, the detailed position and velocity $(\mathbf{x}_i, \mathbf{v}_i)$ of $i = 1$ to N charged particles, so that we may compute the detailed charge and current distribution as

$$\rho_e(\mathbf{x}, t) = \sum_i q_i \delta[\mathbf{x} - \mathbf{x}_i(t)], \tag{11.11}$$

$$\mathbf{j}_e(\mathbf{x}, t) = \sum_i q_i \mathbf{v}_i \delta[\mathbf{x} - \mathbf{x}_i(t)], \tag{11.12}$$

where q_i and the three-dimensional Dirac delta function $\delta[\mathbf{x} - \mathbf{x}_i(t)]$ give the charge and the spatial position $\mathbf{x}_i(t)$ of particle i. We may calculate the latter as a function of time t from the relativistic generalization of Newton's laws:

$$\frac{d\mathbf{x}_i}{dt} = \mathbf{v}_i, \qquad \frac{d\mathbf{p}_i}{dt} = \mathbf{F}_i, \tag{11.13}$$

where

$$\mathbf{p}_i = \gamma_i m_i \mathbf{v}_i \qquad \text{where} \qquad \gamma_i \equiv (1 - v_i^2/c^2)^{-1/2}, \tag{11.14}$$

with m_i being the rest mass. In equation (11.13), the force $\mathbf{F}_i$ usually contains as its largest term the electromagnetic contribution of the Lorentz force:

$$\mathbf{F}_{ei} = q_i \left(\mathbf{E} + \frac{\mathbf{v}_i}{c} \times \mathbf{B} \right). \tag{11.15}$$

The electromagnetic field and the particles can do work on each other. Since the magnetic force is perpendicular to $\mathbf{v}_i$, the rate of the electromagnetic field doing work on the charged particles, $\sum_i \mathbf{v}_i \cdot \mathbf{F}_{ei}$, arises, per unit volume at $\mathbf{x}$, solely from the term

$$\sum_i \delta[\mathbf{x} - \mathbf{x}_i(t)] \mathbf{v}_i \cdot (q_i \mathbf{E}) = \mathbf{j}_e \cdot \mathbf{E}. \tag{11.16}$$

Taking the dot product of $\mathbf{E}$ with equation (11.4), we may express this quantity as

$$\mathbf{j}_e \cdot \mathbf{E} = -\frac{1}{4\pi} \frac{\partial \mathbf{E}}{\partial t} \cdot \mathbf{E} + \frac{c}{4\pi} (\nabla \times \mathbf{B}) \cdot \mathbf{E}.$$

Using the rule for the triple scalar product that crosses and dots may be interchanged, we may write $(\nabla \times \mathbf{B}) \cdot \mathbf{E}$ as $\nabla_B \cdot (\mathbf{B} \times \mathbf{E})$, where ∇_B denotes that the operator acts only on $\mathbf{B}$. We may remove the subscript by allowing ∇ to act on both terms if we add a term $\nabla_E \cdot (\mathbf{E} \times \mathbf{B}) = (\nabla \times \mathbf{E}) \cdot \mathbf{B}$. Thus we have in net

$$\mathbf{j}_e \cdot \mathbf{E} = -\frac{\partial}{\partial t}\left(\frac{E^2}{8\pi}\right) - \nabla \cdot \mathbf{S} + \frac{c}{4\pi}(\nabla \times \mathbf{E}) \cdot \mathbf{B},$$

where

$$\mathbf{S} \equiv \frac{c}{4\pi}(\mathbf{E} \times \mathbf{B}) \tag{11.17}$$

represents the *Poynting vector* and corresponds to a flux of energy associated with the electromagnetic field. To deal with the term $c(\nabla \times \mathbf{E}) \cdot \mathbf{B}$, we note that the dot product of $\mathbf{B}$ with equation (11.2) yields the identification that this term equals $-\mathbf{B} \cdot (\partial \mathbf{B}/\partial t)$; i.e., we now have

$$\frac{\partial}{\partial t}\left(\frac{E^2}{8\pi} + \frac{B^2}{8\pi}\right) + \nabla \cdot \mathbf{S} = -\mathbf{j}_e \cdot \mathbf{E}, \tag{11.18}$$

which is Poynting's theorem: *The explicit time-rate of change of the energy density contained in the electromagnetic field, $(E^2 + B^2)/8\pi$, plus the divergence of the electromagnetic energy flux, $\mathbf{S}$, equals the negative of the volumetric rate of work done on the matter by the electromagnetic field, $\mathbf{j}_e \cdot \mathbf{E}$.* Equation (11.18) represents, of course, another example of the generic conservation law, equation (2.6).

DISTINCTION BETWEEN MACROSCOPIC AND MICROSCOPIC FIELDS

The above discussion does not distinguish between (microscopic) radiation fields and (macroscopic) mean fields. To make this distinction, we suppose the N particles can be subdivided into A species, with all particles in the a-th subset having identical charges q_a and masses m_a. We further suppose that we can perform a meaningful averaging so that the a-th species has a *smoothly varying* number density $n_a(\mathbf{x}, t)$ (instead of a sum of delta functions), and mean velocity field $\mathbf{u}_a(\mathbf{x}, t)$ (local average of the individual particle velocities $\mathbf{v}_i$ in the subset a). The *macroscopic* charge density is then given by

$$\langle \rho_e \rangle = \sum_a q_a n_a, \tag{11.19}$$

whereas the macroscopic current density can be computed from

$$\langle \mathbf{j}_e \rangle = \sum_a q_a n_a \mathbf{u}_a. \tag{11.20}$$

The mean velocity of each species a in turn must be obtained, in principle, by solving a Boltzmann transport equation to obtain the distribution of particles of individual velocities $\mathbf{v} = \mathbf{u} + \mathbf{w}$ in the subset a, when $\mathbf{F}$ on the left-hand side [see equation (8.2)] contains the contribution of equation (11.15).

The sum of the zeroth moment of all charged species in such a formulation will automatically satisfy the conservation of total charge, equation (11.8), even if individual species a can appear or disappear (in the form of source or sink terms on the right-hand sides of the relevant rate equations). The bulk velocity $\mathbf{u}$ of the entire material system will now be given by

$$\rho\mathbf{u} \equiv \sum_a m_a n_a \mathbf{u}_a, \qquad (11.21)$$

where the sum extends over uncharged species ($q_a = 0$) as well as charged, and where the mass density ρ is defined by

$$\rho \equiv \sum_a m_a n_a. \qquad (11.22)$$

Although the differences between the mean velocities $\mathbf{u}_a$ of the individual species a will generally turn out small, for the effects of electromagnetism, we cannot ignore them (as we did in Chapter 8), because it is precisely the small differences between the positive and negative charges that usually give rise to nonzero macroscopic values of the current density $\mathbf{j}_e$. For example, if we had only one positive ions of charge $+e$ and electrons of charge $-e$, with charge neutrality valid on a macroscopic scale, $n_+ = n_- \equiv n_e$, equation (11.20) would give

$$\langle \mathbf{j}_e \rangle = n_e e(\mathbf{u}_+ - \mathbf{u}_-). \qquad (11.23)$$

We will elaborate more on this topic in Volume II when we discuss the equations of plasma physics and magnetohydrodynamics, in which $\langle \rho_e \rangle$ and $\langle \mathbf{j}_e \rangle$ act as the matter sources for macroscopic electric and magnetic fields, $\langle \mathbf{E} \rangle$ and $\langle \mathbf{B} \rangle$, that vary relatively slowly in space and time. (These fields can also have a macroscopic Poynting flux, $\langle \mathbf{S} \rangle = [c/4\pi][\langle \mathbf{E} \rangle \times \langle \mathbf{B} \rangle]$, that carries away energy, not in the form of electromagnetic radiation, but in the form of *hydromagnetic waves*.) For the current discussion, we wish to concentrate on the rapidly fluctuating (radiation) fields, $\delta\mathbf{E} \equiv \mathbf{E} - \langle \mathbf{E} \rangle$ and $\delta\mathbf{B} \equiv \mathbf{B} - \langle \mathbf{B} \rangle$, that arise from the rapidly varying (microscopic) charge and current distributions, $\delta\rho_e \equiv \rho_e - \langle \rho_e \rangle$ and $\delta\mathbf{j}_e \equiv \mathbf{j}_e - \langle \mathbf{j}_e \rangle$, because of the particulate granularity of the actual world.

Having made these preliminary remarks, we henceforth drop the angular brackets and deltas acting on the relevant variables, and trust to the context of the discussion to distinguish whether we are speaking of macroscopic or microscopic fields. We now move to a consideration of the electromagnetic-wave solutions of Maxwell's equations in a vacuum. Before we begin our formal development, however, we comment that nontrivial solutions are possible in a vacuum because according to Faraday's law of induction, time-varying magnetic fields can give rise to electric fields, while according to Maxwell's generalization of Ampere's law, time-varying electric fields can generate magnetic fields. Therefore, *continuously varying* electric and magnetic fields can sustain each other indefinitely, *even when there are no material charges and currents to act as sources for these fields!* This surprising prediction of Maxwell's equation—the propagation of light through a vacuum—proved hard for physicists to swallow at the end of the nineteenth century, and so they reinvented Aristotole's concept of the aether (which had also appeared in various bogus theories of heat) to give a material medium for supporting the oscillations of light when it travels from the distant stars to us. Notice that no such difficulty exists if we accept Newton's viewpoint that light consists of *corpuscles*. Today, we have become familiar with the concept of light as *photons*; so we have lost the classical sense of surprise that light can propagate in a vacuum. In any case, Michelson and Morley's empirical disproof of an all-pervasive aether that provides a substrate for the undulation of light facilitated the eventual acceptance of the theory of special relativity (as invented by Lorentz, Poincaré, and Einstein). Particle physics has given birth to the concept of a quantum vacuum seething with virtual particles and even of false vacuums that can generate the real particles of the *entire universe*; so perhaps an aether under sophisticated guise still resides in humanity's deepest thoughts about the subject.

WAVE SOLUTIONS IN THE VACUUM

In a classical vacuum, where $\rho_e = 0$ and $\mathbf{j}_e = 0$, equations (11.6) and (11.7) read

$$\frac{\partial \mathbf{E}}{\partial t} = c\mathbf{\nabla} \times \mathbf{B} \quad \text{and} \quad \frac{\partial \mathbf{B}}{\partial t} = -c\mathbf{\nabla} \times \mathbf{E}. \qquad (11.24)$$

If we take the time derivative of the first equation and combine it with the curl of the second equation, we obtain

$$\frac{\partial^2 \mathbf{E}}{\partial t^2} = -c^2 \mathbf{\nabla} \times (\mathbf{\nabla} \times \mathbf{E}).$$

Using the triple vector product to expand out the right-hand side, together with the application of the constraint $\mathbf{\nabla} \cdot \mathbf{E} = 0$, we obtain the result that $\mathbf{E}$ satisfies the *homogeneous wave equation*:

$$\frac{\partial^2 \mathbf{E}}{\partial t^2} - c^2 \nabla^2 \mathbf{E} = 0. \qquad (11.25)$$

Similarly, if we take the time derivative of the second relation in equation (11.24) and combine it with the curl of the first relation, together with the constraint $\nabla \cdot \mathbf{B} = 0$, we get (from the symmetry between $\mathbf{E}$ and $\mathbf{B}$):

$$\frac{\partial^2 \mathbf{B}}{\partial t^2} - c^2 \nabla^2 \mathbf{B} = 0. \tag{11.26}$$

Consider waves that have spatial variations only in the z direction, so that we may write

$$\frac{\partial^2}{\partial t^2} - c^2 \nabla^2 = \frac{\partial^2}{\partial t^2} - c^2 \frac{\partial^2}{\partial z^2} = \left(\frac{\partial}{\partial t} + c \frac{\partial}{\partial z} \right) \left(\frac{\partial}{\partial t} - c \frac{\partial}{\partial z} \right). \tag{11.27}$$

If we define $\zeta_\pm \equiv z \pm ct$, we easily show that the wave operator now transforms into

$$\frac{\partial^2}{\partial t^2} - c^2 \nabla^2 = -4c^2 \frac{\partial^2}{\partial \zeta_- \partial \zeta_+},$$

where the partial derivative with respect to ζ_+ is taken at constant ζ_- and vice versa for ζ_-. Thus the most general solutions to equations (11.25) and (11.26) consist of arbitrary functions of ζ_- plus arbitrary functions of ζ_+, i.e., of plane waves propagating at the speed of light c toward positive and negative values of z:

$$\mathbf{E} = \mathbf{E}_+(z - ct) + \mathbf{E}_-(z + ct), \tag{11.28}$$

$$\mathbf{B} = \mathbf{B}_+(z - ct) + \mathbf{B}_-(z + ct), \tag{11.29}$$

where $\mathbf{E}_\pm$ and $\mathbf{B}_\pm$ are arbitrary vector functions of their arguments. The constraints $\nabla \cdot \mathbf{E} = 0 = \nabla \cdot \mathbf{B}$ require that these functions satisfy

$$\hat{\mathbf{z}} \cdot \mathbf{E}'_\pm = 0 = \hat{\mathbf{z}} \cdot \mathbf{B}'_\pm, \tag{11.30}$$

where the superscript prime denotes differentiation with respect to the appropriate argument. The above equation implies that $\mathbf{E}$ and $\mathbf{B}$ must both be *transverse* to the direction $\pm \hat{\mathbf{z}}$ of wave propagation. On the other hand, the original equations (11.24) require that $\mathbf{E}_\pm$ and $\mathbf{B}_\pm$ themselves are mutually perpendicular and equal in magnitude:

$$\mathbf{B}_+ = \hat{\mathbf{z}} \times \mathbf{E}_+ \qquad \text{and} \qquad \mathbf{B}_- = -\hat{\mathbf{z}} \times \mathbf{E}_-, \tag{11.31}$$

in such a way that $\mathbf{E}_\pm \times \mathbf{B}_\pm$ always points in the direction $\pm \hat{\mathbf{z}}$ of wave propagation [which explains the convention used to define the Poynting vector in equation (11.17)].

Our discussion shows that knowledge of the behavior of $\mathbf{E}$ in a *radiation field* gives complete knowledge also of $\mathbf{B}$; consequently, we may focus our attention exclusively on $\mathbf{E}$. In particular, we notice that we have said nothing so far about the possible relationship between the two components

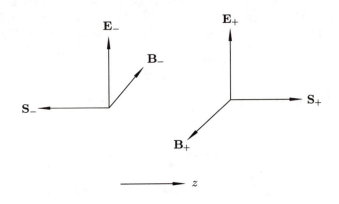

FIGURE 11.1
Relationship between electric vector **E**, magnetic vector **B**, and wave vector **k**
for plane waves propagating in the positive and negative z directions.

of **E**, E_x and E_y, that are mutually perpendicular to the propagation direction in z (see Figure 11.1). This comment motivates our discussion in
Chapter 12 of the possible polarization states of classical electromagnetic
radiation and the Stokes parameters for describing them.

12

Spectral Decomposition and Stokes Parameters

Reference: Chandrasekhar, *Radiative Transfer*, pp. 24–35.

In this chapter we wish to decompose an arbitrarily varying plane wave propagating in a vacuum into different frequencies and different polarization states. We consider the spectral decomposition first, and we begin by reviewing the elements of Fourier transform theory.

FOURIER TRANSFORM THEORY

For any (complex) function $F(x)$ in L_2-space such that

$$\int_{-\infty}^{\infty} |F(x)|^2 \, dx \qquad \text{is finite,} \tag{12.1}$$

where $|F|^2 \equiv FF^*$, there exists a Fourier transform pair such that

$$f(k) = \frac{1}{2\pi} \int_{-\infty}^{\infty} F(x)e^{-ikx} \, dx, \qquad F(x) = \int_{-\infty}^{\infty} f(k)e^{ikx} \, dk. \tag{12.2}$$

The fundamental relation (12.2) between the pair follows from the delta-function representations:

$$\int_{\infty}^{\infty} e^{-i(k-k')x} \, dx = 2\pi\delta(k - k'), \qquad \int_{\infty}^{\infty} e^{ik(x-x')} \, dk = 2\pi\delta(x - x'). \tag{12.3}$$

If we accept the validity of equation (12.3), which Lighthill has put on a rigorous basis, we no longer need Lebesgue integration to discuss Fourier transform theory. When we convolve two functions

$$F \star G(x) \equiv \int_{-\infty}^{\infty} F(\xi)G(x - \xi) \, d\xi, \tag{12.4}$$

its Fourier transform is given by the Faltung theorem as

$$\text{Fourier Transform of } F \star G = 2\pi f(k)g(k). \tag{12.5}$$

The placement of the 2π's is a little awkward to remember, which has led Lighthill to recommend the alternative definitions

$$f(\nu) = \int_{-\infty}^{\infty} F(t)e^{i2\pi\nu t}\, dt, \qquad F(t) = \int_{-\infty}^{\infty} f(\nu)e^{-i2\pi\nu t}\, d\nu. \qquad (12.6)$$

In our discussions, we will use whichever form yields the most convenient representation, with deference given to the conventional treatment if a conflict arises.

The relations (12.2) have simple multidimensional generalizations. For example, in three dimensions,

$$f(\mathbf{k}) = \frac{1}{(2\pi)^3} \int F(\mathbf{x})e^{-i\mathbf{k}\cdot\mathbf{x}}\, d^3x, \qquad F(\mathbf{x}) = \int f(\mathbf{k})e^{i\mathbf{k}\cdot\mathbf{x}}\, d^3k, \qquad (12.7)$$

where the integrals are understood to extend from $-\infty$ to ∞ in all three dimensions.

SPECTRAL DECOMPOSITION

Consider a plane electromagnetic wave propagating in the $+z$ direction. If the wave train is of finite length (so it is L_2 integrable), it has a Fourier decomposition

$$\mathbf{E}_+(z - ct) = \int_{-\infty}^{\infty} \mathbf{e}_+(k)e^{ik(z-ct)}\, dk, \qquad (12.8)$$

where the product $kc \equiv \omega \equiv 2\pi\nu$, the radian frequency. For what follows, we find it convenient to allow $\mathbf{e}_+$ to be complex, in which case we recover the physical electric field by taking the real part of equation (12.8). In general, then, $\mathbf{e}_+$, which must be transverse to $\hat{\mathbf{z}}$, can be written

$$\mathbf{e}_+(k) = \hat{\mathbf{x}}\mathcal{E}_x e^{i\phi_x} + \hat{\mathbf{y}}\mathcal{E}_y e^{i\phi_y}, \qquad (12.9)$$

where $\mathcal{E}$ and ϕ are real constants (dependent on k)—the amplitude and phase.

The Fourier component that contributes to the real part of equation (12.8) therefore reads

$$\mathbf{E}_k = \hat{\mathbf{x}}\mathcal{E}_x \cos[k(z - ct) + \phi_x] + \hat{\mathbf{y}}\mathcal{E}_y \cos[k(z - ct) + \phi_y], \qquad (12.10)$$

where $\mathcal{E}_x$, $\mathcal{E}_y$, ϕ_x, ϕ_y are generally functions of k. Thus an arbitrarily polarized monochromatic wave propagating in a definite direction is characterized by *four* additional parameters rather than the one, I_ν, that we have so far used in developing the equation of transfer. It would obviously be

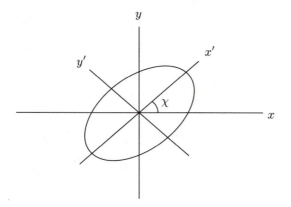

FIGURE 12.1
Geometry for elliptically polarized wave propagating in the $+z$ direction.

inconvenient to try to develop a theory of radiative transfer utilizing quantities with such disparate dimensions as electric fields and angular phases. A better approach follows the development of Stokes, who in 1852 found an apt set of four parameters to describe the propagation of polarized light. This occurred four years before Maxwell wrote his first paper on electromagnetism, but after Young had demonstrated that light had to have two independent transverse modes of oscillation of *something*, which Maxwell showed later, of course, to be the electric vector **E**.

STOKES PARAMETERS FOR AN ELLIPTICALLY POLARIZED MONOCHROMATIC WAVE

Imagine sitting at a fixed position z, which we may take to be $z = 0$ without any loss of generality. Equation (12.10) now becomes

$$\mathbf{E}_k = \hat{\mathbf{x}}\mathcal{E}_x \cos(\omega t - \phi_x) + \hat{\mathbf{y}}\mathcal{E}_y \cos(\omega t - \phi_y), \qquad (12.11)$$

where $\omega \equiv kc = 2\pi\nu$ is the radian frequency. As a function of time, the tip of the electric vector $\mathbf{E}_k$ in equation (12.11) will trace out an ellipse. In general, the principal axes of this ellipse will have a tilt angle χ with respect to x-y coordinate system (see Figure 12.1), i.e.,

$$\begin{pmatrix} \hat{\mathbf{x}}' \\ \hat{\mathbf{y}}' \end{pmatrix} = \begin{pmatrix} \cos\chi & \sin\chi \\ -\sin\chi & \cos\chi \end{pmatrix} \begin{pmatrix} \hat{\mathbf{x}} \\ \hat{\mathbf{y}} \end{pmatrix}. \tag{12.12}$$

We define the zero of time so that the electric vector $\mathbf{E}_k$ lies along the x' direction at $t = 0$; i.e., we have a principal axes decomposition,

$$\mathbf{E}_k = \hat{\mathbf{x}}' E_1 \cos\omega t + \hat{\mathbf{y}}' E_2 \sin\omega t. \tag{12.13}$$

If we time average the square of equations (12.11) and (12.13), we obtain the identification:

$$\mathcal{E}_x^2 + \mathcal{E}_y^2 = E_1^2 + E_2^2 = \text{constant} \equiv \mathcal{E}_0^2. \tag{12.14}$$

We can satisfy the last part of the above equation by defining an ellipticity angle β so that

$$E_1 = \mathcal{E}_0 \cos\beta, \qquad E_2 = \mathcal{E}_0 \sin\beta. \tag{12.15}$$

Collecting terms in equations (12.11), (12.12), (12.13), and (12.15), we get for the coefficient of $\hat{\mathbf{x}} \cos\omega t$,

$$\mathcal{E}_x \cos\phi_x = \mathcal{E}_0 \cos\beta \cos\chi; \tag{12.16}$$

of $\hat{\mathbf{x}} \sin\omega t$,

$$\mathcal{E}_x \sin\phi_x = -\mathcal{E}_0 \sin\beta \sin\chi; \tag{12.17}$$

of $\hat{\mathbf{y}} \cos\omega t$,

$$\mathcal{E}_y \cos\phi_y = \mathcal{E}_0 \cos\beta \sin\chi; \tag{12.18}$$

and of $\hat{\mathbf{y}} \sin\omega t$,

$$\mathcal{E}_y \sin\phi_y = \mathcal{E}_0 \sin\beta \cos\chi. \tag{12.19}$$

The Stokes parameters are defined by the different sums of quadratic products of the above:

$$I = \mathcal{E}_x^2 + \mathcal{E}_y^2 = \mathcal{E}_0^2, \tag{12.20}$$

$$Q = \mathcal{E}_x^2 - \mathcal{E}_y^2 = \mathcal{E}_0^2 \cos 2\beta \cos 2\chi, \tag{12.21}$$

$$U = 2\mathcal{E}_x\mathcal{E}_y \cos(\phi_y - \phi_x) = \mathcal{E}_0^2 \cos 2\beta \sin 2\chi, \tag{12.22}$$

$$V = 2\mathcal{E}_x\mathcal{E}_y \sin(\phi_y - \phi_x) = \mathcal{E}_0^2 \sin 2\beta. \tag{12.23}$$

Notice that for a 100 percent elliptically polarized monochromatic wave, the three parameters $\mathcal{E}_0$, β, χ suffice to give essentially the same information as the four parameters $\mathcal{E}_x$, $\mathcal{E}_y$, ϕ_x, ϕ_y because, apart from an arbitrary zero for time (or phase), only the phase difference,

$$\Delta\phi \equiv \phi_y - \phi_x, \tag{12.24}$$

counts in any practical measurement. Thus it must be true that for such a wave, the four Stokes parameters are not independent, but satisfy an additional relationship among them, which we easily discover to be

$$I^2 = Q^2 + U^2 + V^2. \tag{12.25}$$

This relationship holds only for 100 percent polarized waves (see below).

From laboratory measurements of $\mathcal{E}_x$, $\mathcal{E}_y$, and $\Delta\phi$, we may determine, in principle, the Stokes parameters I, Q, U, and V from equations (12.20), (12.21), (12.22), and (12.23). From knowledge of the latter, we may then deduce the polarization angle χ and the ellipticity parameter β through the relations:

$$\tan 2\chi = U/Q, \qquad \sin 2\beta = V/I. \tag{12.26}$$

Light for which $V = 0$ ($\beta = 0$ or $\pm\pi/2$) is said to be *linearly polarized*; whereas light for which $Q = U = 0$ (in which case $V^2 = I^2$) is said to be *circularly polarized*, with the polarization being left-circularly polarized or right-circularly polarized depending on whether V is positive or negative ($\beta = \pm\pi/4$). [Consult Figure 12.1 together with equations (12.13), (12.15), and (12.23). Considerable room for confusion on conventions exists here; so if you need to measure circular polarization, always check on the sign conventions being used by other people.]

STOKES PARAMETERS FOR PRACTICAL MEASUREMENTS

Even the frequency of decameter-wave radio radiation amounts to tens of megahertz, 10^7 cycles per second. Clearly, then, the practical measurement of electromagnetic wave trains involves taking a time average over many cycles. Moreover, we usually have to use a finite bandwidth in frequency and therefore must deal with a range of wavenumbers. The absolute phases ϕ_x and ϕ_y of the individual Fourier components may vary over time and with wavenumber, but for 100 percent elliptically polarized light, the phase difference, equation (12.24), will be preserved, as will the angles χ and β (if they vary slowly enough with changing k). If we denote the average over time and finite bandwidth by angular brackets, the measurable quantities may then be denoted by

$$\bar{I} = \langle \mathcal{E}_x^2 + \mathcal{E}_y^2 \rangle = \langle \mathcal{E}_0^2 \rangle, \tag{12.27}$$

$$\bar{Q} = \langle \mathcal{E}_x^2 - \mathcal{E}_y^2 \rangle = \bar{I} \cos 2\beta \cos 2\chi, \tag{12.28}$$

$$\bar{U} = 2\langle \mathcal{E}_x \mathcal{E}_y \rangle \cos \Delta\phi = \bar{I} \cos 2\beta \sin 2\chi, \tag{12.29}$$

$$\bar{V} = 2\langle \mathcal{E}_x \mathcal{E}_y \rangle \sin \Delta\phi = \bar{I} \sin 2\beta. \tag{12.30}$$

Generally, observed light from celestial sources will not be 100 percent elliptically polarized, because radiation reaching the observer will generally originate from a variety of regions of different polarizations and different wave phases. Consider therefore a beam consisting of a mixture of many independent streams of elliptically polarized light, $\mathbf{E}_k = \sum_n \mathbf{E}_k^{(n)}$, with different $\mathbf{E}_k^{(n)}$ possessing no permanent phase relationships with each other. Quadratic products of the electric vectors involving cross terms from different streams will vanish upon averaging (under the assumption that the different streams are *incoherent*); thus the Stokes parameters have the additive properties:

$$\bar{I} = \sum_n \bar{I}^{(n)}, \qquad \bar{Q} = \sum_n \bar{Q}^{(n)}, \qquad \bar{U} = \sum_n \bar{U}^{(n)}, \qquad \bar{V} = \sum_n \bar{V}^{(n)}.$$
(12.31)

On the other hand, when we compute the squares of $\bar{I}$, $\bar{Q}$, $\bar{U}$, and $\bar{V}$, Schwartz's inequality yields (for an explicit demonstration, consult Chandrasekhar 1950):

$$\bar{I}^2 \geq \bar{Q}^2 + \bar{U}^2 + \bar{V}^2.$$
(12.32)

In this case, the four Stokes parameters $\bar{I}$, $\bar{Q}$, $\bar{U}$, and $\bar{V}$ really do require four independent measurements.

Because of the inequality (12.32), we may always decompose light from any celestial source into two parts: a completely unpolarized part ("natural light" in which $\bar{Q}_u = \bar{U}_u = \bar{V}_u = 0$) and 100 percent elliptically polarized light (in which $\bar{I}_p = [\bar{Q}_p^2 + \bar{U}_p^2 + \bar{V}_p^2]^{1/2}$). In explicit form, for a measured set, $\bar{I}, \bar{Q}, \bar{U}, \bar{V}$,

$$\bar{Q}_p = \bar{Q}, \qquad \bar{U}_p = \bar{U}, \qquad \bar{V}_p = \bar{V},$$
(12.33)

with

$$\bar{I}_p = (\bar{Q}_p^2 + \bar{U}_p^2 + \bar{V}_p^2)^{1/2},$$
(12.34)

and

$$\bar{I}_u = \bar{I} - \bar{I}_p.$$
(12.35)

The ratio $\bar{I}_p/\bar{I}$ gives the fractional polarization.

RADIATIVE TRANSFER OF POLARIZED LIGHT

A simple multiplicative factor allows us to convert $|\mathbf{E_k}|^2$, where $\mathbf{k}$ is an arbitrary propagation vector, into the specific intensity I_ν. We assume that this has been done (see later chapters), and that I_ν, Q_ν, U_ν, and V_ν are the corresponding values for the Stokes parameters $\bar{I}$, $\bar{Q}$, $\bar{U}$, and $\bar{V}$, in the units of monochromatic specific intensity. Following Chandrasekhar, we may now formulate the equation of radiative transfer for polarized light in terms of a vector equation containing the four components

$$\vec{I}_\nu = \begin{pmatrix} I_\nu^+ \\ I_\nu^- \\ U_\nu \\ V_\nu \end{pmatrix},$$ (12.36)

where

$$I_\nu^+ \equiv \frac{1}{2}(I_\nu + Q_\nu),$$ (12.37)

$$I_\nu^- \equiv \frac{1}{2}(I_\nu - Q_\nu),$$ (12.38)

represent the intensities measured with linear polarization filters oriented in two mutually orthogonal directions (x and y). The vector equation of (steady) transfer now reads

$$\hat{\mathbf{k}} \cdot \mathbf{\nabla} \vec{I}_\nu = \rho\kappa_\nu(\vec{S}_\nu - \vec{I}_\nu),$$ (12.39)

where $\vec{S}_\nu$ is a source function with four components.

The earliest problem to be treated in this fashion involves *Rayleigh scattering*, which explains the polarization produced by scattering of sunlight in the Earth's atmosphere. In other areas of astrophysics, scattering by free electrons or dust grains plays an analogous role. For Rayleigh scattering, we need to replace $\phi_\nu(\hat{\mathbf{k}}, \hat{\mathbf{k}}')I_\nu$ by $\overset{\leftrightarrow}{\mathbf{R}} \cdot \vec{I}_\nu$ in the scattering integral where (see Problem Set 3 and the comments below)

$$\overset{\leftrightarrow}{\mathbf{R}} = \frac{3}{8\pi} \begin{pmatrix} \cos^2\Theta & 0 & 0 & 0 \\ 0 & 1 & 0 & 0 \\ 0 & 0 & \cos\Theta & 0 \\ 0 & 0 & 0 & \cos\Theta \end{pmatrix}.$$ (12.40)

In the above, Θ represents the angle through which incident radiation propagating in direction $\mathbf{k}'$ is scattered into $\mathbf{k}$ (i.e., $\cos\Theta = \hat{\mathbf{k}}' \cdot \hat{\mathbf{k}}$), and the coordinate system has been chosen in accordance with Figure 12.2. To remember which of the diagonal elements $\overset{\leftrightarrow}{\mathbf{R}}$ contains the $\cos^2\Theta$ and which the 1, consider the case of right-angle scattering, $\cos\Theta = 0$. Maxwell's equations have as free solutions only transverse waves. Radiation possessing two senses of polarization leaving the source and scattered through 90° will have the polarization component with electric vector originally oscillating along the line of sight removed from the beam; i.e., electric charges forced to oscillate along the line of sight contribute no observed electromagnetic radiation. This explains the presence of the $\cos^2\Theta$ in the first row and

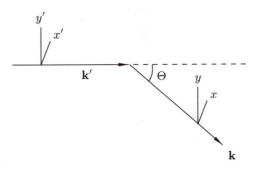

FIGURE 12.2
Geometry for Rayleigh scattering.

first column of $\overleftrightarrow{\mathbf{R}}$. In more detail, if we start with unpolarized light, with equal mixtures of radiation linearly polarized in the x' and y' directions,

$$\vec{I}_\nu \propto \begin{pmatrix} 1/2 \\ 1/2 \\ 0 \\ 0 \end{pmatrix},$$

then once-scattered light will have the polarization properties,

$$\overleftrightarrow{\mathbf{R}} \cdot \vec{I}_\nu \propto \frac{3}{16\pi} \begin{pmatrix} \cos^2 \Theta \\ 1 \\ 0 \\ 0 \end{pmatrix},$$

which constitutes the usual statement of Rayleigh's law (see Problem Set 3). For right-angle scattering, $\Theta = \pi/2$, the scattered light will become completely linearly polarized in the y-direction (equal to the y'-direction) because light polarized in the x' direction, which lies antiparallel to the new direction of propagation, cannot get scattered into the beam. Equivalently, we may say that the x-direction lies along the original direction of propagation, which did not contain any *longitudinally* polarized electromagnetic waves. Notice, finally, that Rayleigh scattering introduces no net circular polarization if none existed originally.

Finally, we comment that $\overleftrightarrow{\mathbf{R}}$ takes the nice diagonal form given in equation (12.40) only because we have used different coordinate systems to represent $\vec{I}_\nu$ before and after the scattering process. For use in equation

(12.39), we want the decomposition of $\vec{I_\nu}$ with respect to a single coordinate system. This requires us to know how $\vec{I_\nu}$ and $\overset{\leftrightarrow}{\mathbf{R}}$ transform under rotations of the coordinate axes, a task carried out in Chandrasekhar (1950, pp. 34–43); see the discussion there.

SOME QUALITATIVE ASTROPHYSICAL APPLICATIONS

Detection of a concentric pattern of polarization vectors in an extended region can indicate that the light comes via scattering from a central point source (see Figure 12.3 on page 130). This technique has helped astronomers discover the true origin of diffuse radiation in circumstances as diverse as nebular condensations near the strong near-infrared source IRc2 in the Orion molecular cloud, and patches of faint light near certain radio galaxies. Scattering by an unresolved distribution of matter cannot lead to a net sense of polarization unless that distribution occurs non-isotropically (or non-uniformly) with respect to the bright illuminating object. For example, reflection of unpolarized light from a young star by a surrounding flattened disk of gas and dust, tilted at a finite angle with respect to the observer, will cause the light to acquire a net linear polarization that lies perpendicular to the (unresolved) long axis of the projected disk (which scatters more of the central light than the short axis). For additional applications, some of which can get quite quantitative, see the astrophysical literature.

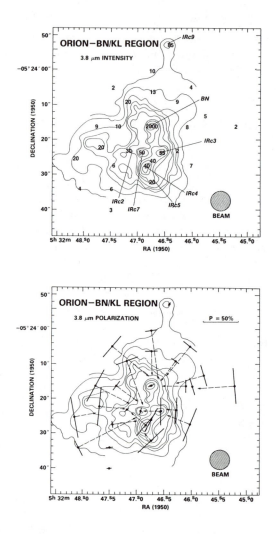

FIGURE 12.3

The upper map shows the infrared intensity map at $3.8\,\mu$m of the
Becklin-Neugebauer/Kleinmann-Low region in Orion. It is not easy to identify
which bright spots correspond to locations of possible protostars. However, the
polarization map below singles out only two positions of intrinsic luminosity:
the object labeled IRc2 (now known to be a source of an intense protostellar
wind) and BN (long suspected to be a relatively high-mass star in the process
of forming). The fingering of these suspects (dashed arrows) comes from the
concentric patterns of polarization vectors expected to accompany
single-particle scattering off dust-laden material surrounding a central
illuminating source. Evidently all the other bright spots in the upper diagram
(IRc 3 through 7) correspond to infrared reflection nebulae. (From M.W.
Werner, H.L. Dinerstein, and R.W. Capps 1983, *Ap. J. (Letters)*, **265**, L13.)

13

Retarded and Lienard-Wiechert Potentials

References: Landau and Lifshitz, *Classical Theory of Fields*, Chapters 8 & 9; Jackson, *Classical Electrodynamics*, 1st ed., pp. 183–188, 268–277.

In this chapter we develop the background material needed to treat classical radiation theory. In particular, we begin by assuming that the charge and current distributions are given, and that we need to find the formal solution of Maxwell's equations with known material sources and sinks. This solution, in terms of retarded potentials, constitutes standard material in many physics texts (for two excellent examples, consult the listed references); so our discussion will highlight only the central points.

SCALAR AND VECTOR POTENTIALS

We begin by replacing $\mathbf{E}$ and $\mathbf{B}$ in Maxwell's equations by an equivalent formulation in terms of electromagnetic potentials. Equation (11.3) implies that we may introduce a vector potential $\mathbf{A}$ such that

$$\mathbf{B} = \boldsymbol{\nabla} \times \mathbf{A}, \tag{13.1}$$

where $\mathbf{A}$ is arbitrary up to an additive gradient of any scalar function $\psi(\mathbf{x}, t)$. If we substitute equation (13.1) into equation (11.2), we may deduce

$$\mathbf{E} = -\boldsymbol{\nabla}\phi - \frac{1}{c}\frac{\partial \mathbf{A}}{\partial t}, \tag{13.2}$$

where the scalar potential ϕ has the same extent of arbitrariness as does $\mathbf{A}$. The nonuniqueness in the definitions of $\mathbf{A}$ and ϕ can be formalized in that a *gauge transformation*,

$$\mathbf{A}' = \mathbf{A} + \boldsymbol{\nabla}\psi, \qquad \phi' = \phi - \frac{1}{c}\frac{\partial \psi}{\partial t}, \tag{13.3}$$

where $\psi(\mathbf{x}, t)$ is completely arbitrary, leaves the physical variables $\mathbf{E}$ and $\mathbf{B}$ unchanged. In particular, we possess the freedom to choose ψ so as to

131

make the computation of $\mathbf{A}$ and ϕ as convenient as possible. For radiation problems, it is conventional to choose ψ so that $\mathbf{A}$ and ϕ satisfy the *Lorentz gauge*,

$$\boldsymbol{\nabla} \cdot \mathbf{A} + \frac{1}{c} \frac{\partial \phi}{\partial t} = 0. \tag{13.4}$$

Equation (13.4) is a scalar equation in a double sense: (1) up to an additive free-wave solution, it fixes the *one* degree of freedom in the choice of ψ and (2) it constitutes a *Lorentz invariant* in the formalism of special relativity because $[\mathbf{A}, i\phi]$ form the components of a *four-vector*. [We use the (older) convention in which $x_4 = ict$ represents the fourth component of a four-vector whose spatial components are $(x_1, x_2, x_3) \equiv \mathbf{x}$. An alternative convention, preferred by general relativists, uses $x_0 = ct$ as the zeroth component of the analogous four-vector. This alternative requires a distinction between so-called *covariant* and *contravariant* components, which provides an unnecessary mathematical nuisance for problems that involve the effects of only special relativity.]

When equations (13.1) and (13.2) are substituted into equations (11.1) and (11.4), we obtain

$$-\nabla^2 \phi - \frac{1}{c} \frac{\partial}{\partial t} (\boldsymbol{\nabla} \cdot \mathbf{A}) = 4\pi \rho_e,$$

$$\boldsymbol{\nabla} \times (\boldsymbol{\nabla} \times \mathbf{A}) = \frac{4\pi}{c} \mathbf{j}_e - \frac{1}{c} \boldsymbol{\nabla} \left(\frac{\partial \phi}{\partial t} \right) - \frac{1}{c^2} \frac{\partial^2 \mathbf{A}}{\partial t^2}.$$

With the help of equation (13.4), we may now write the above relations as the *inhomogeneous wave equation*:

$$\Box^2 \begin{pmatrix} \phi \\ \mathbf{A} \end{pmatrix} = -4\pi \begin{pmatrix} \rho_e \\ \mathbf{j}_e/c \end{pmatrix}, \tag{13.5}$$

where $\Box^2$ is the d'Alembertian operator,

$$\Box^2 \equiv \nabla^2 - \frac{1}{c^2} \frac{\partial^2}{\partial t^2}. \tag{13.6}$$

Our definition, which makes the d'Alembertian the four-dimensional analog of the Laplacian, differs by a minus sign from that adopted in some texts.

GREEN'S FUNCTION SOLUTION

Equation (13.5) has a formal particular solution given by the Green's function technique. Each component of equation (13.5) reads

$$\Box^2 \psi(\mathbf{x}, t) = -4\pi S(\mathbf{x}, t). \tag{13.7}$$

The Green's function technique takes care of the inhomogeneous term S by solving the related equation:

$$\Box^2 G(\mathbf{x}, t; \mathbf{x}', t') = -4\pi\delta(\mathbf{x} - \mathbf{x}')\delta(t - t'), \tag{13.8}$$

where we hold $(\mathbf{x}', t')$ fixed while $(\mathbf{x}, t)$ vary. If the solution of equation (13.8) exists, then we may obtain a particular solution of (13.7) from

$$\psi(\mathbf{x}, t) = \int G(\mathbf{x}, t; \mathbf{x}', t') S(\mathbf{x}', t')\, d^3x'dt', \tag{13.9}$$

where the integration extends over all spacetime. Verification of the result (13.9) follows trivially by applying the d'Alembertian operator to the above and making use of equation (13.8). To the particular function (13.9) we may add any solution of the homogeneous wave equation that we described in Chapter 11; in what follows, however, we are interested only in that part of the radiation field which results from the known material sources.

The standard route to the solution of equation (13.8) makes use of Fourier transforms (see Jackson 1962). Less systematic but more revealing is the physical derivation given below (taken from Landau and Lifshitz 1951). Consider a nonmoving and isotropic source of radiation at the origin of our coordinate system. (For definiteness, think of a fluctuating source of charge.) Waves sent out by this emitter satisfy the inhomogeneous spherically symmetric wave equation:

$$\frac{1}{r^2}\frac{\partial}{\partial r}\left(r^2\frac{\partial\phi}{\partial r}\right) - \frac{1}{c^2}\frac{\partial^2\phi}{\partial t^2} = -4\pi Q(t)\delta(\mathbf{r}). \tag{13.10}$$

Away from the origin $\mathbf{r} = 0$, the right-hand side vanishes, and the most general solution of the homogeneous wave equation reads

$$\phi = \frac{1}{r}\left[f_+(t - r/c) + f_-(t + r/c)\right], \tag{13.11}$$

where the factor $1/r$ is needed to take care of the spherical divergence.

For a particular solution, we may ignore the term $f_-(t + r/c)$ that represents *ingoing* waves. Moreover, we choose the functional form of f_+ so that it gives the correct solution near the origin. As $r \to 0$, derivatives with respect to r dominate those with respect to t; thus equation (13.10) goes over to Coulomb's law

$$\nabla^2\phi = -4\pi Q(t)\delta(\mathbf{r}),$$

which has the well-known solution $\phi = Q(t)/r$, valid for $r \to 0$. Consequently, in equation (13.11), we identify $f_+(t) = Q(t)$, so that the desired particular solution reads

$$\phi = \frac{1}{r}Q(t - r/c). \tag{13.12}$$

If we apply the preceding argument to each point $\mathbf{x}'$ treated as a separate origin of waves in equation (13.8), we obtain the *retarded* Green's function solution:

$$G_{\text{ret}} = \frac{\delta[t' - (t - |\mathbf{x} - \mathbf{x}'|/c)]}{|\mathbf{x} - \mathbf{x}'|}. \tag{13.13}$$

When we substitute G_{ret} into equation (13.9), the integration over t' effectively extends from $-\infty$ up to t, since the argument of the delta function in equation (13.13) cannot equal zero anywhere in space for $t' > t$.

If you are not convinced by the preceding argument, we provide next a direct verification that G_{ret} does constitute a solution of equation (13.8). Notice first the Poisson identity

$$\nabla^2 \left(\frac{1}{|\mathbf{x} - \mathbf{x}'|} \right) = -4\pi\delta(\mathbf{x} - \mathbf{x}'). \tag{13.14}$$

If we now write equation (13.13) in the short-hand notation $G_{\text{ret}} = R^{-1}\delta$, where $R \equiv |\mathbf{x} - \mathbf{x}'|$ and δ is the delta function with the appropriate (retarded) argument, $t' - (t - R/c)$, we obtain

$$\Box^2 G_{\text{ret}} = \delta\nabla^2(R^{-1}) + R^{-1}\Box^2\delta + 2\nabla(R^{-1}) \cdot \nabla\delta. \tag{13.15}$$

The first term of the right-hand side reproduces the right-hand side of equation (13.8); the second term upon differentiation becomes

$$\Box^2\delta = \frac{1}{c^2}\delta''(\nabla R \cdot \nabla R - 1) + \frac{1}{c}\delta'\nabla^2 R,$$

where superscript primes here indicate differentiation with respect to the (retarded) argument of the delta function. Standard (e.g., Cartesian tensor) manipulations reveal

$$\nabla R = \frac{\mathbf{x} - \mathbf{x}'}{|\mathbf{x} - \mathbf{x}'|}, \qquad \nabla^2 R = \frac{2}{R}.$$

Thus we obtain

$$\Box^2\delta = \frac{2}{cR}\delta',$$

which, with $\nabla(R^{-1}) = -R^{-2}\nabla R$ and $\nabla\delta = c^{-1}\delta'\nabla R$, implies that the second and third terms on the right-hand side of equation (13.15) exactly cancel. We have now demonstrated that G_{ret} does indeed satisfy equation (13.8). (Q.E.D.)

If G_{ret} represents an acceptable Green's function, we might also argue *mathematically* for its *advanced* counterpart:

$$G_{\text{adv}} \equiv \frac{\delta[t' - (t + |\mathbf{x} - \mathbf{x}'|/c)]}{|\mathbf{x} - \mathbf{x}'|}. \tag{13.16}$$

Hence we might think that the most general Green's function consists of a combination of the retarded and advanced parts:

$$G = \text{linear combination of } G_{\text{ret}} \text{ and } G_{\text{adv}}. \qquad (13.17)$$

However, the contributions of G_{adv} to equation (13.9) come, anywhere in space, from $t' > t$, i.e., from the *future*. Classical causality prevents the future from influencing the present. For this reason, Sommerfeld advocated the concept of a "radiation condition" to reject the advanced term G_{adv}. Modern trends in physics have blurred to some degree the absolute distinction between future and past, and the timeless genius of Feynman found a meaningful way to make use of the advanced potential in both classical and quantum electrodynamics.

RETARDED POTENTIALS

If we follow the conventional development and retain only G_{ret}, equation (13.9) applied to equation (13.5) yields

$$\begin{pmatrix} \phi \\ \mathbf{A} \end{pmatrix} = \int_V \begin{bmatrix} \rho_e(\mathbf{x}', t') \\ \mathbf{j}_e(\mathbf{x}', t')/c \end{bmatrix} \frac{d^3 x'}{|\mathbf{x} - \mathbf{x}'|}, \qquad (13.18)$$

where t' is now specifically the *retarded time*,

$$t' \equiv t - |\mathbf{x} - \mathbf{x}'|/c, \qquad (13.19)$$

with $t - t'$ representing the interval between emission of an electromagnetic signal at $\mathbf{x}'$ and reception at $\mathbf{x}$.

THE LIENARD-WIECHERT POTENTIAL

The integration in equation (13.18) occurs over only the volume V where charges and currents are found. When can the integration be performed exactly? Let us consider a single charge moving on some trajectory $\mathbf{r}(t)$:

$$\rho_e = q\delta[\mathbf{x} - \mathbf{r}(t)], \qquad \mathbf{j}_e = q\mathbf{v}\delta[\mathbf{x} - \mathbf{r}(t)], \qquad (13.20)$$

where q and $\mathbf{v} \equiv d\mathbf{r}/dt$ are the electric charge and velocity of the particle. Except for the motion of the particle, the problem resembles that of the Green's function method. Motivated by this comment, we adopt the trick of writing

$$\begin{bmatrix} \rho_e(\mathbf{x}', t') \\ \mathbf{j}_e(\mathbf{x}', t')/c \end{bmatrix} = \int \begin{bmatrix} q \\ q\mathbf{v}(\tau)/c \end{bmatrix} \delta[\mathbf{x}' - \mathbf{r}(\tau)]\delta(\tau - t')\, d\tau. \qquad (13.21)$$

When we substitute equation (13.21) into equation (13.18), we obtain upon integration over $\mathbf{x}'$,

$$\begin{bmatrix} \phi(\mathbf{x}, t) \\ \mathbf{A}(\mathbf{x}, t) \end{bmatrix} = \int \begin{bmatrix} q \\ q\mathbf{v}(\tau)/c \end{bmatrix} \frac{\delta(\tau - t')}{|\mathbf{x} - \mathbf{r}(\tau)|}\, d\tau, \qquad (13.22)$$

where t' is now given by the particle retarded time,

$$t' = t - R(\tau)/c, \qquad (13.23)$$

and we have now denoted

$$\mathbf{R}(\tau) \equiv \mathbf{x} - \mathbf{r}(\tau). \qquad (13.24)$$

Equation (13.22) states that the retarded potentials of a single particle may be calculated as an integral over its past history. For $v < c$, i.e., for motion in a vacuum, only one instant in a particle's past history has a light cone that reaches a given location in spacetime $(\mathbf{x}, t)$. The argument of the delta function in equation (13.22) vanishes when $\tau = t'$; thus we are motivated to transform integration variables from τ to τ':

$$\tau' \equiv \tau - t' = \tau - t + R(\tau)/c.$$

In this transformation,

$$d\tau' = [1 + \dot{R}(\tau)/c]\, d\tau.$$

With $R^2(\tau) = \mathbf{R}(\tau)\cdot\mathbf{R}(\tau)$, we obtain $2R(\tau)\dot{R}(\tau) = 2\mathbf{R}(\tau)\cdot\dot{\mathbf{R}}(\tau) = -2\mathbf{R}(\tau)\cdot\mathbf{v}(\tau)$, where $\mathbf{v} = \dot{\mathbf{r}}(\tau)$. Consequently, we have

$$\frac{d\tau}{R(\tau)} = \frac{d\tau'}{R(\tau) - \mathbf{R}(\tau)\cdot\mathbf{v}(\tau)/c}.$$

The difference in the denominator arises physically because the interval $d\tau$ over which the particle emits two successive signals does not equal the interval over which an observer fixed at $\mathbf{x}$ receives these signals if the particle moves toward or away from the observer.

Substituting the above considerations into equation (13.22), we obtain

$$\begin{bmatrix} \phi(\mathbf{x}, t) \\ \mathbf{A}(\mathbf{x}, t) \end{bmatrix} = \int \begin{bmatrix} q \\ q\mathbf{v}(\tau)/c \end{bmatrix} \frac{\delta(\tau')\, d\tau'}{R(\tau) - \mathbf{R}(\tau)\cdot\mathbf{v}(\tau)/c}.$$

The integration is performed by setting $\tau' = 0$, i.e., by finding the value of the retarded time τ so that

$$\tau + R(\tau)/c = t. \tag{13.25}$$

The retarded potentials are now given by the simple formulae:

$$\phi = \left[\frac{q}{R - \mathbf{R} \cdot \mathbf{v}/c}\right]_{\text{ret}}, \qquad \mathbf{A} = \left[\frac{q\mathbf{v}/c}{R - \mathbf{R} \cdot \mathbf{v}/c}\right]_{\text{ret}}, \tag{13.26}$$

where the subscript "ret" on the square brackets indicates that the quantities inside are to be evaluated at the retarded time τ that satisfies equation (13.25). The solutions (13.26) are known as the *Lienard-Wiechert potentials*.

Notice for a nonmoving charge, $\mathbf{v} = 0$, that ϕ and $\mathbf{A}$ reduce to the Coulomb potentials: $\phi = q/R$ and $\mathbf{A} = 0$. Thus the first of the Lienard-Wiechert potentials, ϕ, in equation (13.26) merely states the effect of motion on Coulomb's law, allowing for the fact that it takes a finite time for information about a charge's displacement to reach an observer. A nonzero vector potential $\mathbf{A}$ also arises because a moving charge, which corresponds to an electric current, generates a magnetic field. For example, when an electron oscillates with a displacement amplitude small in comparison to the distance to the observer (so that the observer resides in the *wave zone*; see Chapter 14), the electric field at the position of the observer attempts to follow the charge's position as it appeared one light travel time back. Apart from a static electric field pointing toward the mean position of the electron, then, the observer will also experience an oscillatory perturbational electric field. By Faraday's law of induction, an oscillatory magnetic field will accompany this oscillatory electric field, and the combination of oscillating electric and magnetic fields constitutes the launching of an electromagnetic wave in accordance to Maxwell's description of light. In later chapters, we shall use the results—equation (13.18) for a distribution of charges and equation (13.26) for a single charge—to develop mathematically the physical ideas underlying this verbal description of radiation theory in the context of classical electrodynamics. For now, we merely note that the expression (13.26) results from Maxwell's equations (a relativistically correct theory; indeed, *the* theory that historically motivated relativity considerations) without any approximations concerning the magnitude of $\mathbf{v}$ compared to c. Therefore, Lienard-Wiechert potentials apply to topics such as synchrotron radiation from cosmic rays gyrating relativistically in interstellar magnetic fields. To demonstrate this fact explicitly, we next rederive equation (13.26) from the theory of special relativity.

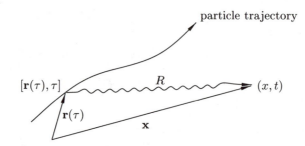

FIGURE 13.1
Information concerning the motion at time τ of a charge q reaches an observer
at time t only after the passage of a light travel time $R(\tau)/c$, where $R(\tau)$ equals
the separation distance at time τ.

ALTERNATIVE DERIVATION USING SPECIAL RELATIVITY

Consider an elementary charge q located in (retarded) spacetime at $[\mathbf{r}(\tau), \tau]$.
(See Figure 13.1.) An observer at spacetime $(\mathbf{x}, t)$ is causally connected to
the charge (through its emitted radiation) if equation (13.25) is satisfied,
i.e., with $R(\tau)$ being given by

$$R(\tau) = c(t - \tau).$$

In the frame of rest of the particle (denoted by superscript primes), Coulomb's
law applies:

$$\phi' = \frac{q}{R}, \qquad \mathbf{A}' = 0, \tag{13.27}$$

at the field point. The four-potential $\vec{A}' \equiv (\mathbf{A}', i\phi')$ satisfies the Lorentz
gauge (13.4). We wish to find $\vec{A} = (\mathbf{A}, i\phi)$ in the lab frame. To do this,
we need to find a covariant four-vector which reduces to $[0, ie/c(t - \tau)]$
in the frame instantaneously at rest with respect to the particle. Such a
four-vector can be constructed from two other four-vectors:

$$\vec{R} \equiv \vec{x} - \vec{r}, \tag{13.28}$$

where $\vec{x} = (\mathbf{x}, ict)$ and $\vec{r} = [\mathbf{r}(\tau), ic\tau]$ are spacetime coordinates, and $\vec{u}$ is
the normalized four-velocity:

$$\vec{u} \equiv \frac{d\vec{r}}{ds} = (\gamma \mathbf{v}/c, i\gamma), \tag{13.29}$$

where γ is the usual Lorentz factor $(1 - v^2/c^2)^{-1/2}$ and ds is the displacement element in spacetime,

$$ds^2 = c^2 d\tau^2 - |d\mathbf{r}|^2. \tag{13.30}$$

By inspection, the desired four-vector can be written as the four-potential

$$\vec{A} \equiv -\frac{q\vec{u}}{\vec{R} \cdot \vec{u}}, \tag{13.31}$$

where the dot product of two four-vectors $\vec{R}$ and $\vec{u}$ in Minkowski space is defined as

$$\vec{R} \cdot \vec{u} \equiv \mathbf{R} \cdot \mathbf{u} - R_4 u_4, \tag{13.32}$$

and has the same numerical value in all inertial frames. In the lab frame, we have

$$\vec{R} \cdot \vec{u} = \gamma \mathbf{R} \cdot \mathbf{v}/c - \gamma c(t - \tau) = \gamma(\mathbf{R} \cdot \mathbf{v}/c - R).$$

Hence equation (13.31) implies

$$\phi = \frac{q}{R - \mathbf{R} \cdot \mathbf{v}/c}, \qquad \mathbf{A} = \frac{q\mathbf{v}/c}{R - \mathbf{R} \cdot \mathbf{v}/c},$$

which reproduce the Lienard-Wiechert potentials, equation (13.26).

14

The Wave Zone and Dipole Radiation

In this chapter we apply the solutions for the retarded potentials derived in the previous chapter to nonrelativistic problems in classical radiation theory. We deal with the situation involving relativistic motions in Chapter 16.

THE WAVE ZONE

For most astronomical applications, the separation x between observer and the center of the source distribution is much greater than either the dimension L typical for the source or the wavelength λ of the spectral regime of interest. By the source here, we mean that part of the system which emits coherently, so that we need to sum the contributions of the electric and magnetic fields (or potentials), rather than sum their absolute squares after the fact of the detailed microscopic calculation. Thus, by L, we are speaking typically of the distance over which an electron travels in a vibrational period, rather than of the dimension of an extended astronomical object such as an emission nebula or a galaxy. Occasionally, however, an entire macroscopic object, such as a rotating, magnetized, neutron star, might radiate electromagnetic waves coherently (see Problem Set 4), and we then might be more interested in investigating the properties of the *near zone*. In any case, the wave-zone approximation consists of the assumptions:

$$x \equiv |\mathbf{x}| \gg L, \quad \text{and} \quad x \gg \lambda. \tag{14.1}$$

Notice that we have not said anything at this point about the relative sizes of L and λ; this will come later when we discuss *multipole expansions*.

The assumption $x \gg \lambda$ allows us to concentrate our attention on variations that arise from the effects of time retardation. More precisely, a particular Fourier component Q_ν varies with time as $\exp(i2\pi\nu t)$; so if $x \gg \lambda = c/\nu$, we obtain the ordering

$$\frac{1}{x}Q_\nu \ll \frac{\nu}{c}Q_\nu \sim \frac{1}{c}\frac{\partial Q_\nu}{\partial t}, \tag{14.2}$$

within factors of 2π. Inside the volume V, the integration variable $\mathbf{x}'$ in equation (13.18) has typical magnitude L. For $x \gg L$ then, we may expand

$$|\mathbf{x} - \mathbf{x}'| \equiv (x^2 + x'^2 - 2\mathbf{x} \cdot \mathbf{x}')^{1/2} \approx x\left(1 - \frac{\mathbf{x} \cdot \mathbf{x}'}{x^2}\right) = x - \hat{\mathbf{k}} \cdot \mathbf{x}',$$

where $\hat{\mathbf{k}}$ is the unit vector in the (propagation) direction toward the observer:

$$\hat{\mathbf{k}} \equiv \mathbf{x}/x. \tag{14.3}$$

In this approximation, the observer's distance from the radiator $|\mathbf{x} - \mathbf{x}'|$ equals the observer's distance x from the origin (center of volume V containing the coherent radiation) minus the radiator's projected distance $\hat{\mathbf{k}} \cdot \mathbf{x}'$ from the origin along the line of sight. This formula forms a familiar aspect of the small-angle approximation to astronomers. The corresponding retarded time given in equation (13.19) now has the simplified expression,

$$t' = t - \frac{1}{c}(x - \hat{\mathbf{k}} \cdot \mathbf{x}'), \tag{14.4}$$

so that the gradient of t' becomes

$$\nabla t' = -\frac{\hat{\mathbf{k}}}{c}. \tag{14.5}$$

Consider the simplification now possible for equation (13.18). Since $x \gg L$, we may ignore the term $\hat{\mathbf{k}} \cdot \mathbf{x}'$ everywhere except in the time dependence. The last exception results from the high frequencies of common electromagnetic waves [see equation (14.2)], whose time variations are large relative to the variations of distance (divided by c) to the coherently emitting parts of the source distribution (i.e., astronomers can usually observe radiation only in the regime where this statement holds true). Hence, in the wave zone, we have the approximate expressions

$$\begin{bmatrix} \phi(\mathbf{x}, t) \\ \mathbf{A}(\mathbf{x}, t) \end{bmatrix} = \frac{1}{x}\int_V \begin{bmatrix} \rho_e(\mathbf{x}', t') \\ \mathbf{j}_e(\mathbf{x}', t')/c \end{bmatrix} d^3x', \tag{14.6}$$

where t' is given by equation (14.4).

ELECTROMAGNETIC RADIATION FIELDS

The evaluation of $\mathbf{E}$ and $\mathbf{B}$ depends on the computation for various space-time derivatives of $\mathbf{A}$ and ϕ (see Chapter 13). For example,

$$\nabla\phi = -\frac{\nabla x}{x^2}\int_V \rho_e\, d^3x' + \frac{1}{x}\int_V (\nabla t')\frac{\partial\rho_e}{\partial t'}\, d^3x' = -\hat{\mathbf{k}}\left(\frac{1}{x}\phi + \frac{1}{c}\frac{\partial\phi}{\partial t}\right),$$

where we have made use of equation (14.5). The first term on the extreme right-hand side of equation (14.5) is much smaller than the second by equation (14.2); consequently, in the wave zone,

$$\nabla\phi = -\frac{\hat{\mathbf{k}}}{c}\frac{\partial\phi}{\partial t}. \tag{14.7}$$

This result suggests that the application of the del operator ∇ in the wave zone is equivalent to the application of the operator $-(\hat{\mathbf{k}}/c)\partial/\partial t$. Direct calculation verifies the mnemonic value of this heuristic result (which comes from the properties of electromagnetic wave propagation in a vacuum). Thus

$$\nabla\cdot\mathbf{A} = -\frac{\hat{\mathbf{k}}}{c}\cdot\frac{\partial\mathbf{A}}{\partial t}, \qquad \nabla\times\mathbf{A} = -\frac{\hat{\mathbf{k}}}{c}\times\frac{\partial\mathbf{A}}{\partial t}. \tag{14.8}$$

Because of the Lorentz gauge, equation (13.4), we may replace derivatives of ϕ by derivatives of $\mathbf{A}$; thus equation (14.7) can be written

$$\nabla\phi = \hat{\mathbf{k}}\left(-\frac{1}{c}\frac{\partial\phi}{\partial t}\right) = \hat{\mathbf{k}}(\nabla\cdot\mathbf{A}) = \hat{\mathbf{k}}\left(-\frac{\hat{\mathbf{k}}}{c}\cdot\frac{\partial\mathbf{A}}{\partial t}\right), \tag{14.9}$$

if we use equation (13.4). We now straightforwardly evaluate the magnetic and electric fields. For $\mathbf{B}$, we use

$$\mathbf{B} = \nabla\times\mathbf{A} = \frac{1}{c}\left(\frac{\partial\mathbf{A}}{\partial t}\times\hat{\mathbf{k}}\right), \tag{14.10}$$

which recovers the result that $\mathbf{B}$ for the radiation field is perpendicular to the propagation direction $\hat{\mathbf{k}}$. With $\mathbf{A}$ given by equation (14.6), equation (14.10) also shows that the radiation $\mathbf{B}$ only declines as $1/x$, instead of the much steeper power associated with quasi-static magnetic fields far from their material sources. The electric field is given by equation (13.2):

$$\mathbf{E} = -\nabla\phi - \frac{1}{c}\frac{\partial\mathbf{A}}{\partial t} = \hat{\mathbf{k}}\left(\frac{\hat{\mathbf{k}}}{c}\cdot\frac{\partial\mathbf{A}}{\partial t}\right) - \frac{1}{c}\frac{\partial\mathbf{A}}{\partial t} = \left(\frac{1}{c}\frac{\partial\mathbf{A}}{\partial t}\times\hat{\mathbf{k}}\right)\times\hat{\mathbf{k}}, \tag{14.11}$$

which with equation (14.10) shows that $\mathbf{E} = \mathbf{B}\times\hat{\mathbf{k}}$, recovering the relation that both radiation fields are transverse to the direction of propagation $\hat{\mathbf{k}}$, with $|\mathbf{E}| = |\mathbf{B}|$. Poynting's vector becomes

$$\mathbf{S} = \frac{c}{4\pi}(\mathbf{E}\times\mathbf{B}) = \frac{c}{4\pi}(\mathbf{B}\times\hat{\mathbf{k}})\times\mathbf{B} = \frac{B^2}{4\pi}c\hat{\mathbf{k}}. \tag{14.12}$$

The differential power radiated into a solid angle $d\Omega$ centered about the direction $\hat{\mathbf{k}}$ equals

$$dP = \mathbf{S} \cdot (\hat{\mathbf{k}} x^2 \, d\Omega) = \frac{(Bx)^2}{4\pi} c \, d\Omega. \tag{14.13}$$

Since $B \propto 1/x$, the differential power $dP/d\Omega$, when averaged over time, is independent of x for a given propagation direction. The time average of $dP/d\Omega$ equals the frequency-integrated specific intensity $I(\hat{\mathbf{k}}, \mathbf{x})$.

DIPOLE RADIATION

Equations (14.10), (14.11), and (14.12) yield general formulae to obtain radiation properties in the wave zone. In the following discussion, we specialize these results to the important approximation of *dipole radiation*; in Chapter 15, we will consider the more formal problem of a general *multipole expansion*.

The dipole approximation results by simplifying equation (14.4) for the retarded time even further:

$$t' \approx t - x/c, \tag{14.14}$$

which amounts to the neglect of all phase differences between different parts of the (coherently) emitting source, and is valid if the size of the source L is utterly negligible with respect to the wavelength λ of the emitted light. (Recall that L typically equals the distance traveled by a radiating charge in an oscillation period, whereas λ equals the distance traveled by light over the same period.) The wave-zone solution for the retarded vector potential in the dipole approximation reads [see equation (14.6)]:

$$\mathbf{A}(\mathbf{x}, t) = \frac{1}{cx} \int_V \mathbf{j}_e(\mathbf{x}', t - x/c) \, d^3 x'.$$

In particular, for a nonrelativistic distribution of moving charges,

$$\mathbf{A}(\mathbf{x}, t) = \frac{1}{cx} \sum_a q_a \mathbf{v}_a(t - x/c) = \frac{\dot{\mathbf{d}}}{cx}, \tag{14.15}$$

where $\mathbf{d}$ is the dipole moment of the charge distribution,

$$\mathbf{d} \equiv \sum_a q_a \mathbf{x}_a, \tag{14.16}$$

evaluated at the retarded time (14.14).

Differentiating equation (14.15) in time, we obtain

$$\frac{\partial \mathbf{A}}{\partial t} = \frac{1}{cx} \ddot{\mathbf{d}} \quad \Rightarrow \quad \mathbf{B} = \frac{1}{c^2 x} \ddot{\mathbf{d}} \times \hat{\mathbf{k}},$$

which implies, in turn,

$$\frac{dP}{d\Omega} = \frac{1}{4\pi c^3}|\ddot{\mathbf{d}} \times \hat{\mathbf{k}}|^2 = \frac{|\ddot{\mathbf{d}}|^2}{4\pi c^3}\sin^2\theta,$$

where θ equals the angle between $\ddot{\mathbf{d}}$ (at the retarded time) and $\hat{\mathbf{k}}$. The angular distribution of radiated power has a maximum perpendicular to the dipole axis (say, due to the oscillation of a single charge) and vanishes over the poles (where there is no generation of a time-varying component of $\mathbf{E}$ perpendicular to the line of sight). If we integrate over all solid angles, with $d\Omega = 2\pi\sin\theta\,d\theta$, to obtain the total radiated power, we obtain the famous Larmor formula:

$$P = \frac{2|\ddot{\mathbf{d}}|^2}{3c^3}. \tag{14.17}$$

SOME SIMPLE EXAMPLES

1. Electron-Electron Collisions

Electrons have a constant charge-to-mass ratio, $q_a/m_a = -e/m_e$ independent of a; therefore

$$\dot{\mathbf{d}} = -\frac{e}{m_e}\sum_a m_a \mathbf{v}_a = -\frac{e}{m_e}\mathbf{P},$$

where $\mathbf{P}$ is the total momentum of the pair of colliding electrons. If the pair can be considered an isolated system, $\ddot{\mathbf{d}} = -(e/m_e)\dot{\mathbf{P}} = 0$, no dipole radiation will emerge. Each electron moves exactly opposite to its partner, so their individual contributions to the time-varying electric fields at large distances cancel in the dipole approximation. *Bremsstrahlung* ("braking radiation") must therefore be dominated by electric-dipole radiation from (nonrelativistic) electron-ion collisions. Thermal ions are typically nearly equally numerous as free electrons, and the relative velocity between electrons and ions differs by only a factor $\sim \sqrt{2}$ from electron-electron encounters.

2. Thomson Model of an Atom

According to Thomson, electrons have equilibrium positions in an atom like raisins in a raisin pudding. When perturbed slightly away from this state, they will vibrate about their equilibrium positions like a harmonic oscillator, with some characteristic radian frequency ω_0 (see Goldstein 1959 for the general theory of small-amplitude oscillations about equilibria). Relative to the electrically neutral unperturbed state, the resulting dipole moment associated with the displacement $\mathbf{x}_1$ equals $\mathbf{d} = -e\mathbf{x}_1$, and the total radiated power equals $P = (2e^2/3c^3)|\ddot{\mathbf{x}}_1|^2$, where $\mathbf{x}_1$ satisfies, to zeroth order, the harmonic-oscillator solution, $\mathbf{x}_1 = \mathbf{x}_0 \cos \omega_0 t$. Thus

$$P = \frac{2e^2}{3c^3} x_0^2 \omega_0^4 \cos^2 \omega_0 t;$$

and the average power radiated over one cycle equals

$$\bar{P} = \frac{e^2 \omega_0^4}{3c^3} x_0^2.$$

The correspondence principle requires that $\bar{P}$ be given by $A_{21}\hbar\omega_{21}$, where A_{21} is the Einstein A associated with the emission process, and ω_{21} represents another way to write the resonance (line) frequency ω_0. On the other hand, we may heuristically identify half the displacement amplitude of the oscillation $\mathbf{x}_0/2$ with the expectation value of the position vector $\mathbf{r}$ in the overlap of the wave functions of the initial and final states involved in the quantum transition

$$\mathbf{x}_0/2 \to \langle 2|\mathbf{r}|1\rangle;$$

hence we expect the coefficient for spontaneous emission A_{21} to be given by a formula like

$$A_{21} = \frac{4e^2\omega_{21}^3}{3c^3\hbar}|\langle 2|\mathbf{r}|1\rangle|^2. \tag{14.18}$$

In a future chapter, we shall find that this formula is indeed accurate if states 2 and 1 are nondegenerate so that their statistical weights equal unity.

3. Classical Theory of Absorption and Radiation Damping

Consider radiation propagating past an electron bound in a Thomson atom. We represent the electric field by the real part of

$$\mathbf{E} = \mathbf{E}_0 e^{i(\mathbf{k}\cdot\mathbf{x}-\omega t)}. \tag{14.19}$$

Without loss of generality, we assume $\mathbf{E}_0$ to be real. We also assume $k|\mathbf{x}_1| \ll 1$, so that the spatial excursion of the electron (of order 1 Å) can be ignored in the context of the relatively smooth spatial variation of $\mathbf{E}$ (which occurs on a wavelength $\gg 1$ Å). We may then set $\mathbf{x} = 0$ in equation (14.19). We also assume that the electron has only nonrelativistic motions $|\mathbf{v}| \ll c$ (i.e., kinetic energies much less than 0.5 MeV), so that we may ignore $-e(\mathbf{v}/c) \times \mathbf{B}$ in comparison to $-e\mathbf{E}$ in the Lorentz force exerted by the electromagnetic wave (with $|\mathbf{B}| = |\mathbf{E}|$) on the electron. The steady-state response of the electron to the presence of the electromagnetic wave has a solution given by the real part of

$$\mathbf{x}_1 = \mathbf{x}_0 e^{i\omega t}. \tag{14.20}$$

To compute $\mathbf{x}_0$, we need to write down the electron's equation of motion. In this equation of motion, we need to include a radiative damping force proportional to $\mathbf{v} \propto \dot{\mathbf{x}}_1$:

$$\ddot{\mathbf{x}}_1 = -\omega_0^2 \mathbf{x}_1 - \gamma \dot{\mathbf{x}}_1 - \frac{e}{m_e} \mathbf{E}_0 e^{-i\omega t}, \tag{14.21}$$

where γ is a radiation reaction coefficient. To determine γ, we note that according to Larmor's formula (14.17), the radiative rate of loss of energy by the electron, averaged over one cycle, equals

$$\langle \frac{dW}{dt} \rangle = -\langle P \rangle = -\frac{2e^2}{3c^3} \langle |\ddot{\mathbf{x}}_1|^2 \rangle, \tag{14.22}$$

where

$$\langle |\ddot{\mathbf{x}}_1|^2 \rangle \equiv \frac{1}{\tau} \int_{t-\tau/2}^{t+\tau/2} \ddot{\mathbf{x}}_1 \cdot \ddot{\mathbf{x}}_1 \, dt,$$

with $\tau \equiv 2\pi/\omega$. If we integrate by parts and note that $\dot{\mathbf{x}}_1$ and $\ddot{\mathbf{x}}_1$ are $90°$ out of phase for the real part of the steady-state response, we obtain

$$\langle |\ddot{\mathbf{x}}_1|^2 \rangle = -\frac{1}{\tau} \int_{t-\tau/2}^{t+\tau/2} \dddot{\mathbf{x}}_1 \cdot \dot{\mathbf{x}}_1 \, dt.$$

Hence, we may write

$$\langle \frac{dW}{dt} \rangle = \langle \mathbf{F}_{RR} \cdot \dot{\mathbf{x}}_1 \rangle,$$

where we identify the radiation reaction force as

$$\mathbf{F}_{RR} \equiv \frac{2e^2}{3c^3} \dddot{\mathbf{x}}_1. \tag{14.23}$$

Historically, Abraham was the first to propose equation (14.23) for the classical radiation reaction force (the damping of a charge's motion which arises because of the emission of radiation). However, can a force dependent on the *third* time-derivative of the particle's displacement have physical meaning? What does it mean to raise the order of Newton's equation of motion by one? (See the discussion by Landau and Lifshitz 1951.) For simple harmonic motion, we can avoid the difficulty by replacing $\dddot{\mathbf{x}}_1$ by $-\omega^2 \dot{\mathbf{x}}_1$. We may then write the radiation reaction force as

$$\mathbf{F}_{RR} = -m_e \gamma \dot{\mathbf{x}}_1,$$

where

$$\gamma \equiv \frac{2e^2\omega^2}{3m_e c^3}. \tag{14.24}$$

In particular, note that the ratio of the damping frequency to the oscillation frequency, γ/ω, equals $(2/3)kr_e \ll 1$, for wavelengths much greater than the classical radius of the electron, $r_e \equiv e^2/m_e c^2 = 2.82 \times 10^{-13}$ cm. In this limit, radiation damping has a well-defined notion.

If we solve the equation of motion for the damped harmonic oscillator, equation (14.21), we obtain for the steady-state displacement amplitude [see equation (14.21)]:

$$\mathbf{x}_0 = \frac{(e/m_e)\mathbf{E}_0}{(\omega^2 - \omega_0^2) + i\omega\gamma}.$$

The complex denominator implies that, in the presence of radiative damping, the response $\mathbf{x}_1$ is slightly out of phase with respect to the imposed field $\mathbf{E}$. When we perform the average indicated in equation (14.22), we now obtain

$$\langle\frac{dW}{dt}\rangle = -\frac{e^2}{3c^3}\omega^4 \mathbf{x}_0 \cdot \mathbf{x}_0^* = -\left(\frac{e^2}{3c^3}\right)\left[\frac{(e^2/m_e^2)E_0^2}{(\omega^2 - \omega_0^2)^2 + \omega^2\gamma^2}\right].$$

In the middle step above, we have used a superscript asterisk to denote complex conjugation.

The average radiative flux incident on the electron equals

$$\langle S \rangle = \frac{c}{8\pi}E_0^2.$$

We define the total cross section σ_{sca} as the ratio of the scattered power $-\langle dW/dt \rangle$ to the incident flux $\langle S \rangle$, i.e.,

$$\langle\frac{dW}{dt}\rangle = -S\sigma_{\text{sca}}.$$

With γ given by equation (14.24), we obtain

$$\sigma_{\text{sca}} = \left(\frac{4\pi e^2}{m_e c}\right)\left[\frac{\gamma\omega^2}{(\omega^2 - \omega_0^2)^2 + \omega^2\gamma^2}\right]. \tag{14.25}$$

SOME LIMITING CASES OF INTEREST

As we have seen above, usually $\gamma \ll \omega$. If in addition, we have

(a) $\omega \ll \omega_0$ (Rayleigh scattering by bound electron):

$$\sigma_{\text{sca}} = \left(\frac{8\pi}{3}r_e^2\right)\left(\frac{\omega^4}{\omega_0^4}\right). \tag{14.26}$$

Notice the ω^4 dependence of the total scattering cross section.

(b) $\omega_0 = 0$ (Thomson scattering by free electron):

$$\sigma_{\text{sca}} = \frac{8\pi}{3} r_{\text{e}}^2. \tag{14.27}$$

Notice that the Thomson scattering cross section has no dependence on the radiation frequency (it is grey). This result holds only for radiation where the photon energy is much less than the rest energy of the electron ~ 0.5 MeV.

(c) $\omega \approx \omega_0$ (resonant scattering of line radiation):

$$\sigma_{\text{sca}} = \left(\frac{\pi e^2}{m_{\text{e}} c}\right) \mathcal{L}, \tag{14.28}$$

where $\mathcal{L}$ is the Lorentz profile:

$$\mathcal{L} = \frac{1}{\pi} \left[\frac{(\gamma/4\pi)}{(\nu - \nu_0)^2 + (\gamma/4\pi)^2} \right], \tag{14.29}$$

and $\nu \equiv \omega/2\pi$, $\nu_0 \equiv \omega_0/2\pi$. If we compare the formula (14.28) with its quantum analog, equation (10.4), we see a one-to-one correspondence, except that the quantum-mechanical expression contains the extra factor f_{12}. Hence, relative to the classical formula, we may think of f_{12} as an "effective number" of oscillators, which explains its name as the "oscillator strength."

15

Multipole Radiation and Thermal *Bremsstrahlung*

In this chapter we discuss two topics: (1) radiation mechanisms beyond the dipole approximation and (2) application of the dipole radiation formula to the problem of thermal *Bremsstrahlung*. We begin with multipole radiation, but instead of taking the approach of a systematic expansion (see Jackson 1962, Chapter 16), we shall specialize the discussion to the recovery of the *electric quadrupole* and *magnetic dipole* formulae (see Landau and Lifshitz 1951, pp. 215–217). These formulae suffice for most considerations beyond the dipole approximation.

MULTIPOLE RADIATION

In Chapter 14, we obtained the dipole approximation by keeping only the first term $t_0 = t - x/c$ in the more accurate expression for the retarded time,

$$t' = t_0 + t_1 \qquad \text{where} \qquad t_1 = \hat{\mathbf{k}} \cdot \mathbf{x}'/c. \tag{15.1}$$

If we substitute equation (15.1) into equation (14.6) and expand for $t_1 \ll t_0$, we obtain for the vector potential

$$\mathbf{A} = \mathbf{A}_0 + \mathbf{A}_1 + \cdots, \tag{15.2}$$

where $\mathbf{A}_0$ corresponds to the dipole approximation and where $\mathbf{A}_1$ represents the next-order correction:

$$\mathbf{A}_1 = \frac{1}{cx} \int_V t_1 \frac{\partial \mathbf{j}_e}{\partial t_0} (\mathbf{x}', t_0) \, d^3 x' = \frac{\hat{\mathbf{k}}}{c^2 x} \cdot \frac{\partial}{\partial t} \int_V \mathbf{x}' \mathbf{j}_e(\mathbf{x}', t - x/c) \, d^3 x'. \tag{15.3}$$

For a nonrelativistic distribution of moving charges,

$$\mathbf{A}_1 = \frac{\hat{\mathbf{k}}}{c^2 x} \cdot \frac{d}{dt} \sum_a q_a(\mathbf{x}_a \mathbf{v}_a), \tag{15.4}$$

where the particle position $\mathbf{x}_a$ and velocity $\mathbf{v}_a$ are to be evaluated at the zeroth-order retarded time $t_0 = t - x/c$. As usual, it pays to express the dyadic $\mathbf{x}_a \mathbf{v}_a$, where $\mathbf{v}_a = d\mathbf{x}_a/dt$, as the sum of symmetric and antisymmetric tensors. We begin by writing

$$\mathbf{x}_a \mathbf{v}_a = \frac{d}{dt}(\mathbf{x}_a \mathbf{x}_a) - \mathbf{v}_a \mathbf{x}_a.$$

If we add $\mathbf{x}_a \mathbf{v}_a$ to both sides of the above and divide by 2, we will have accomplished the desired decomposition. Our reward comes when we dot the resulting expression by $\hat{\mathbf{k}}$:

$$\hat{\mathbf{k}} \cdot \mathbf{x}_a \mathbf{v}_a = \frac{1}{2}\left[\hat{\mathbf{k}} \cdot \frac{d}{dt}(\mathbf{x}_a \mathbf{x}_a) + (\mathbf{x}_a \times \mathbf{v}_a) \times \hat{\mathbf{k}}\right],$$

where we have used the formula for the triple vector product to replace $\hat{\mathbf{k}} \cdot (\mathbf{x}_a \mathbf{v}_a - \mathbf{v}_a \mathbf{x}_a)$ by the last term on the right-hand side.

The above manipulations motivate us to define the *magnetic moment* $\mathbf{M}$ associated with the collection of moving charges as

$$\mathbf{M} \equiv \sum_a \frac{q_a}{2c}(\mathbf{x}_a \times \mathbf{v}_a) = \sum_a \frac{q_a}{2m_a c}\mathbf{L}_a, \tag{15.5}$$

where $\mathbf{L}_a$ is the orbital angular momentum of the a-th charge:

$$\mathbf{L}_a \equiv m_a \mathbf{x}_a \times \mathbf{v}_a. \tag{15.6}$$

We may now write equation (15.4) as

$$\mathbf{A}_1 = \frac{\hat{\mathbf{k}}}{2c^2 x} \cdot \frac{d^2}{dt^2}\left(\sum_a q_a \mathbf{x}_a \mathbf{x}_a\right) + \mathbf{A}_M, \tag{15.7}$$

where $\mathbf{A}_M$ is the vector potential associated with the magnetic dipole,

$$\mathbf{A}_M \equiv \frac{1}{cx}\dot{\mathbf{M}} \times \hat{\mathbf{k}}. \tag{15.8}$$

Without changing equations (14.10) and (14.11) for $\mathbf{B}$ and $\mathbf{E}$, we may add to $\mathbf{A}$ any vector parallel to $\hat{\mathbf{k}}$; thus we may replace equation (15.7) by the expression

$$\mathbf{A}_1 \to \mathbf{A}_Q + \mathbf{A}_M, \tag{15.9}$$

where $\mathbf{A}_Q$ is the vector potential associated with an electric quadrupole,

$$\mathbf{A}_Q \equiv \frac{1}{6c^2 x}\hat{\mathbf{k}} \cdot \ddot{\mathbf{Q}}, \tag{15.10}$$

with $\mathbf{Q}$ being the symmetric and traceless *quadrupole moment*,

$$\mathbf{Q} \equiv \sum_a q_a \left(3\mathbf{x}_a \mathbf{x}_a - |\mathbf{x}_a|^2 \mathbf{I}\right), \tag{15.11}$$

and with **I** being the unit tensor.

Equations (14.10) and (14.11) now give the radiation magnetic and electric fields as

$$\mathbf{B} = \frac{1}{c^2 x} \left[\ddot{\mathbf{d}} \times \hat{\mathbf{k}} + \frac{1}{6c} (\hat{\mathbf{k}} \cdot \dddot{\mathbf{Q}}) \times \hat{\mathbf{k}} + (\ddot{\mathbf{M}} \times \hat{\mathbf{k}}) \times \hat{\mathbf{k}} \right], \qquad (15.12)$$

$$\mathbf{E} = \mathbf{B} \times \hat{\mathbf{k}}. \qquad (15.13)$$

For a system in which the charge-to-mass ratio q_a/m_a equals a constant, the electric dipole moment **d** is proportional to the total linear momentum **P**, and the magnetic dipole moment **M** is proportional to the total orbital angular momentum **L**. Processes which conserve both (e.g., collisions between two electrons) cannot produce radiation by either the electric or magnetic dipole processes. In general, when the electric-dipole term does not equal zero, the electric quadrupole and magnetic dipole fields are smaller by a factor of roughly v_a/c. (The power radiated is smaller by a factor $\sim [v_a/c]^2$.) Since the characteristic frequency $\nu = c/\lambda$ generated by a system of charges moving at speed v_a through a size L will have the same order of magnitude as v_a/L, we see—as anticipated verbally in the last chapter—that a multipole expansion for the radiation field is essentially one in small L/λ. If v_a/c or L/λ cannot be assumed to be small (e.g., in the problem of synchrotron radiation), a formal multipole expansion will possess little practical utility.

BREMSSTRAHLUNG AND THE ROLE OF THE PLASMA PARAMETER

We would now like to calculate the volumetric rate of free-free emission (thermal *Bremsstrahlung*) by an (at least partially) ionized gas of temperature T, and electron and ion charge densities, $-en_e$ and $\sum_i Z_i en_i$. Although the classical trajectory of a free electron encountering an ion has an exact representation in terms of a hyperbola (when we ignore radiation damping for the purpose of computing the source terms in the retarded potentials), the expressions that emerge from such a treatment quickly become quite complex (see Landau and Lifshitz 1951, pp. 207–213). A simpler analysis results if we realize that the typical encounter between an electron and ion in many astrophysical contexts will involve only a small-angle scattering.

The mean spacing L between electrons and ions equals $n_e^{-1/3}$. An interesting measure then consists of the ratio between the mean Coulomb energy $Z_i e^2/L$ and the thermal energy kT:

$$\frac{Z_i e^2 n_e^{1/3}}{kT} = \left(\frac{Z_i}{4\pi} \right) (n_e \lambda_{\text{De}}^3)^{-2/3}, \qquad (15.14)$$

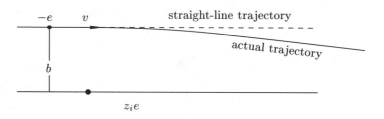

FIGURE 15.1
Approximation of straight-line trajectory to compute the acceleration
associated with *Bremsstrahlung* radiation from electron-ion Coulomb scattering
through small angles.

where λ_{De} is the electron Debye length (see Chapter 7),

$$\lambda_{\mathrm{De}} \equiv \left(\frac{kT}{4\pi n_e e^2} \right)^{1/2} . \tag{15.15}$$

The quantity $n_e\lambda_{\mathrm{De}}^3$ equals the number of free electrons in a Debye cube;
physicists call its reciprocal the *plasma parameter*. If we express T and n_e
in cgs units, $n_e\lambda_{De}^3 \approx 330(T^3/n_e)^{1/2}$. For many astrophysical situations
(but not in the interiors of planets or low-temperature white dwarfs), the
quantity $n_e\lambda_{\mathrm{De}}^3 \gg 1$, so that the expression in equation (15.14), representing the average Coulomb energy compared to the random kinetic energy,
is much less than unity. Hence the plasma behaves on average like an ideal
gas, and Coulomb encounters when they occur typically involve only mild
(small-angle) scatterings.

PERTURBATION TREATMENT

Adopting the assumption that most of the contribution in Coulomb scattering comes from small-angle scatterings, we solve for even the nonradiative
orbit by a perturbation technique. We take the zero-order orbit to be a
straight line (the classical analog of the *Born approximation*). We then calculate the force—and, therefore, the acceleration—that the electron would
experience on this straight line (see Figure 15.1). Then we may proceed to
Fourier analyze the magnitude of the acceleration vector, and therefore the
resulting radiation field. For thermal *Bremsstrahlung*, the total emission is
unpolarized, so we will not bother to keep track of the polarized radiation
that arises from the individual encounters.

In this perturbation scheme, the nonrelativistic equation of motion for the electron yields for the magnitude of the acceleration vector

$$|\ddot{\mathbf{x}}| = \frac{Z_i e^2}{m_e(b^2 + v^2 t^2)}, \tag{15.16}$$

where b is the impact parameter (see Figure 15.1). In the dipole approximation, the electric field observed at a distance r from the electron has magnitude

$$|\mathbf{E}| = \frac{e|\ddot{\mathbf{x}}|}{c^2 r} \sin\theta,$$

where θ equals the angle between the observer's line of sight and the direction of $\ddot{\mathbf{x}}$. Thus we have

$$|\mathbf{E}| = \frac{Z_i e^3 \sin\theta}{m_e c^2 r(b^2 + v^2 t^2)}. \tag{15.17}$$

If we Fourier analyze $|\mathbf{E}|$ using Lighthill's convention [see equation (12.6)], we obtain

$$|\mathbf{E}| = \int_{-\infty}^{\infty} \mathcal{E}_\nu e^{-i2\pi\nu t} \, d\nu.$$

Since $|\mathbf{E}|$ is real, we require $\mathcal{E}_{-\nu} = \mathcal{E}_\nu^*$. Astronomers and physicists prefer to define ν so that it has only positive values; we can accomplish this by writing the above equation as

$$|\mathbf{E}| = \int_0^{\infty} \left(\mathcal{E}_\nu e^{-i2\pi\nu t} + \mathcal{E}_\nu^* e^{i2\pi\nu t} \right) d\nu. \tag{15.18}$$

In either case, $\mathcal{E}_\nu$ can be obtained through the inverse transform as

$$\mathcal{E}_\nu = \frac{Z_i e^3 \sin\theta}{m_e c^2 r} \int_{-\infty}^{\infty} \frac{e^{i2\pi\nu t}}{b^2 + v^2 t^2} \, dt.$$

We may perform the integral by contour integration; the result reads

$$\mathcal{E}_\nu = \left(\frac{Z_i e^3 \sin\theta}{m_e c^2 r} \right) \left(\frac{\pi}{bv} \right) e^{-2\pi|\nu|b/v}. \tag{15.19}$$

The closure (Parseval's) theorem of Fourier transforms implies that we may take the norm of any function either in its original space or in Fourier space. Think of the analogy with finite-dimensional vector spaces, where the sum of the squares of the components in any orthogonal representation gives the norm of the vector. In this example,

$$\int_{-\infty}^{\infty} |\mathbf{E}|^2 \, dt = \int_{-\infty}^{\infty} \mathcal{E}_\nu \mathcal{E}_\nu^* \, d\nu = 2 \int_0^{\infty} \mathcal{E}_\nu \mathcal{E}_\nu^* \, d\nu, \tag{15.20}$$

which follows straightforwardly from the Lighthill transforms relating $|\mathbf{E}|$ and $\mathcal{E}_\nu$ and the delta-function representations,

$$\int_{-\infty}^{\infty} e^{i2\pi(\nu-\nu')t}\, dt = \delta(\nu-\nu'), \qquad \int_{-\infty}^{\infty} e^{-i2\pi\nu(t-t')}\, d\nu = \delta(t-t'). \quad (15.21)$$

Equation (15.20) suggests that we identify $(c/2\pi)\mathcal{E}_\nu\mathcal{E}_\nu^*\, d\nu$ as the time-integrated flux observed between ν and $\nu + d\nu$ (with $\nu > 0$) due to the radiation of our one electron. The energy per Hertz radiated by the electron at frequency ν therefore has the expression

$$\int_0^\pi \frac{c}{2\pi}\mathcal{E}_\nu\mathcal{E}_\nu^* r^2 2\pi \sin\theta\, d\theta = \frac{4}{3}\left(\frac{Z_i^2 e^6}{m_e^2 c^3}\right)\frac{\pi^2}{(bv)^2}e^{-4\pi\nu b/v}.$$

THERMAL BREMSSTRAHLUNG

The total rate of emission by all electrons in a thermal distribution imping-ing on a single ion of charge $Z_i e$ at any impact parameter b equals

$$\int_{v_{\min}}^{\infty} v[n_e f_e(v)\, 4\pi v^2 dv]\left(\frac{4\pi^2 Z_i^2 e^6}{3m_e^2 c^3 v^2}\right)\int_{b_{\min}}^{\infty} e^{-4\pi\nu b/v}\frac{2\pi b\, db}{b^2}. \quad (15.22)$$

In the above, $v_{\min}$ and $b_{\min}$ give minimum cutoff values that are imposed for reasons which will become obvious shortly. We assume that $f_e(v)$ is a Maxwellian, in which case the above expression becomes

$$n_e\left(\frac{2m_e}{\pi kT}\right)^{1/2}\left(\frac{8\pi^3 Z_i^2 e^6}{3m_e^2 c^3}\right) I, \quad (15.23)$$

where I is the double integral

$$I \equiv \int_{x_{\min}}^{\infty} dx\, e^{-x}\int_{\xi_{\min}}^{\infty}\frac{e^{-\xi}}{\xi}\, d\xi,$$

with $\xi_{\min} \equiv 4\pi\nu b_{\min}/v$ and $x_{\min} \equiv m_e v_{\min}^2/2kT$. In the double integral for I, we recognize the inner integral as $E_1(\xi_{\min})$.

What should we choose for $x_{\min}$ and $\xi_{\min}$? Clearly, for free-free emission, we need consider only electrons with sufficient energy to emit at least *one photon* of energy $h\nu$, i.e.,

$$x_{\min} = \frac{h\nu}{kT}. \quad (15.24)$$

For $b_{\min}$, we may choose one of the following two possibilities: (a) To pro-tect our perturbation procedure from large-angle deflections, we can choose $Z_i e^2/b_{\min} = m_e v^2$. (b) To have at least one quantum unit of angular mo-mentum, we can choose $b_{\min} m_e v = \hbar$, a relation that we may also regard as arising from the uncertainty principle. The two expressions for b_{min} have the same value at $v = Z_i e^2/\hbar$. Collecting ideas, we have $b_{min} = Z_i e^2/m_e v^2$ for $v \leq Z_i e^2/\hbar$, while $b_{min} = \hbar/m_e v$ for $v \geq Z_i e^2/\hbar$. Thus

$$\xi_{min} = (h\nu/\chi_i)(x/x_i)^{-3/2} \quad \text{for } x \leq x_i,$$

and

$$\xi_{min} = (h\nu/kT)x^{-1} \quad \text{for } x \geq x_i, \tag{15.25}$$

with $x \equiv m_e v^2/2kT$ and $x_i = \chi_i/kT$, where $\chi_i \equiv (m_e/2)(Z_i e^2/\hbar)^2$ is the ionization energy associated with the ion $Z_i e$.

With the above identifications, we obtain

$$I = \int_{h\nu/kT}^{\infty} e^{-x} E_1(\xi_{min}) \, dx, \tag{15.26}$$

where ξ_{min} is given by equation (15.25). For $h\nu \ll \chi_i$ or kT, $\xi_{min} = (h\nu/\chi_i)(x/x_i)^{-3/2} \ll 1$ in the important part of the integral (15.26). We now approximate $E_1(\xi_{min}) \approx \ln(\gamma/\xi_{min})$ where $\gamma = 0.5772\ldots$ is Euler's constant. Thus for $h\nu/kT \ll 1$,

$$I \approx \ln\left[\gamma(\chi_i/h\nu)x_i^{-3/2}\right] - \frac{3}{2}\gamma \quad \text{since} \quad \int_0^{\infty} e^{-x} \ln x \, dx = -\gamma.$$

Notice the slow logarithmic dependence of I on ν. Free-free emission, equation (15.23), gives almost a completely flat emission spectrum at frequencies where $h\nu/kT \ll 1$. The flat emission spectrum arises because an electron-ion encounter has a very short duration in comparison with the corresponding wave periods. The encounter looks effectively like an impulse, and the Fourier transform of a delta function corresponds to a broad flat spectrum [see the second of the relations in equation (15.21)]. The flat spectrum cannot extend to indefinitely high frequencies, but must have an exponential cutoff as depicted in Figure 15.2, for the reasons that follow.

For $h\nu \gg \chi_i$ or kT, $\xi_{min} = (h\nu/kT)x^{-1}$ is of order unity in the important part of the integral (15.26). We may therefore approximately evaluate $E_1(\xi_{min})$ at the bottom limit of the integral (15.26) where $\xi_{min} = 1$. Thus, for $h\nu/kT \gg 1$,

$$I \approx E_1(1)e^{-h\nu/kT}.$$

The cutoff $e^{-h\nu/kT}$ arises because the Maxwell-Boltzmann distribution contains an exponentially small number of electrons capable of emitting photons of frequency ν when $h\nu \gg kT$.

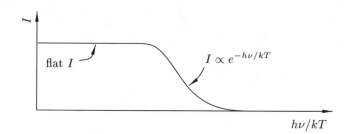

FIGURE 15.2
Behavior of thermal *Bremsstrahlung* integral I with variations of radiation frequency ν.

FREE-FREE EMISSION

Our classical deliberations above have incorporated quantum considerations in a plausible but essentially *ad hoc* fashion. To motivate the writing down of accurate formulae obtained by honest quantum-mechanical computations, we note that our discussion showed I always to have the form $e^{-h\nu/kT}$ times a slowly varying function of ν. This suggests that we replace I by the expression

$$I \rightarrow \frac{4}{\pi\sqrt{3}}\bar{g}_{\text{ff}}(\nu)e^{-h\nu/kT}, \qquad (15.27)$$

where $\bar{g}_{\text{ff}}(\nu)$ is a quantum-mechanical Gaunt factor that varies slowly with ν and is of order unity for most $h\nu/kT$ of practical interest (see, however, Problem Set 4).

Collecting expressions (15.23) and (15.27), we now have that the volumetric rate of emission at frequency ν associated with electrons accelerated by all ions equals

$$\rho j_{\nu}^{\text{ff}} = \sum_{i} n(Z_i)n_{\text{e}} \left(\frac{2m_{\text{e}}}{3\pi kT}\right)^{1/2} \left[\frac{32\pi^2 Z_i^2 e^6}{3m_{\text{e}}^2 c^3}\right]\bar{g}_{\text{ff}}(\nu)e^{-h\nu/kT}, \qquad (15.28)$$

where we have used the notation of radiative transfer for the specific emissivity j_{ν}.

FREE-FREE ABSORPTION COEFFICIENT

To obtain the free-free opacity, we make use of Kirchhoff's law $j_\nu = 4\pi\kappa_\nu B_\nu(T)$, which holds because we have explicitly assumed that the translational degrees of freedom exist in LTE. With $\kappa_\nu = j_\nu/4\pi B_\nu$, we obtain

$$\rho\kappa_\nu^{\text{ff}} = \sum_i n(Z_i)n_e \left(\frac{2m_e}{3\pi kT}\right)^{1/2} \left[\frac{4\pi Z_i^2 e^6}{3m_e^2 ch\nu^3}\right] \bar{g}_{\text{ff}}(\nu)\left(1 - e^{-h\nu/kT}\right).$$

(15.29)

We recognize the factor $1 - \exp(-h\nu/kT)$ as the LTE correction for stimulated emission.

In stellar interiors, most ions come from H^+ and He^{++} in regions where hydrogen and helium are completely ionized. In such regions, the ν^{-3} and $T^{-1/2}$ dependence of $\rho\kappa_\nu^{\text{ff}}$ implies that the associated Rosseland mean (see Chapter 2) satisfies *Kramers's law*:

$$\kappa_R \propto \rho T^{-7/2}.$$

This dependence (satisfied approximately also by bound-free opacities in stellar interiors) plays an important role in the *mass-luminosity* relationship for low-mass stars. (For an elementary discussion, see Shu 1982, Problem 8.1.)

Other important applications arise in the theory of thermal radio continuum emission from $H\textsc{ii}$ regions (and ionized stellar winds), as well as the analysis of X-rays from galactic clusters. In Problem Set 4, we will consider some aspects of how the properties of free-free radiation are used as diagnostics of astrophysical phenomena.

16

Radiation from Relativistically Moving Charges

In Chapter 15 we considered an expansion in terms of multipole moments, which, we commented, really represented an expansion in v/c. Clearly, such an expansion would have limited utility for relativistically moving charges where $v/c \to 1$. To treat such cases, we must begin anew. Fortunately, relativistically moving charges in the cosmos (e.g., cosmic rays) usually move independently of one another; so it suffices to consider their emitted radiation one particle at a time. We may therefore take as our departure point the Lienard-Wiechert potentials,

$$\begin{pmatrix} \phi \\ \mathbf{A} \end{pmatrix} = \frac{q}{(R - \mathbf{R} \cdot \mathbf{v}/c)} \begin{pmatrix} 1 \\ \mathbf{v}/c \end{pmatrix}, \tag{16.1}$$

evaluated at the retarded time τ, where

$$\tau + R(\tau)/c = t, \tag{16.2}$$

with

$$\mathbf{R} = \mathbf{x} - \mathbf{r}(\tau). \tag{16.3}$$

CALCULATION OF $\mathbf{E}$ AND $\mathbf{B}$

The potentials ϕ and $\mathbf{A}$ vary with $\mathbf{x}$ and t in equation (16.1) through the dependences of $\mathbf{R}$ on $\mathbf{x}$ and $\mathbf{r}(\tau)$ and through $\mathbf{v}(\tau) \equiv \dot{\mathbf{r}}$, where τ is defined as a function of $(\mathbf{x}, t)$ through the equations (16.2) and (16.3). Thus, to calculate

$$\mathbf{B} = \nabla \times \mathbf{A}, \tag{16.4}$$

and

$$\mathbf{E} = -\nabla \phi - \frac{1}{c} \frac{\partial \mathbf{A}}{\partial t}, \tag{16.5}$$

we need to be able to compute $\nabla \tau$ and $\partial \tau / \partial t$, where ∇ operates at constant t and $\partial / \partial t$ operates at constant $\mathbf{x}$. The expression for $\partial \tau / \partial t$ holds special interest since it represents the ratio of the intervals of time between the same two events as experienced by the charged particle and by the observer. By differentiating equation (16.2) with respect to t and the identity $R^2 \equiv \mathbf{R} \cdot \mathbf{R}$ with respect to τ, we easily obtain this ratio as

$$\frac{\partial \tau}{\partial t} = \mathcal{K}^{-1}, \qquad \text{with} \qquad \mathcal{K} \equiv 1 + \dot{R}/c = 1 - \frac{\mathbf{R}}{R} \cdot \frac{\mathbf{v}}{c}, \tag{16.6}$$

where $\mathbf{v} = \dot{\mathbf{r}}$ evaluated at the retarded time τ. When $v \to c$, notice that $\mathcal{K}$ can closely approach zero for the observational direction $\mathbf{R}/R$ parallel to $\mathbf{v}$. This fact—the apparent shortening of the duration between events, $dt = \mathcal{K}\, d\tau$, if the charge almost catches up with the electromagnetic signals it sends—underlies, as we shall see, many of the interesting beaming effects associated with relativistic motion.

We leave further formal manipulations for Problem Set 3 and here merely quote the final results:

$$\mathbf{E} = \frac{q}{(R - \mathbf{R} \cdot \mathbf{v}/c)^3} \left\{ \left(1 - \frac{v^2}{c^2} \right) \left(\mathbf{R} - R\frac{\mathbf{v}}{c} \right) + \frac{\mathbf{R}}{c^2} \times \left[\left(\mathbf{R} - R\frac{\mathbf{v}}{c} \right) \times \dot{\mathbf{v}} \right] \right\}, \tag{16.7}$$

$$\mathbf{B} = \frac{\mathbf{R}}{R} \times \mathbf{E}. \tag{16.8}$$

Notice that $\mathbf{B}$ is perpendicular to $\mathbf{E}$ everywhere, even in the near zone.

The individual contributions to $\mathbf{E}$ in equation (16.7) have one of two forms: (1) those proportional to a velocity-corrected displacement vector, $\mathbf{R} - R\mathbf{v}/c$; and (2) those proportional to the acceleration $\dot{\mathbf{v}}$ and, like $\mathbf{B}$, perpendicular to $\mathbf{R}$. The first decline as $1/R^2$ at large R; the second, as $1/R$. As we shall see later, the first corresponds essentially to the (Lorentz-transformed) Coulomb field; the second, to the radiation field associated with the acceleration of a charge.

THE ELECTROMAGNETIC FIELD OF A CHARGE MOVING AT CONSTANT VELOCITY

To obtain a better appreciation of the "Coulomb" part of the electromagnetic field, consider the example $\mathbf{v} = $ constant. The vector displacement $\mathbf{R}(\tau) - R(\tau)\mathbf{v}/c = \mathbf{R}(\tau) - (t - \tau)\mathbf{v}$ then has a simple interpretation as the *present* vector distance (at time t) from the charge to the point of observation (see Figure 16.1). Moreover, as we will prove later, we may express the velocity-corrected distance that enters so prominently in the denominator

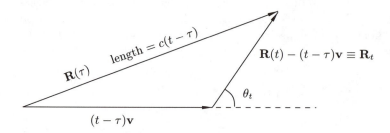

FIGURE 16.1
Interpretation of distances that enter in Lienard-Wiechert potentials when the charge q moves at constant velocity $\mathbf{v}$.

of equation (16.7) in terms of variables that relate only to the *instantaneous* configuration of the system (rather than to quantities that refer to the retarded time):

$$R - \mathbf{R} \cdot \mathbf{v}/c = R_t \left(1 - \frac{v^2}{c^2} \sin^2 \theta_t \right)^{1/2} = \left(|\mathbf{R}_t|^2 - |\mathbf{R}_t \times \frac{\mathbf{v}}{c}|^2 \right)^{1/2}, \quad (16.9)$$

where θ_t is the angle formed between the velocity $\mathbf{v}$ and the present radius vector $\mathbf{R}_t$ from the charge to the observer.

Proof: If we dot $\mathbf{R}_t$ with itself, we obtain

$$R_t^2 = R^2 + R^2 \frac{v^2}{c^2} - 2RR \cdot \frac{\mathbf{v}}{c} = R^2 \left(1 + \frac{v^2}{c^2} - 2\frac{v}{c} \cos \theta \right).$$

The cosine of the angle θ equals the adjacent side divided by the hypotenuse $R(\tau) = c(t - \tau)$ (consult Figure 16.1):

$$\cos \theta = \frac{(t - \tau)v + R_t \cos \theta_t}{c(t - \tau)} = \frac{v}{c} + \frac{R_t}{R} \cos \theta_t. \quad (16.10)$$

Eliminating $\cos \theta$ in equation (16.10) and defining $\beta \equiv v/c$, we now have

$$(1 - \beta^2)R^2 - 2(\beta \cos \theta_t)R_t R - R_t^2 = 0,$$

which we may solve as a quadratic equation for the displacement $R(\tau)$ at the retarded time in terms of present variables:

$$R = \frac{R_t}{(1 - \beta^2)} \left[\beta \cos \theta_t + \left(\beta^2 \cos \theta_t^2 + 1 - \beta^2 \right)^{1/2} \right], \quad (16.11)$$

with the sign before the radical chosen to make R positive. An elementary trigonometric identity now allows us to write

$$R = \frac{R_t}{(1 - \beta^2)} \left[\beta \cos\theta_t + \left(1 - \beta^2 \sin^2\theta_t\right)^{1/2} \right]. \qquad (16.12)$$

On the other hand,

$$\mathbf{R} \cdot \frac{\mathbf{v}}{c} = R\frac{v}{c}\cos\theta = R\beta^2 + R_t\beta\cos\theta_t, \qquad (16.13)$$

according to our derived formula (16.10) for $\cos\theta$. Thus, if we subtract expression (16.13) from R, we get

$$R - \mathbf{R} \cdot \frac{\mathbf{v}}{c} = (1 - \beta^2)R - R_t\beta\cos\theta_t = R_t(1 - \beta^2\sin^2\theta_t)^{1/2},$$

using equation (16.12) in the last step. This result represents the desired equation (16.9). (Q.E.D.)

If we use equation (16.9), equations (16.7) and (16.8) now become, for $\mathbf{v} = $ constant,

$$\mathbf{E} = \frac{q(1 - v^2/c^2)\mathbf{R}_t}{(|\mathbf{R}_t|^2 - |\mathbf{R}_t \times \mathbf{v}/c|^2)^{3/2}}, \qquad \mathbf{B} = \frac{\mathbf{v}}{c} \times \mathbf{E}. \qquad (16.14)$$

If $\mathbf{v}$ were zero, equation (16.14) would merely give the static electric field associated with a stationary charge q. For $\mathbf{v} \neq 0$, equation (16.14) represents the Lorentz transformation of the field to a frame which moves at a velocity $-\mathbf{v}$ relative to the rest frame of the charge. Notice that a charge which moves at a constant velocity relative to an observer always possesses an electric field which points radially away (for positive q) from the *instantaneous* position of the charge. However, in directions where $\mathbf{R}_t$ does not lie perpendicular to $\mathbf{v}$, the magnitude of $\mathbf{E}$ will usually be less than the strict Coulomb value q/R_t^2.

The result that $\mathbf{E}$ points radially away (for a positive charge moving at constant velocity $\mathbf{v}$) from the particle's instantaneous position yields a picturesque description of how electromagnetic radiation gets generated if that charge should suffer sudden deceleration (*Bremsstrahlung*). Imagine that such a charge, moving at constant $\mathbf{v}$ in the past, suddenly comes to rest at the origin at time $t = 0$. At any later time, the electric field for $r < ct$ will point parallel to $\mathbf{r}$, radially away from the origin. No information that the charge has stopped, however, reaches any point $r > ct$; hence, in this region, the electric field points parallel to $\mathbf{R} = \mathbf{r} - \mathbf{v}t$, radially away from the position $\mathbf{v}t$ where the charge *would have been if it had not come to rest*. Across the spherical surface $r = ct$, therefore, the electric field must make make a sudden turn from the direction $\hat{\mathbf{r}}$ to $\hat{\mathbf{R}}$. Within a thin shell of thickness $c\Delta t$ of the expanding wavefront at $r = ct$, then, where Δt represents the small interval of time for the deceleration actually to take place, the electric field must lie almost perpendicular to $\hat{\mathbf{r}}$ in order to connect continuously to the exterior configuration. In other words, the electric field (and induced magnetic field) inside the shell must consist of the transverse radiation field associated with the sudden deceleration.

Another way to view the process for a relativistic charge is as follows. When a relativistic charge moves past an observer (or another charge), the component of $\mathbf{R}_t$ parallel to $\mathbf{v}$ will change sign, with a rapid change of the corresponding longitudinal component of the electric field $\mathbf{E}$ from positive to negative values. Such a field might have little effect on a potential collision partner with finite inertia. At the same time, the measured component of $\mathbf{E}$ (and $\mathbf{B}$) perpendicular to the line of motion will produce a sharp pulse (of a single sign) as the denominator in the expression for $\mathbf{E}$ in equation (16.14) swings through its minimum value, $(1 - v^2/c^2)^{3/2}R_t^3$. Consequently, in its practical electromagnetic interaction with other particles, a fast charge appears to carry with it a cloak of transverse electromagnetic fields. This cloak, which has a Coulombic origin, can be Fourier analyzed as if it were a *virtual* radiation field, leading to a method for dealing with electromagnetic interactions between relativistically moving particles. [See Jackson 1962, Chapter 15, for an account of how the idea of virtual quanta can be used to make semiclassical calculations of relativistic *Bremsstrahlung*, with the collisional partner scattering the virtual radiation fields (i.e., Fourier components which sum to give fields that vary as $1/R^2$ at large R) to cause some of them to become real radiation fields (i.e., Fourier components which sum to give fields that vary as $1/R$ at large R). Quantum-mechanically, we may picture a charge as always surrounded by a cloud of virtual photons, some of which get shaken off to become real photons when the charge is violently accelerated or decelerated.]

RADIATION FIELD IN THE WAVE ZONE

Now that we have examined the nature of the first term inside the square bracket of equation (16.7), let us look at the second term, the real wave field that arises because of particle acceleration. In the wave zone, we may approximate $\mathbf{R} \approx \hat{\mathbf{k}}x$, where x is some mean distance to the charge. With this approximation, equations (16.7) and (16.8) become

$$\mathbf{E} = \frac{q}{c^2 x \mathcal{K}^3} \left\{ \hat{\mathbf{k}} \times [(\hat{\mathbf{k}} - \mathbf{v}/c) \times \dot{\mathbf{v}}] \right\}, \qquad \mathbf{B} = \hat{\mathbf{k}} \times \mathbf{E}. \qquad (16.15)$$

For nonrelativistic motions, $v/c \ll 1$, we recover the Larmor formula $\mathbf{B} = (\ddot{\mathbf{d}} \times \hat{\mathbf{k}})/c^2 x$, where $\mathbf{d} \equiv q\mathbf{r}(\tau)$ is the electric dipole moment associated with our moving charge.

For $\beta \equiv v/c$ close to 1, on the other hand, the radiation fields are strongly beamed in the direction of the motion, $\theta = 0$, where the factor $\mathcal{K}$ in equation (16.6) now reads

$$\mathcal{K} = 1 - \hat{\mathbf{k}} \cdot \mathbf{v}/c = 1 - \beta \cos \theta,$$

and nearly equals zero for $\theta = 0$ and $\beta \approx 1$. Conversely, equation (16.15) implies that the radiation field vanishes in the two directions where the aberration factor $\hat{\mathbf{k}} - \mathbf{v}/c$ (at the retarded time) points parallel (or antiparallel) to the acceleration vector $\dot{\mathbf{v}}$. To appreciate the nature of the beaming, notice that the quantity $\mathcal{K}$ is small (of order $[1 - \beta]$) when $1 - \beta \cos \theta \approx 1 - \beta + \theta^2/2$ is small, i.e., within an angle centered about the prevailing line of motion equal to

$$\theta \approx [2(1 - \beta)]^{1/2} \approx [(1 + \beta)(1 - \beta)]^{1/2} = (1 - \beta^2)^{1/2} \equiv \gamma^{-1},$$

where $\gamma \gg 1$ is the usual Lorentz factor $(1 - \beta^2)^{-1/2}$.

As a special example, consider the case when the acceleration $\dot{\mathbf{v}}$ and the velocity $\mathbf{v}$ vectors are parallel. For example, in *linear accelerators*, we have

$$\mathbf{E} = \frac{q}{c^2 x \mathcal{K}^3} [\hat{\mathbf{k}} \times (\hat{\mathbf{k}} \times \dot{\mathbf{v}})], \qquad \mathbf{B} = -\frac{q}{c^2 x \mathcal{K}^3} [\hat{\mathbf{k}} \times \dot{\mathbf{v}}].$$

The power dP radiated into an area $x^2 d\Omega$ intercepted by the solid angle $d\Omega$ centered about $\hat{\mathbf{k}}$ equals

$$dP = \frac{cB^2}{4\pi} (x^2 d\Omega),$$

i.e.,

$$\frac{dP}{d\Omega} = \frac{c(xB)^2}{4\pi} = \frac{q^2}{4\pi c^3} \left[\frac{|\dot{\mathbf{v}}|^2 \sin^2 \theta}{(1 - \beta \cos \theta)^6} \right]. \tag{16.16}$$

The power defined in equation (16.16) refers to the *received power* (but notice that $|\dot{\mathbf{v}}|$ is calculated in the laboratory frame according to the particle's point of view). The *emitted power* measured relative to the passage of events experienced by the particle's observer is a factor $(d\tau/dt)^{-1} = 1 - \beta \cos \theta$ times the value in equation (16.16) measured by the radiation's observer. (See Jackson 1962, pp. 472–475 and Chapter 17, for a more extended discussion.) In any case, for $\beta \approx 1$, the power perceived by a stationary observer is beamed strongly into two lobes straddling the direction of motion (although no radiation directs itself exactly into $\theta = 0$). For $\beta \approx 0$, the charge radiates broadly toward the two sides (maxima at $\sin^2 \theta = 1$), as illustrated in Figure 16.2. In the instantaneous frame of rest of the particle (if we correct for time dilation, angular aberration, etc.), the radiation pattern looks like the first picture (except it should be expanded in scale, because $\dot{\mathbf{v}}$ is large in such frame [see Chapter 17]); however, relative to an observer with

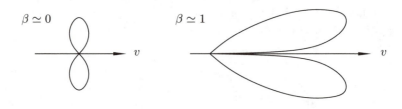

FIGURE 16.2
Angular pattern of radiation for a charge that experiences acceleration $\dot{\mathbf{v}}$ parallel to its velocity $\mathbf{v}$, in the cases of nonrelativistic ($\beta \ll 1$) and relativistic ($\beta \approx 1$) motions. The emission possesses rotational symmetry about an axis that runs along the direction of $\mathbf{v}$.

which the charge possesses a large velocity, a relativistic blueshift (plus aberration, etc.) strengthens the photons emitted in the forward direction, while a relativistic redshift weakens those emitted in the backward direction. Thus the beaming pattern becomes like that depicted in the second picture. Since no intrinsic emission occurs exactly in the forward direction, no amount of Doppler boosting can fill in the deficit at θ equal to zero.

17

Emitted Power and Received Spectrum

Reference: Jackson, *Classical Electrodynamics*, pp. 468–488.

In the last chapter, we finished with a discussion of the angular distribution of radiation from a relativistic charge accelerated in the same direction as its velocity. Here we wish to consider the angular and frequency distribution for an arbitrarily moving relativistic charge. Our discussion of both begins with the formula

$$\frac{dP}{d\Omega} = \frac{c(xE)^2}{4\pi} \equiv \frac{q^2|\mathbf{g}|^2}{4\pi c^3}, \qquad (17.1)$$

where equation (16.15) allows us to identify

$$\mathbf{g} = \frac{1}{\mathcal{K}^3}\left\{\hat{\mathbf{k}} \times \left[\left(\hat{\mathbf{k}} - \frac{\mathbf{v}}{c}\right) \times \dot{\mathbf{v}}\right]\right\}, \qquad \text{with} \qquad \mathcal{K} \equiv 1 - \hat{\mathbf{k}} \cdot \mathbf{v}/c. \qquad (17.2)$$

ANGULAR DISTRIBUTION

If we expand out the triple vector product in equation (17.2), we obtain

$$\mathbf{g} = \frac{1}{\mathcal{K}^3}[(\hat{\mathbf{k}} \cdot \dot{\mathbf{v}})(\hat{\mathbf{k}} - \mathbf{v}/c) - \mathcal{K}\dot{\mathbf{v}}].$$

Dotting the above with itself, we get after collecting terms,

$$g^2 = \frac{1}{\mathcal{K}^4}|\dot{\mathbf{v}}|^2 + \frac{2}{\mathcal{K}^5}(\hat{\mathbf{k}} \cdot \dot{\mathbf{v}})(\dot{\mathbf{v}} \cdot \mathbf{v}/c) - \frac{1}{\mathcal{K}^6}(1 - v^2/c^2)(\hat{\mathbf{k}} \cdot \dot{\mathbf{v}})^2, \qquad (17.3)$$

where we have made use of $\mathcal{K} = 1 - \hat{\mathbf{k}} \cdot \mathbf{v}/c$ to simplify the last term. Construct a coordinate system with $\mathbf{v}$ lying along the z axis, with $\dot{\mathbf{v}}$ lying in the x-z plane making an angle i with respect to $\mathbf{v}$, and with the unit vector $\hat{\mathbf{k}}$ describing the observing direction having (x, y, z) components equal to $(\sin\theta\cos\phi, \ \sin\theta\sin\phi, \ \cos\theta)$. Then,

$$\hat{\mathbf{k}} \cdot \mathbf{v}/c = \beta\cos\theta,$$

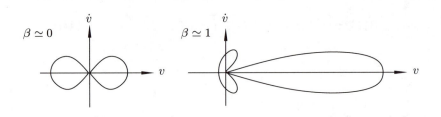

FIGURE 17.1
Angular pattern of radiation in the x-z plane for a charge that experiences acceleration $\dot{\mathbf{v}}$ in the x direction perpendicular to its velocity $\mathbf{v}$ in the z-direction, in the cases of nonrelativistic ($\beta \ll 1$) and relativistic ($\beta \approx 1$) motions.

$$\hat{\mathbf{k}} \cdot \dot{\mathbf{v}} = |\dot{\mathbf{v}}|(\sin\theta\cos\phi\sin i + \cos\theta\cos i),$$

$$\dot{\mathbf{v}} \cdot \mathbf{v}/c = |\dot{\mathbf{v}}|\beta\cos i.$$

With the above formulae and $\mathcal{K} = 1 - \beta\cos\theta$, equation (17.3) gives the angular distribution of the received radiation when i is the angle formed by $\dot{\mathbf{v}}$ and $\mathbf{v}$ at the retarded time τ. For $\dot{\mathbf{v}}$ parallel to $\mathbf{v}$, i.e., for $i = 0$, we easily recover the result derived in the last chapter [see equation (16.16)]:

$$\frac{g^2}{|\dot{\mathbf{v}}|^2} = \frac{\sin^2\theta}{(1 - \beta\cos\theta)^6}. \qquad (17.4)$$

For $\dot{\mathbf{v}}$ perpendicular to $\mathbf{v}$, i.e., for $i = 90°$, we obtain

$$\frac{g^2}{|\dot{\mathbf{v}}|^2} = (1 - \beta\cos\theta)^{-4} - (1 - \beta^2)(1 - \beta\cos\theta)^{-6}\sin^2\theta\cos^2\phi. \qquad (17.5)$$

The angular distributions in the extreme cases of ultrarelativistic and nonrelativistic motions look as in Figure 17.1 for the x-z plane (where $\cos\phi = \pm 1$). Clearly, relativistic motion in this case focuses the front lobe and bends the back lobe forward.

TOTAL EMITTED POWER

To obtain the total *received* power, we integrate $dP/d\Omega$ over all solid angles $d\Omega = \sin\theta\,d\theta d\phi$. If relativistic effects are important, this will not equal the total *emitted* power, which we get by integrating $(dP/d\Omega)(d\tau/dt)^{-1} = (q^2\mathcal{K}g^2/4\pi c^3)$ over the same solid angles. We do the latter here, but similar computations could be performed with equal facility for the total received power.

If we define $\mu \equiv \cos\theta$, we find that we need to perform integrals of the type

$$I_{j+1} \equiv \int_{-1}^{+1} \frac{d\mu}{(1-\beta\mu)^{j+1}} = \frac{(1+\beta)^j - (1-\beta)^j}{j\beta(1-\beta^2)^j},$$

$$J_{j+1} \equiv \int_{-1}^{+1} \frac{\mu\,d\mu}{(1-\beta\mu)^{j+1}} = \frac{1}{j}\frac{dI_j}{d\beta},$$

$$K_{j+1} \equiv \int_{-1}^{+1} \frac{\mu^2\,d\mu}{(1-\beta\mu)^{j+1}} = \frac{1}{j}\frac{dJ_j}{d\beta}.$$

Upon integrating equation (17.3), we then get

$$\int \mathcal{K}g^2\,d\Omega = 2\pi|\dot{\mathbf{v}}|^2 \left\{ I_3 + 2\beta J_4 \cos^2 i \right.$$
$$\left. - (1-\beta^2)\left[K_5 \cos^2 i + \frac{1}{2}(I_5 - K_5)\sin^2 i \right] \right\}.$$

With some algebra and the formulae, $\gamma^{-2} = 1 - \beta^2$ and $d\gamma/d\beta = \gamma^3\beta$, we can derive the following relationships in a straightforward manner:

$$I_3 = 2\gamma^4,\ I_5 = 2\gamma^8(1+\beta^2),\ J_4 = \frac{8}{3}\gamma^6\beta,\ K_5 = \frac{2}{3}\gamma^8(1+5\beta^2).$$

If we write $\cos^2 i = 1 - \sin^2 i$, we may then identify the coefficient of the non-i dependent term inside the curly braces as

$$I_3 + 2\beta J_4 - \gamma^{-2}K_5 = \frac{4}{3}\gamma^6.$$

Similarly, the coefficient of the $\sin^2 i$ term inside the curly braces becomes

$$-\left[2\beta J_4 + \frac{1}{2}\gamma^{-2}(I_5 - 3K_5) \right] = -\frac{4}{3}\gamma^6\beta^2.$$

Collecting expressions, we may finally write the total emitted power as (a result obtained by Lienard in 1898 before the invention of relativity):

$$P_{\text{em}} \equiv \frac{e^2}{4\pi c^3} \int \mathcal{K}g^2 d\omega = \frac{2e^2}{3c^3}\gamma^6[|\dot{\mathbf{v}}|^2 - |\dot{\mathbf{v}} \times \mathbf{v}/c|^2], \qquad (17.6)$$

where $|\dot{\mathbf{v}} \times \mathbf{v}/c| = |\dot{\mathbf{v}}|\beta\sin i$, and $\gamma = (1-\beta^2)^{-1/2}$.

ALTERNATIVE DERIVATION BY SPECIAL RELATIVITY

Equation (17.6) may also be obtained from the arguments of special relativity. In the instantaneous frame of rest of the charge, the total emitted power is given exactly by Larmor's formula for dipole radiation,

$$P_{em} = \frac{2q^2}{3c^3}|\dot{\mathbf{v}}|^2, \tag{17.7}$$

since all of the higher-order multipole moments vanish in a frame where $v/c = 0$. Consider now the Lorentz invariant for the spacetime interval,

$$ds^2 \equiv c^2 d\tau^2 - |d\mathbf{r}|^2 = c^2 \gamma^{-2} d\tau^2.$$

The quantity ds/c equals the interval of *proper time*, which should not be confused with our notation for $d\tau$.

We may form a normalized four-velocity by defining

$$\vec{u} \equiv \frac{d\vec{x}}{ds} = \frac{\gamma}{c}\frac{d\vec{x}}{d\tau} = \gamma\left(\begin{array}{c} \mathbf{v}/c \\ i \end{array}\right),$$

and an unnormalized four-acceleration by

$$\vec{a} \equiv c^2 \frac{d\vec{u}}{ds} = c\gamma \frac{d\vec{u}}{d\tau} = \gamma\left[\begin{array}{c} d(\gamma\mathbf{v})/d\tau \\ icd\gamma/d\tau \end{array}\right]. \tag{17.8}$$

The scalar expression formed by

$$P_{em} \equiv \frac{2q^2}{3c^3}\vec{a}\cdot\vec{a} \tag{17.9}$$

is a Lorentz invariant. In the instantaneous frame of rest of the charge, $\gamma = 1$ and $\vec{a} = (\dot{\mathbf{v}}, 0)$. (Note that $\gamma^{-2} = 1 - \mathbf{v}\cdot\mathbf{v}/c^2$ implies $\gamma^{-3}d\gamma/d\tau = \dot{\mathbf{v}}\cdot\mathbf{v}/c$, which is zero in a frame where $\mathbf{v} = 0$.) Thus equation (17.9) reduces to equation (17.7) in the rest frame of the charge. The total power perceived to be emitted by *any* inertial observer at the instantaneous position of the source forms a Lorentz invariant because it represents an energy divided by a time, and energy and time both transform in the same way (as the fourth component of four-vectors).

In the lab frame, equation (17.9) has the expression,

$$P_{em} = \frac{2q^2}{3c}\left\{\left[\gamma\frac{d}{d\tau}\left(\gamma\frac{\mathbf{v}}{c}\right)\right]^2 - \left(\gamma\frac{d\gamma}{d\tau}\right)^2\right\}.$$

If we square out the first term (i.e., dot the term inside the square bracket with itself), we obtain

$$P_{em} = \frac{2q^2}{3c^2}\left[2\gamma^3\frac{d\gamma}{d\tau}\left(\frac{\dot{\mathbf{v}}}{c}\cdot\frac{\mathbf{v}}{c}\right) + \gamma^4\frac{|\dot{\mathbf{v}}|^2}{c} - \left(\frac{d\gamma}{d\tau}\right)^2\right],$$

where we have used $\gamma^2(-1 + v^2/c^2) = -1$ to simplify the last term on the right-hand side. Writing $d\gamma/d\tau = \gamma^3(\dot{\mathbf{v}} \cdot \mathbf{v}/c^2)$, we obtain

$$P_{\text{em}} = \frac{2q^2}{3c^3}\gamma^6 \left[\left(\dot{\mathbf{v}} \cdot \frac{\mathbf{v}}{c}\right)^2 + \left(1 - \frac{v^2}{c^2}\right)|\dot{\mathbf{v}}|^2 \right].$$

The above result reproduces equation (17.6) when we expand $(\dot{\mathbf{v}} \times \mathbf{v}) \cdot (\dot{\mathbf{v}} \times \mathbf{v}) = \dot{\mathbf{v}} \cdot [\mathbf{v} \times (\dot{\mathbf{v}} \times \mathbf{v})]$. (Q.E.D.)

If we define the components of acceleration parallel and perpendicular to the velocity through

$$va_{\parallel} \equiv |\mathbf{v} \cdot \dot{\mathbf{v}}|, \qquad va_{\perp} \equiv |\mathbf{v} \times \dot{\mathbf{v}}|,$$

with

$$|\dot{\mathbf{v}}|^2 = a_{\parallel}^2 + a_{\perp}^2,$$

we can also write equation (17.6) as

$$P_{\text{em}} = \frac{2q^2}{3c^3}\gamma^4(a_{\perp}^2 + \gamma^2 a_{\parallel}^2). \tag{17.10}$$

For a given acceleration (measured in a laboratory frame, but at the retarded time), a relativistic charge radiates much more energy than does a nonrelativistic one. For the power radiated in the instantaneous frame of rest of the particle numerically to equal equation (17.10), it must be true that the accelerations measured in that frame (where $\gamma' = 1$) satisfy

$$a_{\parallel}' = \gamma^3 a_{\parallel}, \tag{17.11}$$

$$a_{\perp}' = \gamma^2 a_{\perp}. \tag{17.12}$$

The relationships (17.11) and (17.12) do indeed result from carrying out such a Lorentz transformation. (See Rybicki and Lightman 1979, Problem 4.3.)

RECEIVED SPECTRUM

To perform a spectral decomposition of the received angular power (usually, the quantity of interest), take the time integral of equation (17.1) and call it $dW/d\Omega$, the total energy received per steradian in a given propagation direction $\hat{\mathbf{k}}$:

$$\frac{dW}{d\Omega} = \frac{q^2}{4\pi c^3} \int |\mathbf{g}|^2 \, dt = \frac{q^2}{2\pi c^3} \int \mathbf{g}_\nu \cdot \mathbf{g}_\nu^* \, d\nu, \tag{17.13}$$

where $\mathbf{g}_\nu$ is the Lighthill transform of $\mathbf{g}$,

$$\mathbf{g}_\nu \equiv \int_{-\infty}^{+\infty} \mathbf{g} e^{i2\pi\nu t} \, dt. \tag{17.14}$$

But $\mathbf{g}$ is most conveniently expressed as a function of the retarded time τ; therefore, we need to transform variables via

$$dt = (1 - \hat{\mathbf{k}} \cdot \mathbf{v}/c) \, d\tau = \mathcal{K} \, d\tau.$$

Substituting equation (17.2) into equation (17.14), we write

$$\mathbf{g}_\nu = \int_{-\infty}^{\infty} \frac{1}{\mathcal{K}^2} \left\{ \hat{\mathbf{k}} \times \left[\left(\hat{\mathbf{k}} - \frac{\mathbf{v}}{c} \right) \times \dot{\mathbf{v}} \right] \right\} e^{i2\pi\nu[\tau + R(\tau)/c]} \, d\tau. \tag{17.15}$$

We may simplify this expression by using a trick similar to how we expressed wave-zone electromagnetic fields in Chapter 14 solely in terms of time derivatives of the vector potential. With this comment, we are motivated to examine the quantity

$$\frac{d}{d\tau} \left\{ \frac{1}{\mathcal{K}} [\hat{\mathbf{k}} \times (\hat{\mathbf{k}} \times \mathbf{v})] \right\}.$$

Carrying out the differentiation, we find that it equals

$$\frac{1}{\mathcal{K}^2} \left(\hat{\mathbf{k}} \cdot \frac{\dot{\mathbf{v}}}{c} \right) [\hat{\mathbf{k}} \times (\hat{\mathbf{k}} \times \mathbf{v})] + \frac{1}{\mathcal{K}} [\hat{\mathbf{k}} \times (\hat{\mathbf{k}} \times \dot{\mathbf{v}})].$$

Multiplying the last term by $1 = (1 - \hat{\mathbf{k}} \cdot \mathbf{v}/c)/\mathcal{K}$ and factoring out $1/\mathcal{K}^2$, we obtain

$$\frac{1}{\mathcal{K}^2} \left\{ (\hat{\mathbf{k}} \cdot \dot{\mathbf{v}})[\hat{\mathbf{k}} \times (\hat{\mathbf{k}} \times \mathbf{v}/c)] + \hat{\mathbf{k}} \times (\hat{\mathbf{k}} \times \dot{\mathbf{v}}) - (\hat{\mathbf{k}} \cdot \mathbf{v}/c)[\hat{\mathbf{k}} \times (\hat{\mathbf{k}} \times \dot{\mathbf{v}})] \right\}.$$

If we expand out the triple vector products in the first and third terms, we can rewrite them as

$$-(\hat{\mathbf{k}} \cdot \dot{\mathbf{v}})(\mathbf{v}/c) + (\hat{\mathbf{k}} \cdot \mathbf{v}/c)(\dot{\mathbf{v}}) = -\hat{\mathbf{k}} \times \left(\frac{\mathbf{v}}{c} \times \dot{\mathbf{v}} \right).$$

Collecting all the expressions together, we now get the desired result,

$$\frac{d}{d\tau} \left\{ \frac{1}{\mathcal{K}} [\hat{\mathbf{k}} \times (\hat{\mathbf{k}} \times \mathbf{v})] \right\} = \frac{1}{\mathcal{K}^2} \left\{ \hat{\mathbf{k}} \times \left[\left(\hat{\mathbf{k}} - \frac{\mathbf{v}}{c} \right) \times \dot{\mathbf{v}} \right] \right\}, \tag{17.16}$$

which equals part of the integrand in equation (17.15).

If we substitute equation (17.16) into equation (17.15) and integrate by parts under the assumption that we have a finite wave train so that the

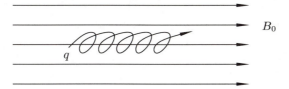

FIGURE 17.2
Helical motion of a charge spiraling in a constant magnetic field $\mathbf{B}_0$.

endpoints give zero, and if we recall that $\partial(\tau + R/c)/\partial\tau = 1 - \hat{\mathbf{k}} \cdot \mathbf{v}/c = \mathcal{K}$, we obtain the desired result:

$$\mathbf{g}_\nu = -i2\pi c\nu \int_{-\infty}^{\infty} \hat{\mathbf{k}} \times (\hat{\mathbf{k}} \times \mathbf{v}/c) e^{i2\pi\nu[\tau + R(\tau)/c]} \, d\tau. \tag{17.17}$$

To be able to use this result, we need to know the detailed time dependence of the particle trajectory $\mathbf{r}(\tau)$ in

$$\mathbf{R} = \mathbf{x} - \mathbf{r}(\tau).$$

In preparation for our discussion of synchrotron radiation in Chapter 19, we now compute this dependence for a charge gyrating in a uniform static magnetic field $\mathbf{B}_0$.

CHARGE TRAJECTORY IN A UNIFORM MAGNETIC FIELD

If we ignore radiation reaction, the relevant (relativistically correct) equation of motion reads

$$\frac{d}{d\tau}(m\gamma\mathbf{v}) = q\left(\frac{\mathbf{v}}{c} \times \mathbf{B}_0\right). \tag{17.18}$$

If we dot the above with $\mathbf{v}$, we can show that $|\mathbf{v}|$ is conserved, so that γ (i.e., the particle energy) is constant. Since the magnetic force is also perpendicular to $\mathbf{B}_0$, we can moreover demonstrate that the component of $\mathbf{v}$ parallel to $\mathbf{B}_0$ is conserved, as must therefore be the magnitude of the component of $\mathbf{v}$ perpendicular to $\mathbf{B}_0$ (since the total speed $|\mathbf{v}|$ is constant). The solution must take therefore, as is well known, the form of a helical motion about a magnetic field line (Figure 17.2). To obtain the angular speed ω_B of rotation about the field line, construct a coordinate system

(1,2,3) such that $\mathbf{B}_0 = B_0 \mathbf{e}_3$ lies in the 3 direction. We may now rewrite equation (17.18) as

$$\frac{d\mathbf{v}}{d\tau} = \mathbf{v} \times (\mathbf{e}_3 \omega_B), \tag{17.19}$$

where ω_B equals

$$\omega_B \equiv qB_0/\gamma mc, \tag{17.20}$$

giving the desired formula for the relativistic gyration frequency. With ω_B equal to a constant, the orbital solution of equation (17.19) assumes the well-known form:

$$\mathbf{r}(\tau) = \mathbf{e}_3 v_3 \tau + \frac{v_{12}}{\omega_B}(\mathbf{e}_1 \cos \omega_B \tau + \mathbf{e}_2 \sin \omega_B \tau), \tag{17.21}$$

where $v_3 = \mathbf{v} \cdot \mathbf{B}_0/B_0$ and $v_{12} \equiv (v_1^2 + v_2^2)^{1/2} = |\mathbf{v} \times \mathbf{B}_0/B_0|$ are constants. The quantity v_{12}/ω_B corresponds to the radius of gyration, and we have chosen the origin and orientation of the (1,2,3) coordinate system so that the 3 and 2 coordinates of the particle equal zero at $\tau = 0$.

Because of relativistic beaming, however, the most convenient decomposition for calculating the Fourier components of the radiation fields is not parallel and perpendicular to the magnetic field axis, but parallel and perpendicular to the velocity vector $\mathbf{v}$ at time τ. If we denote α as the pitch angle of the helix defined through

$$\cos \alpha = \mathbf{v} \cdot \mathbf{B}_0/vB_0, \tag{17.22}$$

the effective radius of curvature of the helical orbit equals

$$\mathcal{R} = v/\omega_B \sin \alpha. \tag{17.23}$$

Notice that $\mathcal{R} = v/\omega_B$ if $\sin \alpha = 1$ (circular orbit), whereas $\mathcal{R} = \infty$ if $\cos \alpha = 1$ (straight line parallel to $\mathbf{B}_0$). In our next undertaking we will make use of these considerations to derive the emission characteristics of synchrotron radiation.

18

Synchrotron Theory: Simple Version

The formal manipulations required for synchrotron theory can get formidable. To arrive at an intuitive feel for the subject, let us consider a simple version of the theory that gets the basics correct in order of magnitude. We then apply this bare-bones theory to the problem of radio galaxies. In Chapter 19 we will derive more accurate expressions for this important radiation process.

TOTAL POWER EMITTED BY A SINGLE ELECTRON

Equation (17.10) gives the total power emitted by a single charge q undergoing arbitrary relativistic motion as

$$P_{\text{em}} = \frac{2q^2}{3c^3}\gamma^4(a_\perp^2 + \gamma^2 a_\parallel^2),$$

where $a_\perp$ and $a_\parallel$ are the accelerations perpendicular and parallel to the velocity vector $\mathbf{v}$ evaluated at the retarded time τ. For a charge in a uniform magnetic field, equation (17.19) implies that the entire (nonradiative) acceleration occurs perpendicular to $\mathbf{v}$, i.e.,

$$a_\perp = \omega_B v \sin \alpha, \qquad a_\parallel = 0,$$

where α is the angle between $\mathbf{v}$ and $\mathbf{B}_0$ and ω_B is the charge's relativistic gyration frequency given by equation (17.20).

It is much easier for cosmic-ray electrons to attain large values of γ than for cosmic-ray protons; so we may concentrate on the synchrotron radiation produced by electrons. The total emitted power now becomes

$$P_{\text{em}} = \frac{2e^2}{3c^3}\gamma^2 \frac{e^2 B_0^2}{m_e^2 c^2} v^2 \sin^2 \alpha = \frac{2}{3} r_e^2 c \gamma^2 \beta^2 B_0^2 \sin^2 \alpha, \tag{18.1}$$

where we have written $r_e \equiv e^2/m_e c^2$ for the classical radius of the electron. If we further write $\sigma_T \equiv 8\pi r_e^2/3$ as the Thomson cross section and

$$U_B \equiv B_0^2/8\pi \qquad (18.2)$$

as the energy density of the static magnetic field, we can put equation (18.1) in the suggestive form

$$P_{em} = 2\beta^2\gamma^2 c\sigma_T U_B \sin^2 \alpha, \qquad (18.3)$$

which looks as if the electron were colliding with a static magnetic field. For this reason, synchrotron radiation also goes by the name *magneto-bremsstrahlung*.

We now assume for simplicity that at each energy $\gamma m_e c^2$, we have an *isotropic* distribution of cosmic-ray electrons. The angle-averaged value of the emission rate per particle for a monoenergetic collection is then given by

$$\langle P_{em} \rangle = \frac{4}{3}\beta^2\gamma^2 c\sigma_T U_B, \qquad (18.4)$$

since

$$\langle \sin^2 \alpha \rangle \equiv \frac{1}{4\pi} \int_0^{2\pi} d\varphi \int_0^\pi (\sin^2 \alpha) \sin \alpha \, d\alpha = \frac{2}{3}.$$

Although the total emitted power P_{em} of a single electron does not equal the total received power P because of possible relativistic motion toward or away from the observer, the average power of an *isotropic* collection of electrons, in the absence of bulk relativistic motion (for which we may make a separate correction), must equal, by the conservation of total energy, the average received power. Thus we may equate $\langle P \rangle$ with $\langle P_{em} \rangle$.

FREQUENCY DISTRIBUTION OF A MONOENERGETIC DISTRIBUTION

Over what range of frequencies is the average received power distributed? At the position of the (average) center of the relativistic circular motion (with $\gamma \gg 1$) of an electron of energy $\gamma m_e c^2$, the radiation appears beamed toward the direction of the observer in a series of pulses spaced in time $2\pi/\omega_B$ apart, but with each pulse lasting only a fraction $\Delta\theta/2\pi \sim \gamma^{-1}/2\pi$ of the orbit cycle. (Consult Chapters 16 and 17, as well as Problem Set 3, and see Figure 18.1. Here, we gloss over the distinct possibility that the average of the beaming might be different than the beaming of the average, on the grounds that such subtleties cannot introduce large order-of-magnitude errors in our derived results.)

A further sharpening of the *received* pulses by a factor γ^{-2} occurs because the interval $\Delta\tau \equiv \tau_2 - \tau_1$ between first and last illumination of

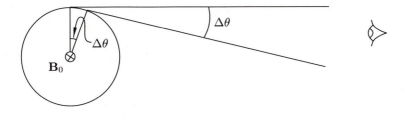

FIGURE 18.1
Beaming of radiation by ultrarelativistic electron circling a magnetic field
pointing into the plane of the paper.

the observer by the charge in each cycle takes longer than the interval
$\Delta t \equiv t_2 - t_1$ between first and last reception by the observer if the charge
moves nearly at the speed of light toward the observer during the beaming.
In particular, we have the approximate relationship

$$\frac{\Delta t}{\Delta \tau} \approx \left(\frac{d\tau}{dt}\right)^{-1} = 1 - \beta \cos\theta \sim 1 - \beta + \Delta\theta^2/2,$$

which $\sim \gamma^{-2}$ if $\Delta\theta \sim \gamma^{-1}$ and $\gamma \gg 1$. Thus the received pulses look
schematically as indicated in Figure 18.2.

If we were to Fourier analyze this pulse shape, we would find appreciable
power at angular frequencies all the way from the fundamental, ω_B, to the

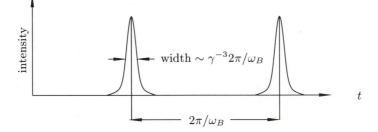

FIGURE 18.2
Temporal pattern of received pulses after accounting for angular beaming of
synchrotron radiation and the difference in intervals between events as
measured at the retarded time τ of emission and at the actual time t of
reception by a distant observer.

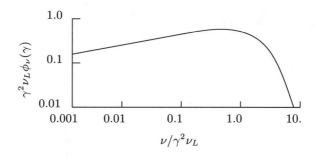

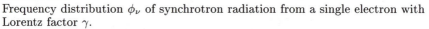

FIGURE 18.3
Frequency distribution ϕ_ν of synchrotron radiation from a single electron with
Lorentz factor γ.

very high harmonic, $\gamma^3 \omega_B \equiv \gamma^2 \omega_L$, where ω_L is defined (purely formally)
as the electron's *Larmor frequency*:

$$\omega_L \equiv \frac{eB_0}{m_e c}. \tag{18.5}$$

We may introduce the standard frequency ν (measured in Hz) in the usual
way as $\nu = \omega/2\pi$; and we write the Fourier decomposition for the averaged
received power per electron with Lorentz factor γ as

$$\langle P_\nu(\gamma) \rangle = \frac{4}{3} \beta^2 \gamma^2 c \sigma_T U_B \phi_\nu(\gamma), \tag{18.6}$$

where we require the integral of $\langle P_\nu(\gamma) \rangle$ over all ν to reproduce equation (18.4). In other words, the frequency distribution function $\phi_\nu(\gamma)$ satisfies the normalization condition

$$\int_0^\infty \phi_\nu(\gamma)\, d\nu = 1, \tag{18.7}$$

with our prior discussion leading us to expect that ϕ_ν will otherwise look
something like Figure 18.3.

Since ϕ_ν represents the frequency distribution of received radiation from
a collection of electrons with a single value of γ, we should formally have
a discrete spectrum (Fourier series rather than Fourier integral) since the
basic motion is periodic with period $2\pi/\omega_B = \gamma/\nu_L$. However,

$$\nu_L \equiv \frac{eB_0}{2\pi m_e c}$$

equals 28 Hz for $B_0 = 10^{-5}$ Gauss (a typical value in extended radio sources such as our own Galaxy and the radio lobes of powerful radio galaxies). The electrons that produce emission at radio frequencies of a few GHz therefore have Lorentz factors $\gamma \sim 10^4$. The spacing $\nu_B = \gamma^{-1}\nu_L$ between successive harmonics is then so narrow as to be negligible in any radio observation that uses a finite bandwidth, and we may effectively treat the radiation as a continuous spectrum. The fact that the cosmic-ray electrons of observed celestial sources will have a distribution of energies rather than a single monoenergetic value only reinforces this conclusion.

FREQUENCY DISTRIBUTION OF A POWER-LAW ENERGY DISTRIBUTION

Let $n(\gamma)\,d\gamma$ equal the number density of cosmic-ray electrons with energies between $m_e c^2 \gamma$ and $m_e c^2(\gamma + d\gamma)$. The synchrotron volume emissivity due to this (isotropic) collection of cosmic-ray electrons is given by

$$\rho j_\nu = \int_1^\infty \langle P_\nu(\gamma)\rangle n(\gamma)\,d\gamma. \tag{18.8}$$

High-energy cosmic-ray protons incident on the Earth have a power-law distribution, and it is believed that cosmic-ray electrons also satisfy a power-law relationship, although the demodulation of the effects of the magnetized solar wind (which tends to sweep out cosmic rays trying to enter the solar system) becomes more problematic for them, at least, at the lower energies. In any case, good reason exists empirically to adopt a distribution of the form

$$n(\gamma)\,d\gamma = n_0 \gamma^{-p}\,d\gamma, \tag{18.9}$$

where the particle exponent p has a typical value ≈ 2.5. We suppose that such an energy spectrum extends in any realistic celestial source only to values of γ larger than some low-energy cutoff corresponding to $\gamma_{\min}$.

If we substitute equations (18.6) and (18.9) into equation (18.8), we cannot formally integrate until we know the functional form of $\phi_\nu(\gamma)$. At this point, we make the grossest approximation of all and replace the actual $\phi_\nu(\gamma)$ by a delta function centered on the highest harmonic frequency $\gamma^2 \nu_L$ with appreciable power:

$$\phi_\nu(\gamma) \rightarrow \delta(\nu - \gamma^2 \nu_L). \tag{18.10}$$

At first sight, equation (18.10) might appear to be a poor representation for the function $\phi_\nu(\gamma)$ depicted in Figure 18.3, but, in fact (to factors of order unity), it does not matter very much what $\phi_\nu(\gamma)$ looks like, provided that the superposition of different energies spans a sufficiently large range

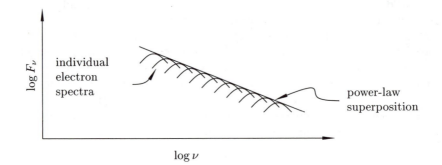

FIGURE 18.4
Frequency dependence $\nu^{-(p-1)/2}$ of synchrotron radiation associated with a power-law superposition γ^{-p} of electrons with Lorentz factor γ.

of γ in our power law. If it does, then the superposition of practically *any* function $\phi_\nu(\gamma)$ will produce a power-law spectrum (see Figure 18.4), and the only thing that counts is for us to get $\phi_\nu(\gamma)$ centered at about the right frequency $(\gamma^2\nu_L)$ and to satisfy the normalization condition (18.7), both of which the replacement (18.10) does do.

For large values of γ, moreover, no great error results by setting β equal to unity everywhere except where it enters in γ. If we transform from γ to the variable $\nu' \equiv \gamma^2\nu_L$, we may now integrate equation (18.8) and obtain

$$\rho j_\nu \sim \frac{2}{3}c\sigma_T n_0 U_B \nu_L^{-1}\left(\frac{\nu}{\nu_L}\right)^{-(p-1)/2}, \qquad (18.11)$$

where the *spectral index* $(p-1)/2 = 0.75$ if $p = 2.5$. Such a spectral index is typical for many (optically thin) radio sources whose (nonthermal) emission is believed to arise by the synchrotron process. In an optically thin source, the monochromatic luminosity L_ν radiated at each frequency ν simply equals the volume emissivity ρj_ν integrated over the source volume V; thus

$$L_\nu = \int_V \rho j_\nu \, dV \propto \nu^{-(p-1)/2} \qquad (18.12)$$

if ρj_ν satisfies equation (18.11).

A nonthermal (synchrotron) spectrum which declines as $\nu^{-0.7}$ or $\nu^{-0.8}$ should be contrasted with a thermal (free-free) *Bremsstrahlung* spectrum which stays nearly flat (or declines only as $\nu^{-0.1}$) when both sources of radiation remain optically thin. In normal galaxies, like the Milky Way system, then, we expect synchrotron emission (from supernova remnants, etc.) to dominate at low frequencies (wavelengths at 50 cm or longer, say), and thermal radiation (from H II regions, etc.) to dominate at high frequencies (wavelengths at 6 cm or shorter, say); however, the exact crossover frequency (or wavelength) depends on the specific galaxy (or the specific part of it) being observed.

APPLICATION TO RADIO GALAXIES

Active galaxies exist in which the synchrotron process generates the entire radio spectrum from decameter to millimeter wavelengths. This conclusion follows both from the measured spectral index and from the polarization properties of the radiation (to be covered in Chapter 19). Such objects constitute the radio galaxies, and in the late 1950s Geoffrey Burbidge pointed out the basic problem of energetics that confronts any attempt to explain the nature of the strongest of such sources, the double-lobed radio galaxies. Suppose we try to estimate how much energy, particle plus magnetic, is needed to produce the observed synchrotron emission. Equation (18.2) gives the magnetic energy density; the cosmic-ray electron energy-density reads

$$U_{\mathrm{CR}} = n_0 \int_{\gamma_{\mathrm{min}}}^{\infty} (\gamma m_e c^2) \gamma^{-p} \, d\gamma = \frac{n_0 m_e c^2}{2 - p} \gamma_{\mathrm{min}}^{-(p-2)}, \qquad (18.13)$$

for $p > 2$. What should we choose for γ_{min}? A conservative procedure involves choosing γ_{min} such that $\gamma_{\mathrm{min}}^2 \nu_{\mathrm{L}}$ yields the lowest radio frequency (typically 10^7 Hz) for which synchrotron radiation was actually observed. If there exists radio emission at yet lower frequencies, there must be cosmic-ray electrons with yet lower γ's to account for them, which would only increase the required U_{CR}.

Now, for given source luminosity L_ν in equation (18.12), minimizing the total energy requirements $(U_{\mathrm{CR}} + U_B)V$, for a homogeneous source model occupying volume V, amounts to minimizing $U_{\mathrm{CR}} + U_B$ for given ρj_ν. A fixed ρj_ν at each ν [i.e., a given $\nu^{(p-1)/2} \rho j_\nu$] implies from equation (18.11), with $\nu_{\mathrm{L}} \propto B_0$, that we want

$$n_0 = C B_0^{-(p+1)/2},$$

where C has a known value. Thus, in equation (18.13), the desired $U_{\mathrm{CR}} \propto B_0^{-(p+1)/2}$ decreases with increasing B_0, while $U_B \propto B_0^2$ in equation (18.2) increases with increasing B_0. The sum reaches a minimum when

$$0 = \frac{d}{dB_0}(U_{\mathrm{CR}} + U_B) = -\frac{(p+1)U_{\mathrm{CR}}}{2B_0} + \frac{2U_B}{B_0},$$

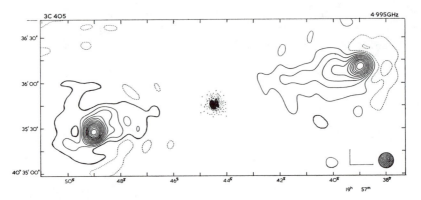

FIGURE 18.5
Radio map of the radio source Cygnus A. (From S. Mitton and M. Ryle 1969, *M.N.R.A.S.*, **146**, 221.)

i.e., when

$$U_{CR} = \frac{4}{p+1}U_B. \tag{18.14}$$

Since the factor $4/(p + 1)$ nearly equals 1 when $p \approx 2.5$, equation (18.14) sometimes goes by the name of the "equipartition requirement." I.e., the minimum energy required to explain the optically thin synchrotron radiation from a radio source places roughly equal amounts of energy in radiating particles (cosmic-ray electrons) as in magnetic fields. If there are unobserved components of energy (e.g., cosmic-ray protons, which in the Galactic environment have ~ 100 times more energy content than the cosmic-ray electrons), or if the magnetic field does not have its optimal strength, the energy requirement would go up correspondingly.

What is the computed minimum energy requirement for a double-lobed radio galaxy like Cygnus A? (See Figure 18.5.) For the given radio emission, a value of $B_0 \approx 10^{-4}$ Gauss is optimal. If the two lobes can be approximated as having the volume equivalent to a sphere of radius $R = 30$ kpc, then we easily compute that the magnetic energy alone

$$E_{mag} = \left(\frac{B_0^2}{8\pi}\right)\frac{4\pi}{3}R^3 \sim 10^{60} \text{ erg,}$$

with a comparable amount contained in fast electrons. This represents an enormous energy, equivalent to the release of more than 10^9 supernovae! (In other sources, $B_0 \sim 10^{-5}$ G and $E_{mag} \sim 10^{58}$ erg are more typical.) Clearly, radio galaxies involve energetics that make pale even the most powerful of stellar phenomena.

Another aspect of the problem deserves mention. Consider the radiative lifetime of a cosmic-ray electron against synchrotron losses. Its energy equation must satisfy

$$m_e c^2 \frac{d\gamma}{d\tau} = -P_{em} = -2\beta^2 \gamma^2 c \sigma_T U_B \sin^2 \alpha.$$

Setting β again to 1, the characteristic time scale for synchrotron losses becomes

$$-\frac{\gamma}{d\gamma/d\tau} = \frac{m_e c/\gamma}{2\sigma_T U_B \sin^2 \alpha}.$$

For $B_0 \approx 10^{-5}$ Gauss, and $\sin \alpha$ of order unity, this lifetime works out to be about 10^7 yr for $\gamma \sim 10^4$. Such a lifetime borders on being embarrassingly short compared to the expected ages of the radio lobes, and the embarrassment becomes more acute if we consider the lifetimes of the higher values of γ needed to account for the higher-frequency synchrotron radiation. Thus some process needs continuously to produce (accelerate to ultrarelativistic speeds) the inferred cosmic-ray electrons in the radio lobes locally (*in situ*). The mechanism now favored from a variety of observational and theoretical arguments focuses on the beamed transport of energy out to the radio lobes from a central "machine" (accreting supermassive black hole?) located in the nucleus of the giant elliptical galaxy that the two radio lobes usually straddle. In support of this picture, astronomers have found that compact radio sources sometimes exist in the nuclei of the central galaxies that have intriguing radio properties (see below).

SYNCHROTRON SELF-ABSORPTION AND COMPACT RADIO SOURCES

The (nonthermal) radio spectrum of the compact radio sources found in the nuclei of galaxies differs as a class from the more extended objects (the radio lobes), in that the former often have *flat* radio spectra (i.e., the radio flux F_ν stays nearly constant with increasing ν rather than falling off, say, as $\nu^{-0.6}$ or ν^{-1}). As we will demonstrate in the next chapter, the opacity for *synchrotron self-absorption*, including the effects of stimulated emission, satisfies

$$\rho \kappa_\nu = K_0 n_0 B_0^{(p+2)/2} \nu^{-(p+4)/2}, \tag{18.15}$$

where K_0 is a constant coefficient. Thus, in a very compact source, we have the possibility that the source will become optically thick at low frequencies, even if it is optically thin at high frequencies (like the HII region problem with synchrotron radiation now replacing thermal *Bremsstrahlung*).

To investigate this possibility, recall from Chapter 3 that the solution of the equation of transfer for a source with uniform properties and no background illumination reads

$$I_\nu = S_\nu (1 - e^{-\tau_\nu}), \tag{18.16}$$

where the synchrotron source function is given by

$$S_\nu \equiv j_\nu / 4\pi\kappa_\nu \propto B_0^{-1/2}\nu^{5/2} \tag{18.17}$$

when we make use of equations (18.11) and (18.15). Notice that the frequency dependence of S_ν does not contain the cosmic-ray energy-distribution index p.

We can derive the result (18.17) for S_ν by another, more revealing, line of reasoning. If we were dealing with thermal radiation, the source function S_ν would equal the Planck function $B_\nu(T)$ (review Chapter 3):

$$S_\nu = \left(\frac{2\nu^2}{c^2}\right)\left(\frac{h\nu}{e^{h\nu/kT} - 1}\right),$$

where the term in the first parenthesis equals a phase-space factor, and the term in the second equals the mean energy of the oscillator capable of emitting radiation of frequency ν (see Chapter 30). The latter quantity equals kT in the (low-frequency) Rayleigh-Jeans limit: $kT \gg h\nu$, when the oscillator contains sufficient thermal energy to radiate many photons of energy $h\nu$ and the correction for stimulated emission is important. When the radiator constitutes a nonthermal (synchrotron) source, we expect the same expression $(2\nu^2/c^2)$ for the phase-space factor, but we anticipate that we must replace kT by the mean energy ϵ of the synchrotron electron capable of emitting radiation of frequency ν. By our previous arguments, we can associate $\epsilon \equiv \gamma m_e c^2$ with $\nu \approx \gamma^2 \nu_{\rm L}$; i.e., we have the association $\gamma \approx (\nu/\nu_{\rm L})^{1/2}$; hence

$$S_\nu \approx \left(\frac{2\nu^2}{c^2}\right)\left(\frac{\nu}{\nu_{\rm L}}\right)^{1/2} m_e c^2 \propto B_0^{-1/2}\nu^{5/2},$$

which recovers equation (18.17). Indeed, if we had bothered to carry along the proportionality constant, we would have gotten an approximate expression for the opacity coefficient K_0.

In equation (18.16), τ_ν equals the monochromatic optical depth through the source,

$$\tau_\nu \equiv \int \rho\kappa_\nu \, ds,$$

which for a uniform spherical source of radius R_s has the average value

$$\tau_\nu = \frac{4}{3}\rho\kappa_\nu R_s, \tag{18.18}$$

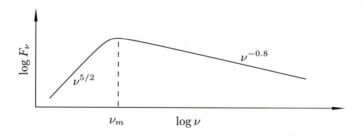

FIGURE 18.6
Expected radio emission from a compact radio source which becomes optically thick at frequencies lower than some transition value ν_m.

because the average distance through the source equals its volume $4\pi R_s^3/3$ divided by its cross-sectional area πR_s^2, or $4R_s/3$. If this uniform spherical source is located at a (Euclidean) distance r, the received radiative flux equals

$$F_\nu = \pi I_\nu \left(\frac{R_s}{r}\right)^2 = \pi I_\nu \theta_s^2, \tag{18.19}$$

where $\theta_s \equiv R_s/r$ is the angular radius of the observed object (assuming $R_s \ll r$). At optically thin frequencies where $\tau_\nu \ll 1$, equation (18.16) has the approximation, $I_\nu \approx S_\nu \tau_\nu = \rho j_\nu/3\pi$, which implies for equation (18.19):

$$F_\nu = \frac{1}{3}\rho j_\nu \theta_s^2. \tag{18.20}$$

This result has the simple interpretation that under optically thin conditions, we ought to be able to see the total volumetric rate of emission $L_\nu = \rho j_\nu (4\pi R_s^3/3)$ after it spreads out over a surface area $4\pi r^2$, which gives equation (18.20) when we identify $\theta_s = R_s/r$.

On the other hand, for optically thick frequencies where $\tau_\nu \gg 1$, equation (18.16) has the approximation, $I_\nu \approx S_\nu$, which implies for equation (18.19),

$$F_\nu = \pi S_\nu \theta_s^2. \tag{18.21}$$

This result also has a simple interpretation, namely, that under optically thick conditions we see only an (angular) surface brightness S_ν distributed on an (angular) circular patch $\pi\theta_s^2$. Since the frequency dependences of ρj_ν and S_ν are given by equations (18.11) and (18.17), respectively, we schematically expect for a homogeneous compact synchrotron source an emergent spectral energy distribution of the form of Figure 18.6. Some compact radio sources do have this shape to their observed radio fluxes, although more complex shapes are also known (probably because a homogeneous source represents an oversimplification).

TRANSITION FREQUENCY

The transition frequency ν_m, where F_ν reaches a maximum as a function of ν and makes a turnover, holds some interest. In the spectral neighborhood of the transition frequency, we ought to be able to use either the optically thick or the optically thin formulae and get order-of-magnitude correct answers. If we apply the optically thick result, equation (18.21), together with equation (18.17), at ν_m, we obtain the identification

$$F_{\nu_m} \propto B_0^{-1/2} \nu_m^{5/2} \theta_s^2, \qquad (18.22)$$

with a known coefficient of proportionality. Since ν_m, F_{ν_m}, and θ_s can be obtained from VLBI measurements, we may use equation (18.22) to deduce the (average) magnetic field strength B_0 in compact radio sources where a turnover in the spectrum due to synchrotron self-absorption occurs. Empirically derived field strengths by this technique range typically from 10^{-1} to 10^{-4} G (not surprisingly), somewhat higher in the compact sources than the extended ones.

Radio astronomers like to define a quantity proportional to the specific intensity called the "brightness temperature" T_b (refer to Problem Set 4 for the motivation):

$$T_b \equiv c^2 I_\nu / 2k\nu^2, \qquad (18.23)$$

where k is the Boltzmann constant. In terms of the flux F_ν [see equation (18.19)] measured from an angularly resolved source,

$$T_b = c^2 F_\nu / 2\pi k\nu^2 \theta_s^2. \qquad (18.24)$$

For compact radio sources with a turnover in the synchrotron spectrum, the maximum brightness temperature $T_b(m)$ occurs roughly at ν_m.

In 1969 Kellermann and Pauliny-Toth noted a strange fact. They found that resolved compact radio sources possess values of $T_b(m)$ which often approached 10^{11} K (indicating a nonthermal origin for the emission), but never exceeded several times this value (e.g., 10^{12} K). They explained this phenomenon in terms of competition from *inverse Compton losses*, a topic that we shall now discuss.

INVERSE COMPTON LOSSES

When an ultrarelativistic electron with Lorentz factor γ encounters a photon, the resulting collision tends to upshift the photon frequency ν roughly to $\gamma^2\nu$, with the exact coefficient dependent on the (Compton) scattering geometry. The result may be viewed as the consequence of two successive Lorentz transformations. For $h\nu' \ll m_e c^2$ in the frame of rest of the electron, Compton scattering behaves as Thompson scattering in preserving the photon frequency ν'. The scattering process also possesses forward-backward symmetry. The mean scattering angle therefore equals 90°. For 90° scattering, the photon ν before collision gets blueshifted by one factor of γ, $\nu' = \gamma\nu$, because of the transformation to the electron's rest frame (to simplify the algebra of the Compton encounter). After collision, the photon ν'' gets blueshifted approximately by another factor of γ, $\nu'' = \gamma\nu' = \gamma^2\nu$, when we transform back to the laboratory frame. (For details, see Rybicki and Lightman 1979.)

In a region where the mean magnetic field $B_0 \sim 10^{-3}$ G, say, the typical Lorentz factor needed to produce a few GHz radiation by Compton scattering of the synchrotron radiation is $\gamma \sim 10^3$. For $\gamma \sim 10^3$, the Compton scattering of $\nu \sim 10^9$ Hz photons will then produce $\gamma^2\nu \sim 10^{15}$ Hz photons, i.e., radiation at optical wavelengths. Thus there arises the distinct possibility that the cosmic-ray electrons, which produce radio waves by the synchrotron process, will Compton scatter those same photons one by one (an intrinsically quantum process) to the optical regime and beyond. In a compact enough source, the radiation densities may be high enough that the Compton process may even dominate over the primary synchrotron process.

To deduce the conditions needed for Compton losses to exceed synchrotron losses, we need the formula for the Compton power P_C produced by an ultrarelativistic electron that encounters photons in a given (isotropic) radiation field of energy density $U_{\rm ph}$. Rybicki and Lightman [see their equation (7.16b)] show that the angle-averaged power equals

$$\langle P_C \rangle = \frac{4}{3}\beta^2\gamma^2 c\sigma_T U_{\rm ph}, \qquad (18.25)$$

if we restrict our attention to photon energies $\ll m_e c^2$. Equation (18.25) bears a strong resemblance to equation (18.4), which gives the synchrotron power emitted by a relativistic electron encountering static magnetic fields of energy density U_B. These two results suggest that the "collisional losses" experienced by a charge passing through electromagnetic fields do not depend sensitively on whether the encounters involve virtual photons or real photons (quasistatic or radiation electromagnetic fields).

If we divide equation (18.25) by equation (18.4), we find that the ratio of Compton losses to synchrotron losses equals

$$\frac{L_C}{L_S} = \frac{U_{\rm ph}}{U_B}, \qquad (18.26)$$

which applies to every single cosmic-ray electron and, therefore, must hold for the entire collection that gives rise to the emitted Compton and synchrotron luminosities, L_C and L_S. (For simplicity, we have assumed a uniform model in which the energy densities U_{ph} and U_B have constant values throughout the entire volume of the source.) The bulk of the radio photons comes out near the peak defined by the turnover frequency ν_m, i.e., $L_S \sim \nu_m F_{\nu_m} 4\pi r^2$. Moreover, for ν_m, we will not make large errors if we adopt the optically thin approximation that the luminosity comes from photons freely streaming at the speed of light c, i.e., that $L_S \sim 4\pi R_S^2 U_{ph} c$. Hence, we may write

$$\frac{U_{ph}}{U_B} \sim \frac{L_S/4\pi R_S^2 c}{B_0^2/8\pi} \sim \frac{\nu_m F_{\nu_m} 4\pi r^2/4\pi R_S^2}{c B_0^2/8\pi}.$$

Thus we obtain

$$\frac{L_C}{L_S} \sim \frac{\nu_m F_{\nu_m}}{\theta_s^2 c B_0^2/8\pi}. \tag{18.27}$$

On the other hand, equation (18.24) applied at $\nu = \nu_m$ yields

$$F_{\nu_m} \propto T_b(m)\nu_m^2 \theta_s^2 \propto T_b(m) F_{\nu_m}^{4/5} B_0^{2/5} \theta_s^{2/5},$$

where we have used equation (18.22) to eliminate ν_m in the last step. Solving the above for F_{ν_m}, we obtain the identification,

$$F_{\nu_m} \propto T_b(m)^5 B_0^2 \theta_s^2.$$

If we substitute this result into equation (18.27), we finally obtain

$$\frac{L_C}{L_S} \propto T_b(m)^5 \nu_m,$$

with a known proportionality constant. Including the proportionality constant into the calculation yields the numerical estimate

$$\frac{L_C}{L_S} \sim \left[\frac{T_b(m)}{10^{12} \text{ K}}\right]^5 \left(\frac{\nu_m}{10^{8.5} \text{ Hz}}\right).$$

This demonstrates that Compton losses remain relatively small for a source with a maximum brightness temperature of the synchrotron radiation $T_b(m) < 10^{12}$ K, but they quickly become catastrophically large if $T_b(m)$ tries to exceed 10^{12} K. Such an excessively bright compact radio source would quickly upshift most of its radio photons to the optical regime (or beyond) through inverse Compton scattering and thereby reduce its radio brightness temperature T_b, presumably to more modest values.

The interesting question then arises whether bright compact radio sources with $T_b(m)$ close to the critical value 10^{12} K have indeed suffered such catastrophic Compton losses. An apparent correlation exists in QSOs such that more radio emission implies more x-ray emission, suggesting that a synchro-Compton mechanism may operate in radio-loud quasars. On the other hand, many of the brightest compact radio sources show evidence of so-called "superluminal expansion," and as Martin Rees and his colleagues have demonstrated, relativistic (bulk) expansion of a compact synchrotron source would much reduce the ratio of Compton to synchrotron losses relative to formulae like those derived above, which apply only when the source remains static in bulk. For further discussion of these points, please see the astrophysical literature.

19

Synchrotron Radiation: Detailed Theory

Reference: Jeffreys and Jeffreys, *Mathematical Physics*, pp. 498–507;
K.C. Westfold 1959, *Ap. J.*, **130**, 241.

In this chapter we provide calculations of the detailed formulae for synchrotron radiation, including polarization properties. We begin with a discussion of the basic mathematical tool needed for the analysis, the asymptotic technique called the *method of stationary phase* for the evaluation of a certain class of integrals.

METHOD OF STATIONARY PHASE

Problems involving the superposition of waves frequently give rise to integrals of the type

$$I \equiv \int_{-a}^{b} A(t) e^{i\Gamma\Phi(t)} \, dt, \tag{19.1}$$

where $A(t)$ and $\Phi(t)$ represent order-unity real functions that vary smoothly with t, and where Γ is a large parameter. For $\Gamma \gg 1$, the rapidly varying total phase $\Gamma\Phi(t)$ implies that contributions of opposite sign tend to cancel in alternating quarter cycles (see Figure 19.1), except where the derivative of Φ vanishes (a point of *stationary phase*).

For simplicity of discussion, let the point of stationary phase $\Phi'(t) = 0$ correspond to $t = 0$, which we shall assume to fall within the range of integration, $-a$ to b. Perform Taylor-series expansions of $A(t)$ and $\Phi(t)$ about $t = 0$:

$$A(t) = A(0) + A'(0)t + \cdots, \qquad \Phi(t) = \Phi(0) + St^2 + \cdots, \tag{19.2}$$

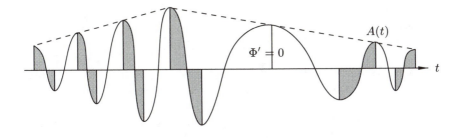

FIGURE 19.1
Contributions to an integral containing a rapidly oscillating integrand come mainly from the neighborhood t of the point where the integrand happens to have a stationary phase, $\Phi'(t) = 0$. Alternate quarter cycles tend to cancel (shaded and light parts in the figure), except for the neighborhood where $\Phi' = 0$.

where $S \equiv \Phi''(0)/2$ may be either positive or negative, depending on the sign of the second derivative of Φ. If we substitute equations (19.2) into (19.1) and keep only the lowest-order terms, we obtain

$$I \approx \Gamma^{-1/2} A(0) e^{i\Gamma\Phi(0)} \int_{-\Gamma^{1/2}a}^{\Gamma^{1/2}b} e^{iS\xi^2} \, d\xi,$$

where we have introduced a new integration variable, $\xi \equiv \Gamma^{1/2} t$. Contour integration allows us to transform the integration from one along the real axis, $\zeta = \xi$, to one along the $\pm 45°$ line, $\zeta = \eta e^{is\pi/4}$, where $s = \text{sgn}(S) = \pm 1$. Moreover, when $\Gamma \gg 1$, we may extend the limits of the integration to $\pm\infty$, in the process introducing only errors that are exponentially small in Γ. Thus we have the approximation

$$I = \Gamma^{-1/2} A(0) e^{i[\Gamma\Phi(0)+s\pi/4]} \int_{-\infty}^{\infty} e^{-|S|\eta^2} \, d\eta = \left(\frac{\pi}{\Gamma|S|}\right)^{1/2} A(0) e^{i[\Phi(0)+s\pi/4]}.$$
$$(19.3)$$

The deformation of the path of integration to transform an oscillatory integrand to an exponentially decaying Gaussian constitutes a standard operation in the more general technique called the *method of steepest descent*. The technique requires modification if $A(t)$ happens to equal zero at $t = 0$ (where $\Phi' = 0$). In this case, we must retain higher-order terms in the expansions (19.2). For example, we easily evaluate the term that arises from $A'(0)t$ by using t^2 as the integration variable rather than t. Especially relevant for the theory of synchrotron radiation is the case when $A'(t)$ also vanishes at $t = 0$, so that the first nonzero contribution gives an integral of the form:

$$\int_{-\infty}^{\infty} \xi^2 e^{iS\xi^2}\, d\xi = -i\frac{d}{dS}\int_{-\infty}^{\infty} e^{iS\xi^2}\, d\xi = \frac{i}{2S}\left(\frac{\pi}{|S|}\right)^{1/2} e^{is\pi/4}. \qquad (19.4)$$

The coefficient in front of this integral will contain a factor $\Gamma^{-3/2}$ (the same exponent as S), which is $\sim \Gamma^{-1}$ smaller than the normal zeroth-order term (19.3).

Even if $A(0) \neq 0$, we see from the above discussion that equation (19.3) represents only the lowest-order term in a series expansion for I in inverse powers of $\Gamma^{1/2}$. We refer to such a series as *asymptotic* (in the sense of Poincaré), because it will generally approach the true solution arbitrarily closely, not in the limit where we take an infinite number of terms (when the series usually diverges), but in the limit $\Gamma \to \infty$ for any *finite* number of terms. Exponentially small terms make no appearance in such a description. With this mathematical preamble, we find ourselves poised to attack the problem of synchrotron radiation.

ORBIT GEOMETRY

Our first step in the development of synchrotron theory involves replacing the helical orbit by an equivalent representation in terms of an instantaneous circular orbit (see Chapter 17). Suppose $\mathbf{r}(\tau)$ near $\tau = 0$ is given by an equivalent circular orbit that takes place in the y'-z' plane:

$$\mathbf{r}(\tau) = \mathcal{R}(\hat{\mathbf{y}}' \cos\varphi + \hat{\mathbf{z}}' \sin\varphi), \qquad (19.5)$$

where $\mathcal{R}$ is given by

$$\mathcal{R} = v/2\pi\nu_B \sin\alpha, \qquad (19.6)$$

and where φ is defined by

$$\varphi \equiv v\tau/\mathcal{R} = (2\pi\nu_B \sin\alpha)\tau. \qquad (19.7)$$

If we differentiate equation (19.5) with respect to τ and divide by c, we get

$$\frac{\mathbf{v}}{c} = \beta(-\hat{\mathbf{y}}' \sin\varphi + \hat{\mathbf{z}}' \cos\varphi). \qquad (19.8)$$

We adopt an (x, y, z) coordinate system, with $\hat{\mathbf{y}} = \hat{\mathbf{y}}'$ and with $\hat{\mathbf{z}}$ pointing in the direction to the observer. We further let θ represent the angle between the observing direction and the velocity vector $\mathbf{v}$ of the particle at $\tau = 0$; thus the (x, y, z) coordinate system can be obtained by applying a rotation of the (x', y', z') system through an angle θ about the $y' = y$ axis:

$$\begin{pmatrix} \hat{\mathbf{z}}' \\ \hat{\mathbf{x}}' \\ \hat{\mathbf{y}}' \end{pmatrix} = \begin{pmatrix} \cos\theta & \sin\theta & 0 \\ -\sin\theta & \cos\theta & 0 \\ 0 & 0 & 1 \end{pmatrix} \begin{pmatrix} \hat{\mathbf{z}} \\ \hat{\mathbf{x}} \\ \hat{\mathbf{y}} \end{pmatrix}. \qquad (19.9)$$

With the transformation (19.9), we can now write equation (19.8) as

$$\frac{\mathbf{v}}{c} = \beta(-\hat{\mathbf{y}}\sin\varphi + \hat{\mathbf{z}}\cos\theta\cos\varphi + \hat{\mathbf{x}}\sin\theta\cos\varphi).$$

If we cross the above with $\hat{\mathbf{k}} \equiv \hat{\mathbf{z}}$, we obtain

$$\hat{\mathbf{k}} \times \frac{\mathbf{v}}{c} = \beta(\hat{\mathbf{x}}\sin\varphi + \hat{\mathbf{y}}\sin\theta\cos\varphi),$$

which becomes, if we cross it again with $\hat{\mathbf{k}} = \hat{\mathbf{z}}$,

$$\hat{\mathbf{k}} \times (\hat{\mathbf{k}} \times \frac{\mathbf{v}}{c}) = \beta(\hat{\mathbf{y}}\sin\varphi - \hat{\mathbf{x}}\sin\theta\cos\varphi). \tag{19.10}$$

Similarly, equation (19.5) has the (x, y, z) vectorial decomposition,

$$\mathbf{r}(\tau) = \mathcal{R}(\hat{\mathbf{y}}\cos\varphi + \hat{\mathbf{z}}\cos\theta\sin\varphi + \hat{\mathbf{x}}\sin\theta\sin\varphi).$$

This result implies that in the wave zone,

$$R = |\mathbf{x}| - \hat{\mathbf{k}} \cdot \mathbf{r} = z - \hat{\mathbf{z}} \cdot \mathbf{r} = z - \mathcal{R}\cos\theta\sin\varphi. \tag{19.11}$$

SYNCHROTRON INTEGRALS

If we substitute equations (19.10) and (19.11) into equation (17.15), we obtain

$$\mathbf{g}_\nu = -i2\pi c\nu e^{i2\pi\nu z/c}$$

$$\times \int_{-\infty}^{\infty} \beta(\hat{\mathbf{y}}\sin\varphi - \hat{\mathbf{x}}\sin\theta\cos\varphi)\exp\left[i2\pi\nu\left(\tau - \frac{\mathcal{R}}{c}\cos\theta\sin\varphi\right)\right]d\tau. \tag{19.12}$$

This expression yields the contribution for all time to the electromagnetic radiation field by our gyrating electron. The orbital period seen at the center of gyration equals $1/\nu_B$, but that observed in the direction $\hat{\mathbf{k}}$ is shorter by a Doppler factor

$$1 - (v\cos\alpha)\cos\alpha/c \approx \sin^2\alpha, \tag{19.13}$$

when $\beta \approx 1$. Thus, per *observed* period at the point of reception, we identify the relevant contribution to the received power in the two orthogonal polarization directions in the plane of the sky as [see equation (17.13)]

$$\begin{bmatrix} dP_\nu(\hat{\mathbf{x}})/d\Omega \\ dP_\nu(\hat{\mathbf{y}})/d\Omega \end{bmatrix} = \frac{e^2\nu_B}{2\pi c^3\sin^2\alpha}(2\pi c\nu)^2$$

$$\times \left| \int_{-1/2\nu_B}^{1/2\nu_B} \begin{pmatrix} \beta\sin\theta\cos\varphi \\ \beta\sin\varphi \end{pmatrix}\exp\left[i2\pi\nu\left(\tau - \frac{\mathcal{R}}{c}\cos\theta\sin\varphi\right)\right]d\tau \right|^2.$$

In equation (19.14), we have separated out the contributions of the $\hat{\mathbf{x}}$ and $\hat{\mathbf{y}}$ components of $\mathbf{g}_\nu$ in order to follow the polarization characteristics of the emitted power.

For $\nu \gg \nu_B$, we have a rapidly varying phase and can expand about $\varphi = v\tau/\mathcal{R} = 0$:

$$\tau - \frac{\mathcal{R}}{c}\cos\theta\sin\varphi = (1 - \beta\cos\theta)\tau + \frac{c^2\beta^3}{6\mathcal{R}^2}\tau^3\cos\theta + \cdots$$

The large parameter Γ of the problem here corresponds to $\gamma \gg 1$ for ultrarelativistic motion. In this limit, $\beta \approx 1$, $\gamma^{-2} \approx 2(1-\beta)$, which implies $\beta \approx 1 - 1/2\gamma^2$, so that

$$\tau - \frac{\mathcal{R}}{c}\cos\theta\sin\varphi = \left[(1-\cos\theta) + \frac{1}{2\gamma^2}\cos\theta\right]\tau + \frac{2\pi}{3}\nu_B^2(\sin^2\alpha)\tau^3\cos\theta,$$

where we have used equation (19.6) with $v \approx c$.

Define the new integration variable,

$$\xi \equiv \gamma\phi = \gamma(2\pi\nu_B\sin\alpha)\tau, \tag{19.15}$$

the dimensionless frequency f_0,

$$f_0 \equiv \frac{\nu}{2\gamma^3\nu_B\sin\alpha}, \tag{19.16}$$

and the angular function Θ,

$$\Theta \equiv \cos\theta + 2\gamma^2(1-\cos\theta). \tag{19.17}$$

Equation (19.14) now yields for the two components of linearly polarized monochromatic power received per unit solid angle in the direction $\hat{\mathbf{k}} = \hat{\mathbf{z}}$:

$$\begin{bmatrix} dP_\nu(\hat{\mathbf{x}})/d\Omega \\ dP_\nu(\hat{\mathbf{y}})/d\Omega \end{bmatrix} = \frac{e^2\nu^2}{2\pi c\gamma^2\nu_B\sin^4\alpha}$$

$$\times \left| \int_{-\xi_0}^{\xi_0} \begin{pmatrix} \sin\theta \\ \xi/\gamma \end{pmatrix} \exp\left[if_0\left(\xi\Theta + \frac{1}{3}\xi^3\cos\theta\right)\right] d\xi \right|^2, \tag{19.18}$$

where $\xi_0 \equiv \pi\gamma\sin\alpha$. The quantity Θ in equation (19.17) is very large (of order γ^2) for all θ except those angles within $\Delta\theta \sim \gamma^{-1}$ of $\theta = 0$. This means that the rapidly varying phase will cancel all contributions except for those within such a beam. For angles $\Delta\theta \sim \gamma^{-1}$ about $\theta = 0$, we may expand and approximate $\Theta \approx (1 + \gamma^2\theta^2)$ as an order-unity quantity for $\gamma\theta$ of order unity. In the same approximation, $\sin\theta$ becomes θ and the integration limits $\pm\xi_0$ can be replaced by $\pm\infty$, so that equation (19.18) reads

$$\begin{bmatrix} dP_\nu(\hat{\mathbf{x}})/d\Omega \\ dP_\nu(\hat{\mathbf{y}})/d\Omega \end{bmatrix} = \frac{e^2\nu^2}{2\pi c\gamma^4\nu_B\sin^4\alpha} \begin{bmatrix} |X(\psi)|^2 \\ |Y(\psi)|^2 \end{bmatrix}, \tag{19.19}$$

where we have defined the integrals

$$
\begin{bmatrix} X(\psi) \\ Y(\psi) \end{bmatrix} \equiv \int_{-\infty}^{\infty} \begin{pmatrix} \psi \\ \xi \end{pmatrix} \exp\left\{ i f_0 \left[\xi(1 + \psi^2) + \frac{\xi^3}{3} \right] \right\} d\xi, \qquad (19.20)
$$

with $\psi \equiv \gamma\theta$.

The integrals in equation (19.20) can be expressed in terms of the modified Bessel functions $K_{1/3}$ and $K_{2/3}$ (see Westfold 1959 and below), but we choose to write the squares of their moduli as the integrals times their complex conjugates:

$$
\begin{bmatrix} |X(\psi)|^2 \\ |Y(\psi)|^2 \end{bmatrix} = \int_{-\infty}^{\infty} d\xi \int_{-\infty}^{\infty} d\eta \begin{pmatrix} \psi^2 \\ \xi\eta \end{pmatrix}
$$
$$
\times \exp\left\{ i f_0 \left[(1 + \psi^2)(\xi - \eta) + \frac{1}{3}(\xi^3 - \eta^3) \right] \right\}. \, (19.21)
$$

Furthermore, if we write $(\xi^3 - \eta^3) = (\xi - \eta)(\xi^2 + \xi\eta + \eta^2)$, we note that the exponent contains a common factor $\xi - \eta$. Motivated by this observation, we rotate the axes of the integration variables by $45°$ by defining

$$
x \equiv \frac{1}{2}(\xi - \eta), \qquad y \equiv \frac{1}{2}(\xi + \eta), \qquad i.e., \qquad \xi = x + y, \qquad \eta = y - x.
$$

The Jacobian of the above transformation equals 2:

$$
\frac{\partial(\xi, \eta)}{\partial(x, y)} = \begin{vmatrix} 1 & 1 \\ -1 & 1 \end{vmatrix} = 2,
$$

with the integration occurring over the whole x-y plane (not to be confused with the plane of the sky). With $\xi^2 + \xi\eta + \eta^2 = x^2 + 3y^2$, we get

$$
\begin{bmatrix} |X(\psi)|^2 \\ |Y(\psi)|^2 \end{bmatrix} = 2 \int_{-\infty}^{\infty} dx \, \exp\left\{ 2 i f_0 x \left[(1 + \psi^2) + \frac{x^2}{3} \right] \right\}
$$
$$
\times \int_{-\infty}^{\infty} \begin{pmatrix} \psi^2 \\ y^2 - x^2 \end{pmatrix} \exp(2 i f_0 x y^2) \, dy.
$$

The inner integrals above have the standard forms discussed at the beginning of this chapter [see equation (19.4)]:

$$
\begin{bmatrix} |X(\psi)|^2 \\ |Y(\psi)|^2 \end{bmatrix} = 2 \int_{-\infty}^{\infty} \left(\frac{\pi}{2 f_0 |x|} \right)^{1/2} e^{i s \pi/4} \begin{pmatrix} \psi^2 \\ i/4 f_0 x - x^2 \end{pmatrix}
$$
$$
\times \exp\left\{ 2 i f_0 x \left[(1 + \psi^2) + \frac{x^2}{3} \right] \right\} dx, \qquad (19.22)
$$

where $s \equiv \mathrm{sgn}(x) = \pm 1$. We interpret the integrals above in a principal value sense in order to avoid the singularity at $x = 0$.

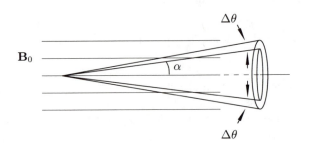

FIGURE 19.2
Conical emission pattern associated with the motion of an ultrarelativistic synchrotron electron.

During an electron's orbit, it beams its radiation into a hollow cone of apex angle $\Delta\theta \sim \gamma^{-1}$ (i.e., $\psi \sim 1$). Figure 19.2 describes the geometry. Since only electrons with local pitch angle nearly equal to α will beam toward the observer at some point in their helical orbits, we may integrate equation (19.19) over all contributing solid angles by writing $d\Omega = 2\pi \sin \alpha d\psi/\gamma$, with the effective integration limits on ψ extendable to $\pm\infty$:

$$\begin{bmatrix} P_\nu(\hat{\mathbf{x}}) \\ P_\nu(\hat{\mathbf{y}}) \end{bmatrix} = \frac{e^2\nu^2}{c\gamma^5\nu_B \sin^3 \alpha} \int_{-\infty}^{\infty} \begin{bmatrix} |X(\psi)|^2 \\ |Y(\psi)|^2 \end{bmatrix} d\psi. \qquad (19.23)$$

If we substitute equation (19.22) into equation (19.23), we may perform the ψ integration first, again in the manner discussed at the beginning of this chapter [see equation (19.4)]:

$$\begin{bmatrix} P_\nu(\hat{\mathbf{x}}) \\ P_\nu(\hat{\mathbf{y}}) \end{bmatrix} = \frac{2e^2\nu^2}{c\gamma^5\nu_B \sin^3 \alpha} \int_{-\infty}^{\infty} \left(\frac{\pi}{2f_0|x|}\right) e^{is\pi/2} \begin{pmatrix} i/4f_0 x \\ i/4f_0 x - x^2 \end{pmatrix}$$
$$\times \exp\left[2if_0\left(x + \frac{x^3}{3}\right)\right] dx.$$

With $f_0 = \nu/2\gamma^3\nu_B \sin \alpha$ and $e^{is\pi/2} = is$, where $s \equiv \mathrm{sgn}(x)$, we obtain

$$\begin{bmatrix} P_\nu(\hat{\mathbf{x}}) \\ P_\nu(\hat{\mathbf{y}}) \end{bmatrix} = \frac{\pi e^2\gamma\nu_B}{c\sin \alpha} \begin{pmatrix} \mathcal{P}_x \\ \mathcal{P}_y \end{pmatrix}, \qquad (19.24)$$

where

$$\begin{pmatrix} \mathcal{P}_x \\ \mathcal{P}_y \end{pmatrix} \equiv -\int_{-\infty}^{\infty} \frac{1}{x^2}\left(\frac{1}{1 + 4if_0 x^3}\right) \exp\left[2if_0\left(x + \frac{x^3}{3}\right)\right] dx. \qquad (19.25)$$

Write $-dx/x^2$ as $d(1/x)$ and perform an integration by parts to make $\mathcal{P}_x$ appear less singular:

$$\mathcal{P}_x = I_- + I_+,$$

where

$$I_\pm \equiv -2if_0 \int_{-\infty}^{\infty} x^{\pm 1} \exp\left[2if_0\left(x + \frac{x^3}{3}\right)\right] dx.$$

The second row of equation (19.25) can now be read off to give

$$\mathcal{P}_y = I_- + 3I_+.$$

With modern computers, we could easily perform the integrations for $I_\pm$ numerically (after turning the contour from $z = x$ to $z = iy$). In fact, we can express $I_\pm$ in terms of already tabulated results:

$$I_+ = \sqrt{3}G\left(\frac{4}{3}f_0\right),$$

$$I_- = -2\sqrt{3}G\left(\frac{4}{3}f_0\right) + \sqrt{3}F\left(\frac{4}{3}f_0\right),$$

where F and G equal the synchrotron functions:

$$F(x) \equiv x \int_x^{\infty} K_{5/3}(y)\,dy, \qquad G(x) \equiv xK_{2/3}(x), \qquad (19.26)$$

and $K_{5/3}$ and $K_{2/3}$ are modified Bessel functions of order 5/3 and 2/3, respectively. Table 19.1 yields the values of $F(x)$ and $G(x)$ given by Westfold's article.

We choose to write the dimensionless frequency that enters into the argument of the various functions as follows:

$$\frac{4}{3}f_0 = \frac{2\nu}{3\gamma^3\nu_B \sin\alpha} \equiv \nu/\nu_c,$$

where ν_c is the characteristic frequency,

$$\nu_c \equiv \frac{3\gamma^2 eB_\perp}{4\pi m_e c}. \qquad (19.27)$$

In the above, we have used $\nu_B = eB_0/2\pi\gamma m_e c$ and expressed

$$B_\perp \equiv B_0 \sin\alpha \qquad (19.28)$$

as the component of $\mathbf{B}_0$ perpendicular to the line of sight (the component that lies in the plane of the sky).

If we collect terms, $\mathcal{P}_x$ and $\mathcal{P}_y$ now acquire the elegant form:

$$\mathcal{P}_x = \sqrt{3}[F(\nu/\nu_c) - G(\nu/\nu_c)], \qquad \mathcal{P}_y = \sqrt{3}[F(\nu/\nu_c) + G(\nu/\nu_c)]. \quad (19.29)$$

x	F	G	x	F	G
0	0	0	1.0	0.655	0.494
0.001	0.213	0.107	1.2	.566	.439
.005	.358	.184	1.4	.486	.386
.010	.445	.231	1.6	.414	.336
.025	.583	.312	1.8	.354	.290
.050	.702	.388	2.0	.301	.250
.075	.772	.438	2.5	.200	.168
.10	.818	.475	3.0	.130	.111
.15	.874	.527	3.5	.0845	.0726
.20	.904	.560	4.0	.0541	.0470
.25	.917	.582	4.5	.0339	.0298
.30	.919	.596	5.0	.0214	.0192
.40	.901	.607	6.0	.0085	.0077
.50	.872	.603	7.0	.0033	.0031
.60	.832	.590	8.0	.0013	.0012
.70	.788	.570	9.0	.00050	.00047
.80	.743	.547	10.0	0.00019	0.00018
0.90	0.694	0.521			

Table 19.1
Values of $F(x)$ and $G(x)$. (From K.C. Westfold 1959, *Ap. J.*, **130**, 241.)

Expressions for the emitted power, $P_\nu^{\mathrm{em}}(\hat{\mathbf{x}})$ and $P_\nu^{\mathrm{em}}(\hat{\mathbf{y}})$, have a multiplicative factor $\sin^2\alpha$ in comparison to the formulae (19.24); thus

$$\begin{bmatrix} P_\nu^{\mathrm{em}}(\hat{\mathbf{x}}) \\ P_\nu^{\mathrm{em}}(\hat{\mathbf{y}}) \end{bmatrix} = \left(\frac{\sqrt{3}}{2}\right) \frac{e^3 B_\perp}{m_e c^2} \left[\frac{F(\nu/\nu_{\mathrm{c}}) - G(\nu/\nu_{\mathrm{c}})}{F(\nu/\nu_{\mathrm{c}}) + G(\nu/\nu_{\mathrm{c}})} \right]. \qquad (19.30)$$

As discussed in Chapter 18, when averaged over an isotropic distribution of cosmic-ray electrons, there exists no difference between the power $\langle P_\nu \rangle$ received as an *average* by all observers on a celestial sphere centered on the source and the average power $\langle P_\nu^{\mathrm{em}} \rangle$ emitted by the electrons. However, observers located at different positions will generally receive different amounts of power for a source model in which the $\mathbf{B}_0$ has a single direction. (Consider, e.g., the case of observers who look at orientations parallel and perpendicular to $\mathbf{B}_0$.) Only for a completely tangled magnetic field configuration (in which case there would be no net polarization) would all observers on the celestial sphere receive the same radiative flux.

POWER-LAW DISTRIBUTION OF ELECTRON ENERGIES

The synchrotron volume emissivity due to a power-law collection of electrons equals

$$\rho \begin{bmatrix} j_\nu(\hat{\mathbf{x}}) \\ j_\nu(\hat{\mathbf{y}}) \end{bmatrix} = \left(\frac{\sqrt{3}}{2} \right) \frac{e^3 B_\perp}{m_e c^2} \int_1^\infty n_0 \gamma^{-p} \begin{bmatrix} F(x) - G(x) \\ F(x) + G(x) \end{bmatrix} d\gamma,$$

with $x \equiv \nu/\nu_c$ implying $\gamma = (2\nu/3\nu_{L\perp})^{1/2} x^{-1/2}$, where $\nu_{L\perp} \equiv eB_\perp/2\pi m_e c$. Making the substitution for γ in terms of x, we get

$$\rho \begin{bmatrix} j_\nu(\hat{\mathbf{x}}) \\ j_\nu(\hat{\mathbf{y}}) \end{bmatrix} = \left(\frac{\sqrt{3}}{2} \right) n_0 \frac{e^3 B_\perp}{m_e c^2} \left(\frac{J_F - J_G}{J_F + J_G} \right) \left(\frac{2\nu}{3\nu_{L\perp}} \right)^{-(p-1)/2}, \qquad (19.31)$$

where J_F and J_G are the pure numbers

$$J_F \equiv \int_0^\infty x^{(p-3)/2} F(x)\, dx = \frac{2^{(p+1)/2}}{p+1} \Gamma\left(\frac{p}{4} + \frac{19}{12} \right) \Gamma\left(\frac{p}{4} - \frac{1}{12} \right), \qquad (19.32)$$

$$J_G \equiv \int_0^\infty x^{(p-3)/2} G(x)\, dx = 2^{(p-3)/2} \Gamma\left(\frac{p}{4} + \frac{7}{12} \right) \Gamma\left(\frac{p}{4} - \frac{1}{12} \right), \qquad (19.33)$$

with Γ being the gamma function. In equations (19.32) and (19.33), we have used $\nu \gg \nu_{L\perp}$ to replace the lower limit on the integral, $3\nu_{L\perp}/2\nu$, by zero.

For optically thin synchrotron emission, the radiation contains only linearly polarized light in the ultrarelativistic limit; circular polarization would arise only for very low frequencies where the γ's for the emitting electrons are not very large compared to unity. Thus the unpolarized portion of the emission corresponds to that part for which $j_\nu(\hat{\mathbf{x}})$ and $j_\nu(\hat{\mathbf{y}})$ are equal (review Chapter 12). In other words, the fractional polarization for each pitch angle α is given by

$$\frac{j_\nu(\hat{\mathbf{y}}) - j_\nu(\hat{\mathbf{x}})}{j_\nu(\hat{\mathbf{y}}) + j_\nu(\hat{\mathbf{x}})} = \frac{J_G}{J_F} = \frac{p+1}{p+7/3}. \qquad (19.34)$$

This value ≈ 72 percent for $p \sim 2.5$; lower polarizations result if the magnetic field does not have a uniform direction, or if there exists Faraday depolarization in the source (see Chapter 20 on plasma effects). In any case, notice that measurements of synchrotron *emission* (as distinct from the *propagation* effects, to be discussed in Chapter 20) yield information only about the component $B_\perp$ perpendicular to the line of sight, and nothing about the component $B_\parallel$ parallel to it.

SYNCHROTRON OPACITY

According to equation (8.24), the single-particle cross section for true absorption involving discrete states is given by

$$\alpha_\nu = \frac{1}{4\pi} h\nu B_{12}\delta(\nu - \nu_{21}), \qquad (19.35)$$

where B_{12} is Einstein's B-coefficient for true absorption, ν_{21} is the frequency of the transition, and δ is the Dirac delta function (an idealization for a more realistic line profile). To make use of this concept, we imagine the radiation of each synchrotron photon as a quantized process involving two electron energy states E_2 and E_1:

$$h\nu = E_2 - E_1 \equiv h\nu_{21}.$$

A complication enters because each frequency ν can be emitted, in principle, by a continuum of transitions between possible upper and lower states. Moreover, we need to correct for the effects of stimulated emission in a non-LTE context.

To proceed, we invoke the relationships among the Einstein rate coefficients for spontaneous emission, stimulated emission, and true absorption. Since we have computed the emissive power of a single synchrotron electron, we begin there. In terms of the Einstein spontaneous transition rate A_{21}, we may write for the average monochromatic power emitted by electrons of energy E_2:

$$P_\nu^{\rm em}(E_2) = h\nu \sum_{E_1} A_{21}\delta(\nu - \nu_{21}).$$

The elementary rate coefficient for the stimulated emission of the same photons satisfies $B_{21} = (c^2/2h\nu^3)A_{21}$. If $N(E_2)\,dE_2 = n(\gamma_2)\,d\gamma_2$ represents the number density of electrons with energy between $E_2 = \gamma_2 m_e c^2$ and $E_2 + dE_2 = (\gamma_2 + d\gamma_2)m_e c^2$, we may now write the (negative) part of the volume absorptivity associated with stimulated emission as

$$-h\nu \sum_{E_2} N(E_2)\,dE_2 \sum_{E_1} B_{21}\delta(\nu - \nu_{21}) = -\frac{c^2}{2h\nu^3}\sum_{E_2} N(E_2)P_\nu^{\rm em}(E_2)\,dE_2.$$

Similarly, since $B_{12} = (g_2/g_1)B_{21}$, the contribution to the true absorption can be written as

$$h\nu \sum_{E_1} n(E_1)\,dE_1 \sum_{E_2} B_{12}\delta(\nu - \nu_{21})$$

$$= \frac{c^2}{2h\nu^3}\sum_{E_2}(g_2/g_1)N(E_2 - h\nu)P_\nu^{\rm em}(E_2)\,dE_2,$$

because $E_1 = E_2 - h\nu$ and $dE_1 = dE_2$ if we keep $h\nu$ constant.

We identify the volume absorptivity (corrected for stimulated emission) in the corresponding frequency interval as [see equation (19.35)]:

$$\rho\kappa_\nu = \frac{c^2}{8\pi h\nu^3} \sum_{E_2} [(g_2/g_1)N(E_2 - h\nu) - N(E_2)]P_\nu^{\text{em}}(E_2).$$

In the continuum limit where p represents the scalar momentum of an electron (not to be confused with the power-law index), we identify $g_2/g_1 = p_2^2\, dp_2/p_1^2\, dp_1 = E_2^2\, dE_2/E_1^2\, dE_1 = E_2^2/(E_2 - h\nu)^2$ for ultrarelativistic particles where $E = pc$. Thus we obtain

$$\rho\kappa_\nu = \frac{c^2}{8\pi h\nu^3} \int \left\{ \left[\frac{E}{(E - h\nu)}\right]^2 N(E - h\nu) - N(E) \right\} P_\nu^{\text{em}}(E)\, dE.$$

The photon energies $h\nu$ (in the radio part of the spectrum typically) radiated by synchrotron electrons are small in comparison with the particle energies (GeV or more typically). For $h\nu \ll E$ (the classical limit), we may approximate the difference as a derivative:

$$\left[\frac{E}{(E - h\nu)}\right]^2 N(E - h\nu) - N(E) = E^2 \left[\frac{N(E - h\nu)}{(E - h\nu)^2} - \frac{N(E)}{E^2}\right]$$

$$\approx -h\nu E^2 \frac{\partial}{\partial E}\left[\frac{N(E)}{E^2}\right].$$

Making the appropriate substitution, we finally obtain

$$\rho\kappa_\nu = -\frac{c^2}{8\pi\nu^2} \int P_\nu^{\text{em}}(E)E^2 \frac{\partial}{\partial E}\left[\frac{N(E)}{E^2}\right]\, dE. \tag{19.36}$$

Notice that Planck's constant has disappeared completely.

For a power-law distribution of energies, $N(E)\, dE = n(\gamma)\, d\gamma = n_0 \gamma^{-p} d\gamma$, we get

$$\rho\kappa_\nu = K_0 n_0 B_\perp^{(p+2)/2} \nu^{-(p+4)/2}, \tag{19.37}$$

where K_0 is the coefficient

$$K_0 \equiv \frac{cr_e}{4\sqrt{3}} \left(\frac{3e}{2\pi m_e c}\right)^{(p+2)/2} \Gamma\left(\frac{3p+2}{12}\right)\Gamma\left(\frac{3p+22}{12}\right), \tag{19.38}$$

with $r_e \equiv e^2/m_e c^2$. The notion of a scalar opacity, however, represents somewhat of an *ad hoc* simplification, since the magnetized medium of relativistic electrons behaves nonisotropically relative to the emission and absorption of synchrotron radiation. See Pacholczyk (1977) for more detailed discussions of these points.

20

Plasma Effects

Reference: Rybicki and Lightman, *Radiative Processes in Astrophysics*, Chapter 8.

In previous chapters we focused on the processes by which electromagnetic radiation is generated. Here we wish to discuss how such waves, once generated, might be modified by propagation through a (partially) ionized and (possibly) magnetized medium (one containing a quasistatic magnetic field $\mathbf{B}_0$). For simplicity, we will refer to such a medium as a magnetized *plasma*. As we will see, plasma effects acquire greater importance at low frequencies; consequently, most of our applications will concern radio astronomy.

WAVE PROPAGATION IN A MAGNETIZED PLASMA

Our basic equations are the Maxwell set:

$$\boldsymbol{\nabla} \cdot \mathbf{E} = 4\pi\rho_{\mathrm{e}}, \qquad \boldsymbol{\nabla} \cdot \mathbf{B} = 0, \tag{20.1}$$

$$\boldsymbol{\nabla} \times \mathbf{E} = -\frac{1}{c}\frac{\partial \mathbf{B}}{\partial t}, \qquad \boldsymbol{\nabla} \times \mathbf{B} = \frac{4\pi}{c}\mathbf{j}_{\mathrm{e}} + \frac{1}{c}\frac{\partial \mathbf{E}}{\partial t}. \tag{20.2}$$

We wish to consider the propagation of a plane monochromatic electromagnetic wave through a plasma threaded by a uniform and static magnetic field $\mathbf{B}_0$, which we consider separately from the radiation fields $\mathbf{E}$ and $\mathbf{B}$:

$$\begin{pmatrix} \mathbf{E} \\ \mathbf{B} \end{pmatrix} = \begin{pmatrix} \mathbf{e}_0 \\ \mathbf{b}_0 \end{pmatrix} e^{i(\mathbf{k}\cdot\mathbf{x}-\omega t)}. \tag{20.3}$$

In response to the presence of this wave, we suppose that, in steady state, the electrically conducting medium acquires a similar periodic structure:

$$\begin{pmatrix} \rho_{\mathrm{e}} \\ \mathbf{j}_{\mathrm{e}} \end{pmatrix} = \begin{pmatrix} \rho_0 \\ \mathbf{j}_0 \end{pmatrix} e^{i(\mathbf{k}\cdot\mathbf{x}-\omega t)}. \tag{20.4}$$

If we substitute equations (20.3) and (20.4) into Maxwell's equations, we obtain the algebraic set

$$i\mathbf{k} \cdot \mathbf{E} = 4\pi\rho_{\mathrm{e}}, \qquad i\mathbf{k} \cdot \mathbf{B} = 0, \tag{20.5}$$

$$i\mathbf{k} \times \mathbf{E} = i\frac{\omega}{c}\mathbf{B}, \qquad i\mathbf{k} \times \mathbf{B} = \frac{4\pi}{c}\mathbf{j}_{\mathrm{e}} - i\frac{\omega}{c}\mathbf{E}. \tag{20.6}$$

The second of equations (20.5) implies that $\mathbf{B}$ and $\mathbf{k}$ are perpendicular to one another; the first of equations (20.6) implies the same for $\mathbf{B}$ and $\mathbf{E}$. We will demonstrate below that $\mathbf{E}$ and $\mathbf{k}$ are also orthogonal, i.e., that $\mathbf{E}$, $\mathbf{B}$, and $\mathbf{k}$ form a mutually orthogonal set of vectors. Moreover, if nontrivial values of ρ_{e} and $\mathbf{j}_{\mathrm{e}}$ arise only in response to the presence of the radiation fields $\mathbf{E}$ and $\mathbf{B}$, then we ultimately have a *homogeneous* set of equations that possesses meaningful solutions only if ω and k satisfy an eigenvalue relationship (the *dispersion relation*, see below).

MODEL OF A COLD ELECTRON PLASMA

To reduce equations (20.5) and (20.6) to a single relation, we adopt the model of a *cold electron plasma*. We assume that thermal electrons constitute the only *mobile* charge carriers that move through the background of a stationary sea of ions; moreover, we assume that the thermal motions of the electrons are nonrelativistic and therefore carry them a negligible distance in comparison with the wavelength of light during one oscillation period of the electromagnetic wave. Without loss of generality, we may then neglect the spatial contribution to the phase variation in $e^{i(\mathbf{k}\cdot\mathbf{x}-\omega t)}$. In this approximation, the (nonrelativistic) equation of motion for any thermal electron reads

$$m_{\mathrm{e}}\dot{\mathbf{v}} = -e\left(\mathbf{E} + \frac{\mathbf{v}}{c} \times \mathbf{B}_0\right), \tag{20.7}$$

where we have anticipated that the contribution of the radiation magnetic field $\mathbf{B}$ to the Lorentz force, $-e\mathbf{v} \times \mathbf{B}/c$, can be ignored in comparison to the contribution of the radiation electric field, $-e\mathbf{E}$, when $v/c \ll 1$, because $\mathbf{B}$ and $\mathbf{E}$ almost have the same magnitude in an electromagnetic wave [see the first of equation (20.6) and the demonstration below that $\omega/k \approx c$ for the regime of practical interest]. We allow, however, the possibility that the *static* magnetic field $\mathbf{B}_0$ may have a non-negligible systematic effect.

With the time variation of all quantities in equation (20.7) assumed to behave as $e^{-i\omega t}$, we require the steady-state velocity response to satisfy

$$-i\omega m_{\mathrm{e}}\mathbf{v} = -e\left(\mathbf{E} + \frac{\mathbf{v}}{c} \times \mathbf{B}_0\right). \tag{20.8}$$

We know that the presence of the term proportional to $\mathbf{B}_0$ tends to make the electron want to gyrate in circles. We can also always decompose an arbitrarily polarized electromagnetic wave, taken for sake of definiteness to be propagating in the z-direction, as a sum of left- and right-handed circularly polarized waves:

$$\mathbf{E}_{\pm} \equiv (\hat{\mathbf{x}} \pm i\hat{\mathbf{y}})\mathcal{E}_0 e^{-i\omega t}. \tag{20.9}$$

The effect of these two differently circularly polarized components on the electron will differ in accordance to whether the rotation of the electric vector takes place in the same or opposite sense as the natural circular motion of the electron in the external magnetic field $\mathbf{B}_0$. For simplicity of discussion, we assume that $\mathbf{B}_0$ points along $\hat{\mathbf{z}}$ (the direction of wave propagation, i.e., the observer's line of sight):

$$\mathbf{B}_0 = B_0\hat{\mathbf{z}}.$$

In practice, this turns out to be an easy assumption to rectify. As we will see shortly, for wave frequencies much greater than the Larmor frequency, the $\mathbf{B}_0$ term has a small effect in comparison with $\mathbf{E}$ in the electron's equation of motion; thus, if we can ignore the electron's motion along the direction of wave propagation in z, a more general orientation of $\mathbf{B}_0$ simply results in the replacement of $B_{\|} \equiv \hat{\mathbf{z}} \cdot \mathbf{B}_0$ everywhere we see B_0 written below.

With $\mathbf{E}$ given by equation (20.9), we may attempt a solution of equation (20.8) of the form

$$\mathbf{v} = v_0(\hat{\mathbf{x}} \pm i\hat{\mathbf{y}})e^{-i\omega t}. \tag{20.10}$$

Substitution of the above into equation (20.8) results in the identification

$$-i\omega m_e v_0 = -e\mathcal{E}_0 \mp ie\frac{v_0}{c}B_0,$$

i.e., if we solve for v_0, we have the steady-state response

$$\mathbf{v} = \frac{-ie\mathbf{E}}{m_e(\omega \mp \omega_{\mathrm{L}})},$$

where ω_{L} is the Larmor frequency

$$\omega_{\mathrm{L}} \equiv eB_0/m_e c.$$

Corresponding to the induced velocity is an electric current,

$$\mathbf{j}_e = -n_e e\mathbf{v} = \frac{in_e e^2 \mathbf{E}}{m_e(\omega \mp \omega_{\mathrm{L}})}. \tag{20.11}$$

The result that $\mathbf{j}_e \propto \mathbf{E}$ represents Ohm's law for this problem. In the absence of collisions with other particles, however, the system suffers no true dissipation (for $\omega \neq \omega_L$), since the time average of the conductive work satisfies

$$\langle \mathbf{j}_e \cdot \mathbf{E} \rangle = \frac{1}{4}(\mathbf{j}_e^* \cdot \mathbf{E} + \mathbf{j}_e \cdot \mathbf{E}^*) = 0$$

if the coefficient of proportionality between $\mathbf{j}_e$ and $\mathbf{E}$—the electrical conductivity—is purely imaginary.

To obtain the charge density ρ_e, we apply the equation governing the conservation of electric charge:

$$0 = \frac{\partial \rho_e}{\partial t} + \boldsymbol{\nabla} \cdot \mathbf{j}_e = -i\omega \rho_e + i\mathbf{k} \cdot \mathbf{j}_e.$$

Thus

$$\rho_e = \frac{\mathbf{k} \cdot \mathbf{j}_e}{\omega} = \frac{in_e e^2 \mathbf{k} \cdot \mathbf{E}}{m_e \omega(\omega \mp \omega_L)}. \tag{20.12}$$

DISPERSION RELATION FOR ELECTROMAGNETIC WAVES

The substitution of equation (20.12) into the first of (20.5) results in the relationship

$$\mathbf{k} \cdot \mathbf{E} = \frac{\omega_{pe}^2}{\omega(\omega \mp \omega_L)} \mathbf{k} \cdot \mathbf{E}, \tag{20.13}$$

where we have defined the square of the electron plasma frequency as

$$\omega_{pe}^2 \equiv 4\pi n_e e^2 / m_e. \tag{20.14}$$

Transposing terms in equation (20.13) to the left-hand side, we may write

$$\epsilon \mathbf{k} \cdot \mathbf{E} = 0, \tag{20.15}$$

where we have defined

$$\epsilon \equiv 1 - \frac{\omega_{pe}^2}{\omega(\omega \mp \omega_L)}. \tag{20.16}$$

If we identify ϵ as the *plasma dielectric constant*, we see that equation (20.15) simply expresses $\boldsymbol{\nabla} \cdot \mathbf{D} = 0$, where $\mathbf{D} \equiv \epsilon \mathbf{E}$ is the displacement vector of the electrodynamics of a continuous medium. Since ϵ will not equal zero for an arbitrary value of ω, we see that equation (20.15) generally requires $\mathbf{k} \cdot \mathbf{E} = 0$; i.e., $\mathbf{k}$ and $\mathbf{E}$ are mutually perpendicular. (Q.E.D.) In turn, equation (20.12) implies that the propagation of transverse electromagnetic waves leads to no charge separation in the plasma, $\rho_e = 0$, and to zero divergence of the electric current, $\boldsymbol{\nabla} \cdot \mathbf{j}_e = 0$.

The substitution of equations (20.11) and (20.12) into the second of equations (20.6) yields

$$\mathbf{k} \times \mathbf{B} = \left[\frac{\omega_{pe}^2}{c(\omega \mp \omega_L)} - \frac{\omega}{c} \right] \mathbf{E} = -\frac{\omega}{c} \epsilon \mathbf{E}. \tag{20.17}$$

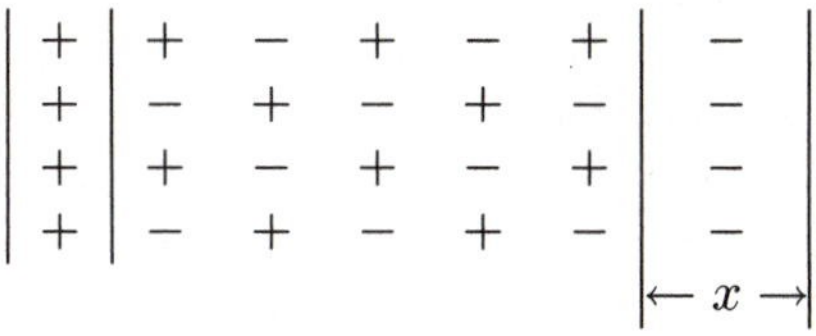

$$\begin{vmatrix} + & + & - & + & - & + & - \\ + & - & + & - & + & - & - \\ + & + & - & + & - & + & - \\ + & - & + & - & + & - & - \end{vmatrix}$$
$$\overset{}{\underset{\longleftarrow x \longrightarrow}{}}$$

FIGURE 20.1
Charge separation leads to restoring electric fields and plasma oscillations.

If we cross the above with $\mathbf{k}$, expand out the triple vector product using $\mathbf{k} \cdot \mathbf{B} = 0$, and invoke the first of equations (20.6) to identify $\mathbf{k} \times \mathbf{E} = \omega \mathbf{B}/c$, we obtain

$$-k^2 \mathbf{B} = -\epsilon \left(\frac{\omega}{c}\right)^2 \mathbf{B},$$

which yields the desired *dispersion relationship*:

$$\omega^2 = c^2 k^2 / \epsilon. \tag{20.18}$$

DYNAMICAL INTERPRETATION OF THE PLASMA FREQUENCY

For many astrophysical situations, ω will be large compared to ω_{L}, and we may first discuss the simpler situation when we totally ignore the latter term in equation (20.16):

$$\epsilon \approx 1 - \frac{\omega_{\mathrm{pe}}^2}{\omega^2}. \tag{20.19}$$

This result has a simple heuristic explanation as follows. Consider in Figure 20.1 displacing the electrons of a piece of the plasma by an amount x with respect to the (stationary) ions. The situation between the charge excesses and deficits at the ends gives the electric field of a capacitor:

$$E_x = 4\pi e n_e x,$$

where $e n_e x$ is the effective surface charge density on the capacitor plates. The equation of motion for a displaced but otherwise free electron inside the capacitor then reads

$$m_e \ddot{x} = -e E_x = -m_e \omega_{\mathrm{pe}}^2 x,$$

which is the familiar equation of motion for a harmonic oscillator. In other words, a free plasma will naturally oscillate at a frequency ω_{pe} in response to *longitudinal* displacements that result in charge separations (see Chapter 29 of Volume II).

Numerically, $\omega_{\mathrm{pe}} = 5.6 \times 10^4 n_e^{1/2}$ rad s^{-1} if n_e is expressed in the units, cm^{-3}, and may not be ignorable in comparison with realistic wave frequencies ω. A passing *transverse* electromagnetic field will produce no charge separation, but it will yield a nonzero (though divergence-free) conduction current $\mathbf{j}_e$. If the wave frequency ω is very large in comparison with the plasma frequency ω_{pe}, the displacement current $\partial\mathbf{E}/\partial t$ will be very much larger than the conduction current $\mathbf{j}_e$. The electromagnetic wave will then propagate much as it does in a vacuum with a phase velocity,

$$\omega/k = c,$$

since ϵ given by equation (20.19) approximately equals unity for $\omega \gg \omega_{\mathrm{pe}}$. On the other hand, if $\omega \approx \omega_{\mathrm{pe}}$, the conduction current tends to cancel the displacement current—see equation (20.17) when we can ignore ω_{L}—and the magnetic field associated with the transverse electromagnetic wave disappears. For ω less than ω_{pe}, equation (20.19) gives a negative value for ϵ, and the phase velocity ω/k in equation (20.18) becomes imaginary; i.e., the electromagnetic wave cannot propagate at all. The wave becomes *evanescent* in the language of the plasma physicist. In this case, since no Ohmic dissipation occurs in the system to provide actual absorption of the wave, the existence of the plasma will reflect the low-frequency radiation that tries to penetrate it. This effect applied to the ionosphere of the Earth, where $n_e \approx 10^6$ cm^{-3}, explains why celestial radio waves with wavelengths longer than about 30 m cannot reach the ground. Conversely, AM broadcasts and other forms of long-range radio communication make use of the ability of the ionosphere to bounce the signal from the transmitter to the receiver to overcome the curvature of the Earth.

PULSAR DISPERSION MEASURE

For $\omega > \omega_{\mathrm{pe}}$, the dielectric "constant" of a plasma ϵ is less than unity, so equation (20.18) implies that the phase velocity of light $\omega/k = c/\epsilon^{1/2}$ exceeds c. This presents no difficulties, since the important speed for physical entities is the *group velocity*, $c_{\mathrm{g}} = \partial\omega/\partial k$ (see Chapter 12 of Volume II). With ϵ given by equation (20.19), we obtain

$$\frac{\partial\omega}{\partial k} = \frac{c}{\epsilon^{1/2}} - \left(\frac{ck\omega_{\mathrm{pe}}^2}{\epsilon^{3/2}\omega^3}\right)\frac{\partial\omega}{\partial k}.$$

If we solve for $\partial\omega/\partial k$ and simplify by identifying ck/ω as $\epsilon^{1/2}$, the group velocity becomes

$$c_{\mathrm{g}} = \frac{\partial\omega}{\partial k} = c\epsilon^{1/2}, \tag{20.20}$$

which is less than c.

We apply equation (20.20) to the propagation of the group of waves that represent the radio pulses from a pulsar. The travel time for pulses of radian frequency ω to get from the pulsar to the telescope equals

$$t_\omega = \int_0^r \frac{ds}{c_g} = \int_0^r \frac{ds}{c} \epsilon^{-1/2}. \tag{20.21}$$

At practical astronomical frequencies for radio observations of the interstellar medium, $\omega \gg \omega_{pe}$; thus we may expand equation (20.19) via the binomial theorem as

$$\epsilon^{-1/2} \approx 1 + \frac{\omega_{pe}^2}{2\omega^2}.$$

If we substitute the above, together with equation (20.14), into equation (20.21), we get

$$t_\omega = \frac{r}{c} + \frac{2\pi e^2}{m_e \omega^2} \mathrm{DM}, \tag{20.22}$$

where r/c is the light travel time at very high (e.g., optical) frequencies, whereas DM is the (pulsar) *dispersion measure*:

$$\mathrm{DM} \equiv \int_0^r n_e \, ds. \tag{20.23}$$

Observation of the increased delay in pulse arrival times that occur at lower frequencies ω yields a measurement of the pulsar dispersion measure DM via equation (20.22). We can then get the average electron density $\langle n_e \rangle$ along the line of sight if we have an estimate for the distance r to the pulsar. Some pulsars show evidence in their radio spectra for 21-cm hydrogen absorption due to the spin-flip hyperfine transition of the ground state of atomic hydrogen. These line features against the continuum emission of the pulsar can often be associated with the velocity of hydrogen gas located in various spiral arms of the Galaxy, whose distances are ascertained by other means. If the velocity of one spiral arm shows up, but the next one out does not, then we may bracket the distance r of the pulsar to lie between those two spiral arms. The mean value for $\langle n_e \rangle \equiv \mathrm{DM}/r$ for Galactic pulsars whose distances can be estimated this way works out to be

$$\langle n_e \rangle \approx 0.03 \text{ cm}^{-3}.$$

Radio astronomers often take this to typify the mean electron abundance in the general interstellar medium and compute fiducial distances to pulsars with unknown values by assuming $r = \mathrm{DM}/(0.03 \text{ cm}^{-3})$.

FARADAY ROTATION

We now consider the situation when we do not ignore the term ω_{L} in equation (20.16). For $\omega \gg \omega_{\mathrm{pe}}$ and ω_{L}, we can approximate equation (20.16) for the two senses of circular polarized radiation by

$$\epsilon_\pm = 1 - \frac{\omega_{\mathrm{pe}}^2}{\omega^2}\left(1 \pm \frac{\omega_{\mathrm{L}}}{\omega}\right).$$

For $\omega \gg \omega_{\mathrm{pe}}$, the dispersion relationship (20.18) can now be expressed as

$$k_\pm = k_0 \mp \Delta k,$$

where k_0 is the part of the wave vector that we have already discussed in connection with pulsar dispersion,

$$k_0 \equiv \frac{\omega}{c} - \frac{\omega_{\mathrm{pe}}^2}{2c\omega}, \tag{20.24}$$

and Δk is the difference associated with the propagation of the two different senses of circular polarization,

$$\Delta k \equiv \frac{\omega_{\mathrm{pe}}^2 \omega_{\mathrm{L}}}{2c\omega^2}. \tag{20.25}$$

In equation (20.25), we modify ω_{L} to mean, in the light of our previous discussion concerning more general field orientations,

$$\omega_{\mathrm{L}} \to eB_\parallel/m_e c, \tag{20.26}$$

where $B_\parallel \equiv \mathbf{B}_0 \cdot \hat{\mathbf{z}}$ and $\hat{\mathbf{z}}$ is the direction of wave propagation.

Consider now an electromagnetic wave train that starts off linearly polarized in the x-direction at the source:

$$\mathbf{E} = \hat{\mathbf{x}}\mathcal{E}_0 e^{-i\omega t} \equiv \frac{1}{2}[(\hat{\mathbf{x}} + i\hat{\mathbf{y}}) + (\hat{\mathbf{x}} - i\hat{\mathbf{y}})]\mathcal{E}_0 e^{-i\omega t}.$$

After propagating a distance z through a magnetized plasma toward the observer, the electric field will behave as

$$\mathbf{E} = \frac{1}{2}\left[(\hat{\mathbf{x}} + i\hat{\mathbf{y}})e^{i(\varphi-\psi)} + (\hat{\mathbf{x}} - i\hat{\mathbf{y}})e^{i(\varphi+\psi)}\right]\mathcal{E}_0 e^{-i\omega t},$$

where we have written

$$\int_0^z k_\pm \, dz = \varphi \mp \psi,$$

with

$$\varphi \equiv \int_0^z k_0 \, dz, \qquad \psi \equiv \int_0^z \Delta k \, dz. \tag{20.27}$$

Factoring out the common phase $e^{i\varphi}$ and expanding $e^{\pm i\psi} = \cos\psi \pm i\sin\psi$, we obtain

$$\mathbf{E} = (\hat{\mathbf{x}}\cos\psi + \hat{\mathbf{y}}\sin\psi)\mathcal{E}_0 e^{i(\varphi - \omega t)}. \tag{20.28}$$

Equation (20.28) has a simple interpretation: radiation that starts linearly polarized in a certain direction ($\hat{\mathbf{x}}$) is rotated by the *Faraday effect* through an angle ψ after propagating a distance z through a magnetized plasma, with ψ given by the formula [see equations (20.25) and (20.27)]:

$$\psi = \frac{2\pi e^3}{m_e^2 c^2 \omega^2}\,\mathrm{RM}, \tag{20.29}$$

where RM is the *rotation measure* and is defined by

$$\mathrm{RM} = \int_0^r n_e B_\parallel\,ds, \tag{20.30}$$

where we have reverted to the $r = z$ and s notation of earlier discussions. We cannot, of course, generally measure the absolute rotation angle ψ, since we do not know the intrinsic polarization direction of the radiation when it started from the source. Instead, in practice we measure the differential rotation between the received polarized radiation at two (or more) frequencies, from which we can then deduce the rotation measure RM. (We suppose that radiation of all relevant frequencies at the source had the same polarization direction, an assumption that holds, for example, in simple models of synchrotron sources.) For measurements toward sources (e.g., pulsars) where the dispersion measure DM is also known, we can derive an estimate of the mean field strength along the line of sight $\langle B_\parallel \rangle = \mathrm{RM}/\mathrm{DM}$. For other directions (e.g., toward extragalactic radio sources such as radio galaxies), an estimate for $\langle B_\parallel \rangle$ can be obtained only if we make some additional assumption about $\langle n_e \rangle$ (e.g., $\langle n_e \rangle = 0.03\,\mathrm{cm}^{-3}$), and the path length through which the radiation traversed such an average electron density. From a variety of such measurements, radio astronomers have concluded that the mean value of $\langle B_\parallel \rangle$ that applies to the diffuse interstellar medium of our Galaxy (where $\langle n_e \rangle \approx 0.03\,\mathrm{cm}^{-3}$) roughly equals 3 microgauss:

$$\langle B_\parallel \rangle = 3\mu\mathrm{G}.$$

MISCELLANEOUS COMMENTS ABOUT FARADAY DEPOLARIZATION, RAZIN EFFECT, ETC.

Apart from the well-established uses for dispersion and rotation measures as astrophysical diagnostics for the propagation medium discussed above, there exist more speculative possibilities for plasma effects. One concerns the role of internal *Faraday depolarization* to account for the relatively low linear polarizations observed in many synchrotron sources. Theoretical uniform source models predict very high polarization fractions, of the order of 70 percent for power-law synchrotron spectra (see Chapter 19). Rarely do actual radio sources approach such high values. One possibility might be a more tangled magnetic field configuration than assumed for the uniform source model, but a thermal plasma mixed in the same volume as the synchrotron electrons gives another possibility. Different portions of the emitting synchrotron volume would then suffer different amounts of Faraday rotation while traversing the volume to propagate toward the observer, resulting in a lower observed polarization level than the theoretical maximum.

If the thermal plasma density is high enough, the synchrotron radiation may even be choked off completely at low frequencies. The basic idea is that the relativistic boosting of synchrotron emission compared to the other radiation processes depends on the electron being fast enough to nearly catch up with the wave crests that it emits, thereby reinforcing the already emitted radiation with fresh crests. The effect is particularly spectacular—giving rise to *Cerenkov radiation* (in which the contribution to the radiation fields no longer comes from a single past time τ, as is does for the vacuum calculation of the Lienard-Wiechert potentials in Chapter 13)— if the electron can actually catch up with the waves it emits, because the phase velocity for electromagnetic wave propagation in the medium is less than c (see the discussion in Jackson [1962] on this point). However, in a plasma, the behavior of the phase velocity ω/k has just the opposite property; it exceeds c. If the amount by which it exceeds c can grow large enough (which depends on ω_{pe} approaching the relevant ω closely enough), then the critical synchrotron assumption fails; that is, the emitting particle's speed $v < c$ cannot closely approach ω/k and thus effectively shuts off the important (relativistic) part of the process. If this interesting hypothesized phenomenon—the *Razin effect*—occurs, a correlation should exist between radio-quiet quasars and other measures of high electron density.

QUANTUM THEORY
OF RADIATION PROCESSES

PART III

QUANTUM THEORY
OF RADIATION PROCESSES

21

Nonrelativistic Quantum Theory of Radiative Processes

With this chapter we begin Part III of this volume, the nonrelativistic quantum theory of radiative processes. We will discuss the problem of radiative *transitions* before tackling the (seemingly more basic) issues of quantized atomic and molecular structure. We adopt this unorthodox strategy under the rationale that prior course work will probably have given the reader of this text more familiarity with the latter topics than with the former, and so the reader will benefit more from an extended discussion of the former than the latter. Even if this is not true, it would be worthwhile to motivate a discussion of atomic and molecular level structure by considering what types of radiative transitions are possible among them.

JUSTIFICATION FOR NONRELATIVISTIC TREATMENT

Before we begin the formal development, we wish to review why, if our interest lies in applications to atoms and molecules, we may generally proceed with a nonrelativistic treatment. For such an order-of-magnitude discussion, it suffices to consider the hydrogen atom within the context of the Bohr model as a prototypical quantized system. In a Bohr hydrogen atom, the electron travels fastest when it lies in the ground electronic state with a size given by the Bohr radius $a_0 = \hbar^2/m_e e^2$ and with a kinetic energy given by $m_e v^2/2 = e^2/2a_0$. These two relations imply that

$$v/c = e^2/\hbar c \equiv \alpha \approx 1/137$$

in the first Bohr orbit, so that $v^2/c^2 \ll 1$ and a nonrelativistic treatment should hold to high approximation. Moreover, the typical radiative transition should involve photon energies of order $\hbar\omega = e^2/2a_0$, or wavenumbers of order $k = \omega/c = (\alpha/2)a_0^{-1}$. Since $ka_0 \ll 1$, multipole expansions should also hold to a high order of approximation for radiative transitions in typical atoms (and molecules).

No coincidence is involved that the same small parameter α governs both statements. Classically, we expect an electron in an atom or molecule to emit radiation with a frequency ν equal to the inverse of its orbital period. In that time the orbiting electron travels a distance $v\nu^{-1} \sim 2\pi a_0$, and the emitted electromagnetic wave travels a distance $c\nu^{-1} = \lambda = 2\pi/k$. Consequently, we expect ka_0 automatically to have the same order of magnitude as v/c. Moreover, if the electron takes many periods to finish radiating (i.e., if the Einstein $A \ll \nu$, as is almost always the case), the launched wavetrain will contain many wavelengths, and the wave frequency will have a well-defined value (i.e., the spectral line will have a small natural width).

For heavy atoms, the effective coupling constant is $Ze^2/\hbar c = Z\alpha$, where Z is the effective nuclear charge, so we see that our approximations will break down when Z becomes of order 10^2. Thus relativistic corrections become important for inner-shell electrons of heavy atoms, but they are usually negligible in all other applications to atoms and molecules.

SINGLE-PARTICLE HAMILTONIAN IN AN ELECTROMAGNETIC FIELD

To go from a classical description to a Schrödinger equation, we need a Hamiltonian formulation of mechanics. A Lagrangian formulation constitutes an intermediate step, where the equation of motion for a particle follows from the Euler-Lagrange equations:

$$\frac{d}{dt}\left(\frac{\partial L}{\partial \dot{\mathbf{x}}}\right) = \frac{\partial L}{\partial \mathbf{x}}, \tag{21.1}$$

with L being the Lagrangian. The nonrelativistic Lagrangian usually equals the kinetic energy, $m|\dot{\mathbf{x}}|^2/2$, minus the potential energy, which we might naively be tempted to write as $-q\phi$. However, in electrodynamics, wherever the charge q and the scalar potential ϕ make appearances, we might expect analogous appearances of $-i/c$ times the current $q\dot{\mathbf{x}}$ and $-i$ times the vector potential $\mathbf{A}$. Hence, we anticipate that, in the presence of electromagnetic fields, the Lagrangian for a charged particle reads

$$L = \frac{1}{2}m|\dot{\mathbf{x}}|^2 - q\phi(\mathbf{x}, t) + q\frac{\dot{\mathbf{x}}}{c} \cdot \mathbf{A}(\mathbf{x}, t). \tag{21.2}$$

Proof: If we differentiate equation (21.2) with respect to $\mathbf{x}$, holding $\dot{\mathbf{x}}$ and t constant, we obtain

$$\frac{\partial L}{\partial \mathbf{x}} = -q\boldsymbol{\nabla}\phi + \frac{q}{c}\boldsymbol{\nabla}(\dot{\mathbf{x}} \cdot \mathbf{A}).$$

Since $\boldsymbol{\nabla}$ does not operate on $\dot{\mathbf{x}}$, we may write equivalently that

$$\frac{\partial L}{\partial \mathbf{x}} = -q\boldsymbol{\nabla}\phi + \frac{q}{c}\left[\dot{\mathbf{x}} \times (\boldsymbol{\nabla} \times \mathbf{A}) + (\dot{\mathbf{x}} \cdot \boldsymbol{\nabla})\mathbf{A}\right].$$

Similarly, if we differentiate equation (21.2) with respect to $\dot{\mathbf{x}}$, holding $\mathbf{x}$ and t constant, we get

$$\frac{\partial L}{\partial \dot{\mathbf{x}}} = m\dot{\mathbf{x}} + \frac{q}{c}\mathbf{A}.$$

Taking the total time derivative of the above yields for equation (21.1),

$$m\ddot{\mathbf{x}} + \frac{q}{c}\left[\frac{\partial \mathbf{A}}{\partial t} + (\dot{\mathbf{x}} \cdot \boldsymbol{\nabla})\mathbf{A}\right] = -q\boldsymbol{\nabla}\phi + \frac{q}{c}[\dot{\mathbf{x}} \times (\boldsymbol{\nabla} \times \mathbf{A}) + (\dot{\mathbf{x}} \cdot \boldsymbol{\nabla})\mathbf{A}].$$

Canceling common terms on the left- and right-hand sides, we obtain the equation of motion for a nonrelativistic charge moving under the influence of Lorentz forces:

$$m\ddot{\mathbf{x}} = q\left(\mathbf{E} + \frac{\dot{\mathbf{x}}}{c} \times \mathbf{B}\right),$$

where $\mathbf{E}$ and $\mathbf{B}$ satisfy

$$\mathbf{B} = \boldsymbol{\nabla} \times \mathbf{A}, \qquad \text{and} \qquad \mathbf{E} = -\boldsymbol{\nabla}\phi - \frac{1}{c}\frac{\partial \mathbf{A}}{\partial t}.$$

The recovery of the correct classical equation of motion demonstrates the validity of the original expression (21.2) (Q.E.D.).

We define the canonical momentum by

$$\mathbf{p} = \frac{\partial L}{\partial \dot{\mathbf{x}}} = m\dot{\mathbf{x}} + \frac{q}{c}\mathbf{A}. \tag{21.3}$$

Notice that $\mathbf{p}$ does not equal the particle momentum $m\dot{\mathbf{x}}$ alone. The Hamiltonian formulation follows by making a Legendre transformation from $\dot{\mathbf{x}}$ to $\mathbf{p}$:

$$H \equiv \dot{\mathbf{x}} \cdot \mathbf{p} - L = m|\dot{\mathbf{x}}|^2 + \frac{q}{c}\dot{\mathbf{x}} \cdot \mathbf{A} - \frac{1}{2}m|\dot{\mathbf{x}}|^2 + q\phi - \frac{q}{c}\dot{\mathbf{x}} \cdot \mathbf{A} = \frac{1}{2}m|\dot{\mathbf{x}}|^2 + q\phi.$$

Eliminating $m\dot{\mathbf{x}}$ in favor of $\mathbf{p}$ by making use of equation (21.3), we finally obtain the (single-particle) Hamiltonian H as a function of $\mathbf{p}$, $\mathbf{x}$, and t:

$$H = \frac{1}{2m}\left|\mathbf{p} - \frac{q}{c}\mathbf{A}(\mathbf{x}, t)\right|^2 + q\phi(\mathbf{x}, t). \tag{21.4}$$

A straightforward exercise demonstrates that Hamilton's equations, $\dot{\mathbf{x}} = \partial H/\partial \mathbf{p}$ and $\dot{\mathbf{p}} = -\partial H/\partial \mathbf{x}$, applied to equation (21.4) also yield the correct nonrelativistic dynamics.

The relativistic generalization of equation (21.4) reads (see Chapter 25):

$$H = [|c\mathbf{p} - q\mathbf{A}(\mathbf{x}, t)|^2 + m^2 c^4]^{1/2} + q\phi(\mathbf{x}, t), \tag{21.5}$$

which, apart from a constant rest energy, mc^2, recovers equation (21.4) in the nonrelativistic limit $|c\mathbf{p} - q\mathbf{A}| \ll mc^2$. Applied to the full form (21.5), Hamilton's equations yield, after some manipulation, the Lorentz formula:

$$\frac{d\mathbf{P}}{dt} = q\left(\mathbf{E} + \frac{\dot{\mathbf{x}}}{c} \times \mathbf{B}\right),$$

where $\mathbf{P}$ is the relativistic momentum:

$$\mathbf{P} \equiv \mathbf{p} - \frac{q}{c}\mathbf{A} = \gamma m\dot{\mathbf{x}},$$

with $\gamma \equiv (1 - |\dot{\mathbf{x}}|^2/c^2)^{-1/2}$. Although the expression (21.5) leads to the relativistically correct equations of motion, it does not by itself constitute a relativistically covariant formulation, since time and space enter asymmetrically in the differentiations. To obtain a relativistically covariant formulation of Hamiltonian mechanics, see Goldstein (1959, Chapter 7).

Because of its large charge-to-mass ratio, our most important applications of the above considerations lie with the electron. For an electron, $q = -e$ and $m = m_e$, we obtain for equation (21.4),

$$H = \frac{1}{2m_e}\left|\mathbf{p} + \frac{e}{c}\mathbf{A}(\mathbf{x}, t)\right|^2 - e\phi(\mathbf{x}, t). \tag{21.6}$$

SEPARATION OF STATIC AND RADIATION FIELDS

In Maxwell's equations, suppose we let ρ_e denote the static charge distribution of everything but the electron whose Hamiltonian is considered in equation (21.6). Apart from these charges, we have only vacuum electromagnetic fields. In other words, we wish to consider a situation in which we let the electron interact only with (a) an electrostatic field associated with an external charge distribution (e.g., the rest of an atom or a molecule), and (b) vacuum radiation fields. Hence, we ignore the possibility of the electron interacting with its *own* electromagnetic field (which requires a special treatment in quantum electrodynamics).

In our restricted program, the first order of business becomes the separation of the static and radiation fields. To do this we must choose a *gauge* in which to perform the calculations. For a fully relativistic treatment [e.g., in which we use equation (21.5) or its Dirac cousin], the Lorentz gauge would be most natural (see Chapter 13):

$$\boldsymbol{\nabla} \cdot \mathbf{A} + \frac{1}{c}\frac{\partial \phi}{\partial t} = 0.$$

For the nonrelativistic theory, however, no question arises of trying to get expressions that hold covariantly in all inertial frames, and we are free to choose some other gauge condition if it provides for convenient calculations. Such a choice is the *Coulomb gauge* to be discussed below:

$$\mathbf{\nabla} \cdot \mathbf{A} = 0. \tag{21.7}$$

To motivate our subsequent development (due to Fermi), we invoke Helmholtz's theorem: any vector $\mathbf{A}$ can be decomposed as the sum of the gradient of a scalar part P and the curl of a vector quantity $\mathbf{Q}$: $\mathbf{A} = \mathbf{\nabla}P + \mathbf{\nabla} \times \mathbf{Q}$. Call $\mathbf{\nabla}P = \mathbf{A}_{\|}$ and $\mathbf{\nabla} \times \mathbf{Q} = \mathbf{A}_{\perp}$, where the symbols $\|$ and $\perp$ refer, respectively, to the *longitudinal* and the *transverse* components of $\mathbf{A}$. In other words, we can always write any vector field as the sum of longitudinal and transverse components, $\mathbf{A} = \mathbf{A}_{\|} + \mathbf{A}_{\perp}$, in such a fashion that $\mathbf{\nabla} \times \mathbf{A}_{\|} = 0$ (because the curl of the gradient of any function P equals zero), and $\mathbf{\nabla} \cdot \mathbf{A}_{\perp} = 0$ (because the divergence of the curl of any vector quantity $\mathbf{Q}$ equals zero).

To apply the above considerations to the vector potential $\mathbf{A}$ of an electromagnetic field, suppose we can make $\mathbf{A}_{\|} = 0$, so that the scalar potential ϕ carries the Coulomb field due to the static charge distribution $\rho_{\mathrm{e}}(\mathbf{x})$. It would be doubly nice if we could simultaneously make the transverse $\mathbf{A} = \mathbf{A}_{\perp}$ [satisfying the Coulomb gauge (21.7)] carry the vacuum radiation field, i.e., make $\mathbf{A}$ represent a *radiation* field.

Proof: Suppose the electromagnetic potentials that we have at hand, $\mathbf{A}$ and ϕ, do not fill the bill. Make a gauge transformation,

$$\mathbf{A}' \equiv \mathbf{A} + \mathbf{\nabla}\psi,$$

$$\phi' \equiv \phi - \frac{1}{c}\frac{\partial \psi}{\partial t},$$

so that $\mathbf{A}'$ and ϕ' have the desired properties. Clearly, the gauge function ψ must be chosen so that $\mathbf{\nabla} \cdot \mathbf{A}' = 0$, i.e., so that $\mathbf{\nabla} \cdot \mathbf{A} + \nabla^2\psi = 0$. If we interpret this constraint as a PDE for ψ, we see that the solution for ψ equals

$$\psi(\mathbf{x}, t) = \frac{1}{4\pi} \int \frac{\mathbf{\nabla}' \cdot \mathbf{A}(\mathbf{x}', t)}{|\mathbf{x} - \mathbf{x}'|}\, d^3x',$$

where $\mathbf{\nabla}'$ operates on the $\mathbf{x}'$ variable. For such a choice of gauge,

$$\nabla^2 \phi' = \nabla^2 \phi - \frac{1}{c}\frac{\partial}{\partial t}(-\mathbf{\nabla} \cdot \mathbf{A}) = -\mathbf{\nabla} \cdot \mathbf{E},$$

since $\mathbf{E} = -\mathbf{\nabla}\phi - c^{-1}\partial \mathbf{A}/\partial t$. But $\mathbf{\nabla} \cdot \mathbf{E} = 4\pi\rho_{\mathrm{e}}$ by one of Maxwell's equations; thus ϕ' satisfies $\nabla^2\phi' = -4\pi\rho_{\mathrm{e}}$, which has the Coulomb solution

$$\phi'(\mathbf{x}) = \int \frac{\rho_{\mathrm{e}}(\mathbf{x}')}{|\mathbf{x} - \mathbf{x}'|}\, d^3x',$$

which is part of what was to be shown.

The other part follows by noting that if $\nabla \cdot \mathbf{A}' = 0$, then $\nabla^2 \mathbf{A}' = -\nabla \times (\nabla \times \mathbf{A}') = -\nabla \times \mathbf{B} = -c^{-1} \partial \mathbf{E} / \partial t$ by another of Maxwell's equations if there are no currents in the problem. But $\mathbf{E} = -\nabla \phi' - c^{-1} \partial \mathbf{A}' / \partial t$, with ϕ' equal to a static potential. Thus, if we differentiate the last equation once in time, we obtain

$$\nabla^2 \mathbf{A}' - \frac{1}{c^2} \frac{\partial^2 \mathbf{A}'}{\partial t^2} = 0, \tag{21.8}$$

i.e., $\mathbf{A}'$ satisfies a wave equation, and therefore can represent a vacuum radiation field. (Q.E.D.)

To simplify the notation in what follows, we drop the primes. However, we should always keep in mind that our gauge choice requires us to be in a special frame, namely, the frame in which the matter forces consist of only an electrostatic field, derivable from a scalar potential $\phi(\mathbf{x})$. For an atom, this means that we do our calculations in the frame of rest of the atomic nucleus.

STRUCTURE AND INTERACTION HAMILTONIANS

In the Coulomb gauge, $\nabla \cdot \mathbf{A} = 0$, we may expand out the nonrelativistic single-electron Hamiltonian (21.6) by dotting $\mathbf{p} + e\mathbf{A}/c$ with itself:

$$H = H_{\text{st}} + H_{\text{int}}, \tag{21.9}$$

where H_{st} represents the static part of the Hamiltonian that governs the structure of the electronic equilibrium state (in the atom, molecule, etc.),

$$H_{\text{st}} = \frac{1}{2m_e} |\mathbf{p}|^2 - e\phi(\mathbf{x}) \equiv H_0, \tag{21.10}$$

and H_{int} represents the perturbational part due to interactions with the radiation field,

$$H_{\text{int}} = H_1 + H_2, \tag{21.11}$$

where H_1 is linear in $\mathbf{A}$ and leads (as we shall see) to one-photon processes,

$$H_1 \equiv \frac{e}{2m_e c} (\mathbf{p} \cdot \mathbf{A} + \mathbf{A} \cdot \mathbf{p}), \tag{21.12}$$

and H_2 is quadratic in $\mathbf{A}$ and leads to two-photon processes,

$$H_2 \equiv \frac{e^2}{2m_e c^2} \mathbf{A} \cdot \mathbf{A}. \tag{21.13}$$

In equation (21.12), we have been careful to distinguish between $\mathbf{p} \cdot \mathbf{A}$ and $\mathbf{A} \cdot \mathbf{p}$ under the anticipation that a quantum treatment will require us to interpret $\mathbf{p}$ as an operator.

RELATIVE IMPORTANCE OF ONE- AND TWO-PHOTON PROCESSES

As a preliminary to our deliberations, we consider order-of-magnitude estimates for the relative importance of the terms H_1 and H_2, using the hydrogen atom again as a prototypical example. Taking the ratio of equation (21.13) to (21.12) yields

$$H_2/H_1 \sim \frac{e^2 A^2/2m_e c^2}{epA/m_e c} = \frac{\alpha^2 a_0 A}{2ev/c} \sim \frac{\alpha a_0 A}{2e}, \qquad (21.14)$$

where $v/c \sim \alpha$, and α and a_0 are again the fine-structure constant and Bohr radius. From $\mathbf{B} = \nabla \times \mathbf{A}$, we may estimate $A = k^{-1}B \sim 2a_0 E/\alpha$, since $B = E$ for a vacuum radiation field, whereas our discussion at the beginning of this chapter showed that the typical wavenumber k for radiation emitted by a hydrogen atom satisfies $k \sim \alpha(2a_0)^{-1}$. Substituting the result $A \sim 2a_0 E/\alpha$ into equation (21.14), we get

$$H_2/H_1 \sim \frac{a_0^2 E}{e},$$

which represents the ratio of the electric field E of the radiation field to the atomic Coulomb electric field e/a_0^2 experienced by the electron. We anticipate that this ratio will usually be very small for conditions of astrophysical interest. To see this, for example, for photon absorption processes, express the energy density in the radiation field $(B^2+E^2)/8\pi = E^2/4\pi$ as the number density of photons n_{ph} times the typical photon energy $\hbar\omega \sim e^2/2a_0$. Hence, $E^2 \sim 2\pi n_{ph}e^2/a_0$, and we can write

$$H_2/H_1 \sim (2\pi n_{ph}a_0^3)^{1/2}.$$

The expression on the right-hand side of the last estimate will be small if the number of photons in a Bohr cube, $n_{ph}a_0^3$, is small compared to unity, i.e., if $n_{ph} \ll a_0^{-3}/2\pi \sim 10^{24}$ cm ^{3}. However, the greatest number of photons a hydrogen atom is likely to see occurs inside stars, say, in regions that contain a blackbody radiation field of temperature $T \sim 10^4$ K. Thus

$$n_{ph} \sim aT^4/kT \sim aT^3/k \sim 10^{14} \text{ cm}^{-3},$$

a large number, but still much smaller than 10^{24} cm^{-3}. This implies that a very small probability exists for absorbing two photons at a time by a hydrogen atom in comparison with the relatively common process of absorbing one photon at a time. In general, then, we may ignore the perturbing effect of H_2 in comparison with that of H_1. Exceptions to this rule do exist, especially if we consider radiative transitions not involving pure absorption. (1) Quantum scattering processes, which intrinsically involve two photons, are contained among the effects of H_2. (2) In a different context, if the spontaneous emission transition probability associated with H_1 is zero (or very small), then the presence of H_2 in the interaction Hamiltonian may have to be taken into account. A famous example concerns the two-photon process by which a hydrogen atom in the $2s$ ($n = 2$, $\ell = 0$) state may radiatively de-excite to the $1s$ (ground) state by emitting two continuum photons with combined frequencies $\nu + \nu' = \nu_{21}$ (see Chapter 24).

Finally, notice that $H_1 \ll H_0$ by the same rough factor $E a_0^2 / e \ll 1$ that $H_2 \ll H_1$. This happy circumstance suggests that we may apply a perturbational approach to the solution of the Schrödinger equation associated with the total Hamiltonian $H = H_0 + H_1 + H_2$. In other words, the radiative effects contained in H_1 (or H_2) usually represent small perturbations on top of the problem of atomic and molecular structure contained in H_0. In the chapters that follow, we exploit the breach opened by this observation.

22

Semiclassical Treatment of Radiative Transitions

Reference: Sakurai, *Advanced Quantum Mechanics*, pp. 34–57.

In this chapter we consider how to give a quantum-mechanical interpretation to the radiation field represented by the vector potential $\mathbf{A}$ in equations (21.12) and (21.13). We will also give a brief account of the time-dependent perturbation theory that describes how interaction with a radiation field allows an atom (or molecule) to radiate when transitions involving only one orbital electron are involved. Our treatment will be semiclassical in that we will regard $\mathbf{A}$ as a vector *function*, rather than as an *operator*, which itself satisfies certain commutation relations. As we will see, such an approach requires some patchwork to make it consistent with the Einstein relations for the emission and absorption of radiation.

CUBIC BOX NORMALIZATION

We begin our analysis by Fourier decomposing $\mathbf{A}$ in a cubic box with sides of length L centered on the origin of our coordinate system:

$$\mathbf{A}(\mathbf{x}, t) = \sum_{\mathbf{k}} \left[\mathbf{a}(\mathbf{k}) e^{i(\mathbf{k} \cdot \mathbf{x} - \omega t)} + \mathbf{a}^*(\mathbf{k}) e^{-i(\mathbf{k} \cdot \mathbf{x} - \omega t)} \right], \qquad (22.1)$$

where the second term, the complex conjugate of the first, is needed to make $\mathbf{A}$ real, and where the fact that $\mathbf{A}$ satisfies $\Box^2 \mathbf{A} = 0$ implies that $\omega = c|\mathbf{k}|$.

In equation (22.1), the sum over $\mathbf{k}$ satisfies the cubic-box boundary conditions that the radiation fields vanish at x, y, $z = \pm L/2$, i.e.,

$$k_x L/2 = n_x \pi, \qquad k_y L/2 = n_y \pi, \qquad k_z L/2 = n_z \pi,$$

with n_x, n_y, n_z equal to 0, ± 1, ± 2, ... (In other words, the sum occurs over n_x, n_y, n_z, each from $-\infty$ to ∞.) In the continuum limit ($L \to \infty$), the sum passes to an integral:

$$\sum_{\mathbf{k}} \cdots \to \frac{V}{(2\pi)^3} \int \cdots d^3k, \qquad (22.2)$$

where $V = L^3$ is the volume of the cubic box, and $d^3k = k^2\, dk\, d\Omega$ for wave propagation within a solid angle $d\Omega$ centered about $\hat{\mathbf{k}}$. Do not worry about the awkward presence of the coefficient V; it will disappear from all important formulae (in which the integrand, being the density of some physical quantity and represented schematically by ..., will contain a canceling factor $1/V$).

QUANTIZATION AND THE CONCEPT OF PHOTONS

The transformation to quantum mechanics occurs with the identification of the canonical momentum $\mathbf{p}$ as the operator $-i\hbar\boldsymbol{\nabla}$:

$$\mathbf{p} = -i\hbar\boldsymbol{\nabla}. \qquad (22.3)$$

With this identification, we note that $\mathbf{p} \cdot \mathbf{A}$, treated as an operator which acts on anything that appears to its right, equals $\mathbf{A}\cdot\mathbf{p}$ since $\boldsymbol{\nabla}\cdot\mathbf{A} = 0$ by our choice of the Coulomb gauge. Thus the one-photon interaction Hamiltonian becomes

$$H_1 = \frac{e}{m_e c}\mathbf{A} \cdot \mathbf{p}. \qquad (22.4)$$

Notice also that $\mathbf{p} = -i\hbar\boldsymbol{\nabla}$ operating on the individual terms $e^{\pm i(\mathbf{k}\cdot\mathbf{x}-\omega t)}$ in equation (22.1) yields $\pm\hbar\mathbf{k}$ times those same functions. This recovery of de Broglie's relationship suggests that $\mathbf{a}e^{i(\mathbf{k}\cdot\mathbf{x}-\omega t)}$ represents a *photon* wave field, with $\hbar\mathbf{k}$ being the associated quantum of momentum. Given the usual convention of signs in Schrödinger's equation, we will see that this term $\propto e^{-i\omega t}$ gives rise to *absorption* processes, whereas its complex conjugate $\propto e^{+i\omega t}$ gives rise to *emission* processes. In terms of the eigenvalues $p = \hbar k$ of the $\mathbf{p}$ operator, we may express equation (22.2) as (see Chapters 1 and 6):

$$\sum_{\mathbf{k}} \cdots \to \frac{V}{h^3} \int \cdots d^3p, \qquad (22.5)$$

where $d^3p = p^2\, dp\, d\Omega$. Neither equation (22.2) nor (22.5) explicitly accounts yet for the presence of photon spin.

To include spin states in the emerging photon concept, we note that the Coulomb gauge $\boldsymbol{\nabla} \cdot \mathbf{A} = 0$ implies $\mathbf{k} \cdot \mathbf{a}(\mathbf{k}) = 0$; i.e., $\mathbf{a}(\mathbf{k})$ has only two components (both perpendicular to $\mathbf{k}$). These correspond classically to two independent (linear or circular) polarization states; quantum mechanically, to two spin states, $\alpha = 1, 2$. Making the photon polarization decomposition, we obtain

$$\mathbf{A}(\mathbf{x}, t) = \sum_{\mathbf{k}, \alpha} \left[\mathbf{e}_\alpha(\hat{\mathbf{k}}) a_\alpha(\mathbf{k}) e^{i(\mathbf{k} \cdot \mathbf{x} - \omega t)} + \text{c.c.} \right], \qquad (22.6)$$

where c.c. means "complex conjugate," where $a_\alpha(\mathbf{k})$ is now a scalar amplitude, and where $\mathbf{e}_1$ and $\mathbf{e}_2$ are two unit vectors perpendicular to $\hat{\mathbf{k}}$. For example, if $\hat{\mathbf{k}}$ lies in the $\hat{\mathbf{z}}$ direction, we could most simply choose the linear polarization states: $\mathbf{e}_1 = \hat{\mathbf{x}}$ and $\mathbf{e}_2 = \hat{\mathbf{y}}$. More suggestively for photon spin, if we choose the circular polarization states: $\mathbf{e}_1 = -(2)^{-1/2}(\hat{\mathbf{x}} + i\hat{\mathbf{y}})$ and $\mathbf{e}_2 = (2)^{-1/2}(\hat{\mathbf{x}} - i\hat{\mathbf{y}})$, we need to be careful to note that $\langle \mathbf{e}_1 | \mathbf{e}_1 \rangle \equiv \mathbf{e}_1^* \cdot \mathbf{e}_1 = 1 = \mathbf{e}_2^* \cdot \mathbf{e}_2 \equiv \langle \mathbf{e}_2 | \mathbf{e}_2 \rangle$, $\langle \mathbf{e}_2 | \mathbf{e}_1 \rangle \equiv \mathbf{e}_2^* \cdot \mathbf{e}_1 = 0 = \mathbf{e}_1^* \cdot \mathbf{e}_2 \equiv \langle \mathbf{e}_1 | \mathbf{e}_2 \rangle$, while $\mathbf{e}_1 \times \mathbf{e}_2 = i\hat{\mathbf{z}} = i\hat{\mathbf{k}}$ (with the appropriate complex conjugate relations).

THE HAMILTONIAN OF THE RADIATION FIELD

For the radiation electromagnetic fields,

$$\mathbf{E} = -\frac{1}{c} \frac{\partial \mathbf{A}}{\partial t}, \qquad \mathbf{B} = \nabla \times \mathbf{A},$$

we easily demonstrate

$$\mathbf{E} = \sum_{\mathbf{k}, \alpha} \left[\mathbf{E}_\alpha(\mathbf{k}) e^{i(\mathbf{k} \cdot \mathbf{x} - \omega t)} + \text{c.c.} \right], \qquad (22.7)$$

$$\mathbf{B} = \sum_{\mathbf{k}, \alpha} \left[\mathbf{B}_\alpha(\mathbf{k}) e^{i(\mathbf{k} \cdot \mathbf{x} - \omega t)} + \text{c.c.} \right], \qquad (22.8)$$

where $\mathbf{E}_\alpha(\mathbf{k}) = ik a_\alpha(\mathbf{k}) \mathbf{e}_\alpha$ with $\omega/c = k$, and $\mathbf{B}_\alpha(\mathbf{k}) = ik a_\alpha(\mathbf{k})(\hat{\mathbf{k}} \times \mathbf{e}_\alpha)$. Given $\mathbf{E}$ and $\mathbf{B}$, we may compute the energy in the radiation field as

$$H_{\text{rad}} = \frac{1}{8\pi} \int_V \left(|\mathbf{E}|^2 + |\mathbf{B}|^2 \right) d^3 x. \qquad (22.9)$$

Write $|\mathbf{E}|^2$ as equation (22.7) times itself written with primed variables, and $|\mathbf{B}|^2$ as equation (22.8) times itself written with primed variables. To evaluate equation (22.9), note now the relationships,

$$\int_V \mathbf{e}_\alpha(\hat{\mathbf{k}}) \cdot \mathbf{e}_{\alpha'}^*(\hat{\mathbf{k}}') e^{\pm i(\mathbf{k} \cdot \mathbf{x} - \omega t)} e^{\mp i(\mathbf{k}' \cdot \mathbf{x} - \omega' t)} d^3 x = L^3 \delta_{\alpha \alpha'} \delta_{\mathbf{k}, \mathbf{k}'},$$

$$\int_V \mathbf{e}_\alpha(\hat{\mathbf{k}}) \cdot \mathbf{e}_{\alpha'}(\hat{\mathbf{k}}') e^{\pm i(\mathbf{k} \cdot \mathbf{x} - \omega t)} e^{\pm i(\mathbf{k}' \cdot \mathbf{x} - \omega' t)} d^3 x = s_\alpha L^3 \delta_{\alpha \alpha'} \delta_{-\mathbf{k}, \mathbf{k}'} e^{\mp 2i\omega t},$$

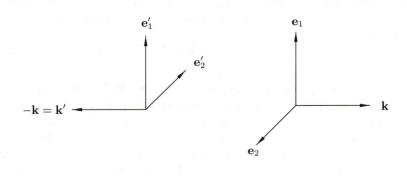

FIGURE 22.1
Geometry for polarization vectors.

where $s_1 = +1$ and $s_2 = -1$ on the basis of Figure 22.1. In the above equations, the Kronecker delta $\delta_{\mathbf{k},\mathbf{k}'}$ equals 1 if the sets of triplet numbers (n_x, n_y, n_z) and (n'_x, n'_y, n'_z) are equal, and it equals 0 otherwise.

With the above formulae, we may now obtain for equation (22.9):

$$H_{\text{rad}} = \frac{L^3}{2\pi} \sum_{\mathbf{k},\alpha} k^2 |a_\alpha(\mathbf{k})|^2, \tag{22.10}$$

where the oscillatory terms that depend on $e^{\mp 2i\omega t}$ vanish upon summation. For the record, we recognize equation (22.10) as a discrete application of the Fourier closure (Parseval's) theorem:

$$H_{\text{rad}} = \frac{1}{8\pi}(2V) \sum_{\mathbf{k},\alpha} \left[|\mathbf{E}_\alpha(\mathbf{k})|^2 + |\mathbf{B}_\alpha(\mathbf{k})|^2 \right],$$

where the factor of 2 comes from the cross products between the terms and their complex conjugates.

RELATIONSHIP TO PHOTON OCCUPATION NUMBER

From a photon point of view, we ought to be able to get the total energy in the radiation field by summing over all photons:

$$H_{\text{rad}} = \sum_{\mathbf{k},\alpha} \hbar\omega \mathcal{N}_\alpha(\mathbf{k}), \tag{22.11}$$

where $\mathcal{N}_\alpha(\mathbf{k})$ is the photon occupation number for polarization state α and propagation vector $\mathbf{k}$. If we compare equation (22.11) with equation (22.10), we may identify $|a_\alpha(\mathbf{k})|$ as

$$|a_\alpha(\mathbf{k})| = \left[\frac{2\pi\hbar\omega\mathcal{N}_\alpha(\mathbf{k})}{L^3 k^2}\right]^{1/2} = c\left[\frac{h\mathcal{N}_\alpha(\mathbf{k})}{V\omega}\right]^{1/2}, \tag{22.12}$$

where we have written $h = 2\pi\hbar$, $V = L^3$, and $k = \omega/c$.

The expression (22.12) makes sense for absorption phenomena, but not entirely for emission phenomena. The absorption of photons must clearly be proportional to the number of photons $\mathcal{N}_\alpha$ available for absorption. The same holds true also for stimulated emission, but the *spontaneous emission* of photons should be able to take place even in the total absence of pre-existing photons. To include the possibility of creating a photon where none existed before, we must replace the term $\mathcal{N}_\alpha$ wherever it appears in emission processes by the term $1 + \mathcal{N}_\alpha$. Such an *ad hoc* procedure proves necessary only because we have formulated the radiation problem in terms of a semiclassical description. In a fully quantized treatment, where a_α and a_α^* would have an interpretation in terms of *annihilation* and *creation* operators (consult the Sakurai reference given at the beginning), the inclusion of spontaneous emission would occur automatically from the outset, and would not have to be tacked on almost as an afterthought.

Quantization of the electromagnetic field ("second quantization"), as first carried out by Dirac, would demonstrate explicitly that when an atom or molecule emits or absorbs a photon, an additional photon appears in or disappears from the state vector that represents the radiation field. The replacement of $\mathcal{N}_\alpha$ by $1 + \mathcal{N}_\alpha$ in the semiclassical treatment of emission processes simulates the relevant effects. In the classical limit when $\mathcal{N}_\alpha \gg 1$, the emission of an additional photon represents a negligible perturbation to the large number of photons already present in the radiation field. For this reason, and because classical physics contains the independent concepts of particles and waves, we found no difficulty in Chapters 15, 18, and 19 in computing the spontaneous emission of *Bremsstrahlung* and synchrotron radiation by an electron energetically capable of emitting a continuous wave train (many photons). It is only in the limit when we need to consider a single photon (or a few), which contains simultaneously *both* aspects of particle and wave, that the methods of classical physics fail us, and we need to entertain a quantum treatment in which we cannot ignore the 1 in $1 + \mathcal{N}_\alpha$. Apart from this feature, there exists nothing in the concepts of true absorption, stimulated emission (the negative of true absorption), or spontaneous emission that make them intrinsically quantum-mechanical rather than classical.

With these remarks, the single-photon interaction Hamiltonian becomes

$$H_1 = \sum_{\mathbf{k},\alpha} \left[H_\alpha^{\mathrm{abs}}(\mathbf{k})e^{-i\omega t} + H_\alpha^{\mathrm{em}}(\mathbf{k})e^{i\omega t} \right], \qquad (22.13)$$

where

$$H_\alpha^{\mathrm{abs}}(\mathbf{k}) = \frac{e}{m_{\mathrm{e}}} \left[\frac{h}{V\omega} \mathcal{N}_\alpha(\mathbf{k}) \right]^{1/2} e^{i\mathbf{k}\cdot\mathbf{x}} \, \mathbf{e}_\alpha(\hat{\mathbf{k}}) \cdot \mathbf{p}, \qquad (22.14)$$

$$H_\alpha^{\mathrm{em}}(\mathbf{k}) = \frac{e}{m_{\mathrm{e}}} \left\{ \frac{h}{V\omega} \left[1 + \mathcal{N}_\alpha(\mathbf{k}) \right] \right\}^{1/2} e^{-i\mathbf{k}\cdot\mathbf{x}} \, \mathbf{e}_\alpha(\hat{\mathbf{k}}) \cdot \mathbf{p}. \qquad (22.15)$$

In the continuum limit, we record, for future use,

$$H_1 = \sum_{\alpha=1}^{2} \frac{V}{(2\pi)^3} \int \left[H_\alpha^{\mathrm{abs}}(\mathbf{k})e^{-i\omega t} + H_\alpha^{\mathrm{em}}(\mathbf{k})e^{i\omega t} \right] d^3k. \qquad (22.16)$$

We remark that the modification to include spontaneous emission makes H_1 no longer real, but this causes no difficulties, because the substitution $\mathbf{p} = -i\hbar\boldsymbol{\nabla}$ makes H_1 an operator, and it is important only that H_1 operating on wave functions gives real observables. We will now consider the effects of radiative perturbations on an electron in a simple atom. (For the rest of this chapter, we will speak of an atom, but we could just as easily speak of [electronic transitions in a] molecule.)

TIME-DEPENDENT PERTURBATION THEORY

The nonrelativistic time-evolution for the wave function ψ of an electron in a simple atom is governed by Schrödinger's equation,

$$H\psi = i\hbar\frac{\partial\psi}{\partial t}, \qquad (22.17)$$

where $H = H_0 + H_1$ with H_0 given by

$$H_0 = -\frac{\hbar^2}{2m_{\mathrm{e}}}\nabla^2 - e\phi(\mathbf{x}), \qquad (22.18)$$

and H_1 by equation (22.13). In ignoring the effects of H_2, we restrict our attention to processes that allow the atom to interact with photons only one at a time. The zeroth-order Hamiltonian H_0 is associated with the structure of the stationary states of the atom and has eigenfunctions of the form

$$\psi_j(\mathbf{x}, t) = \varphi_j(\mathbf{x})e^{-iE_j t/\hbar}. \qquad (22.19)$$

The energy E_j of the j-th eigenstate is an eigenvalue of the time-independent Schrödinger equation

$$H_0\varphi_j = E_j\varphi_j. \tag{22.20}$$

Since H_0 is a Hermitian operator, the entire set of eigenfunctions for *bound* and *free* states forms a complete basis for representing any wave function possible for the atomic system. Thus we may expand the perturbed wave function ψ as follows:

$$\psi(\mathbf{x}, t) = \sum_j c_j\varphi_j(\mathbf{x})e^{-iE_jt/\hbar}. \tag{22.21}$$

The dependences of the coefficients c_j on time t are to be determined by the requirement that ψ satisfies equation (22.17). The homogeneous nature of this equation implies that ψ contains an arbitrary scaling factor, which we choose to satisfy the usual normalization condition

$$\langle\psi|\psi\rangle = 1, \tag{22.22}$$

where we have introduced the Dirac bra-ket notation:

$$\langle\psi_a|\psi_b\rangle \equiv \int \psi_a^*\psi_b\, d^3x. \tag{22.23}$$

The eigenfunctions form an orthonormal set in the sense that

$$\langle\varphi_\ell|\varphi_j\rangle = \delta_{\ell j}, \tag{22.24}$$

where $\delta_{\ell j}$ is the Kronecker delta. (We may want to use a different normalization that produces Dirac delta functions instead of Kronecker deltas for the free states when we take the limit $V \to \infty$.) The orthogonality of φ_ℓ and φ_j when $\ell \neq j$ is easy to demonstrate when the states are nondegenerate, with $E_\ell \neq E_j$.

Proof: Consider the expression

$$\langle\varphi_\ell|H_0|\varphi_j\rangle.$$

If we use H_0 to right-operate on $|\varphi_j\rangle$, we obtain $E_j|\varphi_j\rangle$; if we left-operate on $\langle\varphi_\ell|$ (allowable because H_0 is Hermitian), we obtain $E_\ell\langle\varphi_\ell|$, with E_j and E_ℓ both being real numbers. Thus $\langle\varphi_\ell|H_0|\varphi_j\rangle$ equals both $E_j\langle\varphi_\ell|\varphi_j\rangle$ and $E_\ell\langle\varphi_\ell|\varphi_j\rangle$, which is not possible for $E_\ell \neq E_j$ unless $\langle\varphi_\ell|\varphi_j\rangle = 0$. (Q.E.D.)

The proof breaks down for degenerate eigenstates in which $E_\ell = E_j$. In this case, we need to rely on the analog of the *Gram-Schmidt* procedure in finite-vector spaces to demonstrate that we can always construct an orthonormal set of basis eigenfunctions. In what follows, we will assume that this task has been done.

When we substitute equation (22.21) into equation (22.17), and when we use equation (22.20) to eliminate the zeroth-order terms, we find that the perturbation part satisfies

$$\sum_j c_j H_1 \varphi_j e^{-iE_j t/\hbar} = i\hbar \sum_j \dot{c}_j \varphi_j e^{-iE_j t/\hbar}. \qquad (22.25)$$

We wish to solve this equation subject to the initial condition that the system lies in a well-defined eigenstate i at $t = 0$:

$$\psi(\mathbf{x}, 0) = \varphi_i(\mathbf{x}). \qquad (22.26)$$

Left-multiply equation (22.25) by the bra-function $\varphi_f^* e^{iE_f t/\hbar}$ of a possible final eigenstate f. Integrate the expression over all $\mathbf{x}$, denoting the result by Dirac's bra-ket notation:

$$\sum_j e^{i\omega_{fj} t} c_j \langle \varphi_f | H_1 | \varphi_j \rangle = i\hbar \dot{c}_f, \qquad (22.27)$$

where we have made use of equation (22.24) to simplify the right-hand side. In equation (22.27) we have denoted

$$\omega_{fj} \equiv (E_f - E_j)/\hbar. \qquad (22.28)$$

As we will see shortly, the value of j that counts is $j = i$; therefore the transition frequency defined this way will be positive for absorption processes where $E_f > E_i$; for emission processes where $E_f < E_i$, we would get a factor $e^{-i\omega_{if} t}$ inside the integral in equation (22.27). These signs dictate our earlier identifications of the absorption and emission parts of the interaction Hamiltonian (where we required ω to be positive).

We note that the initial condition, equation (22.26), may be written as $c_j = \delta_{ji}$ at $t = 0$. Motivated by this observation, we follow Dirac and solve by iteration; i.e., we assume as the zeroth iterate that $c_j = \delta_{ji}$ for *all* t, and substitute this guess into the left-hand side of equation (22.27) to obtain for all possible final-state coefficients

$$c_f(t) = -i\hbar^{-1} \int_0^t \langle \varphi_f | H_1 | \varphi_i \rangle e^{i\omega_{fi} t} \, dt. \qquad (22.29)$$

If this result is not accurate enough, we could always carry out further iterations. In fact, we will see that this simple procedure leads to results that are quite satisfactory for most of the physical applications contemplated in this book (see, however, Chapter 24).

THE ABSORPTION OF PHOTONS

Consider a single Fourier component $H_\alpha^{\text{abs}}(\mathbf{k})e^{-i\omega t}$ of the absorption part of the interaction Hamiltonian [see equation (22.14)]. We denote by $c_f(\mathbf{k}, t)$ the value that results from equation (22.29) when $H_\alpha^{\text{abs}}e^{-i\omega t}$ replaces H_1:

$$c_f(\mathbf{k}, t) = -\hbar^{-1}\langle\varphi_f|H_\alpha^{\text{abs}}(\mathbf{k})|\varphi_i\rangle \left[\frac{e^{i(\omega_{fi}-\omega)t} - 1}{\omega_{fi} - \omega}\right].$$

Quantum probabilities are proportional to the absolute squares of wave functions; thus consider the quantity

$$|c_f(\mathbf{k}, t)|^2 = c_f^*(\mathbf{k}, t)c_f(\mathbf{k}, t) = \hbar^{-2}|\langle\varphi_f|H_\alpha^{\text{abs}}(\mathbf{k})|\varphi_i\rangle|^2 \frac{\sin^2[(\omega - \omega_{fi})t/2]}{[(\omega - \omega_{fi})/2]^2},$$
(22.30)

where we have used a half-angle trigonometric formula to write

$$[e^{i(\omega_{fi}-\omega)t}-1][e^{-i(\omega_{fi}-\omega)t}-1] = 2\{1-\cos[(\omega-\omega_{fi})t]\} = 4\sin^2[(\omega-\omega_{fi})t/2].$$

ABSORPTION TRANSITION PROBABILITY

Equation (22.30) gives the probability of absorption of a photon of radian frequency $\omega = ck$ not necessarily equal to ω_{fi} and with a certain propagation direction $\hat{\mathbf{k}}$ and polarization state α. To obtain the total probability $\mathcal{P}_{if}$ of having a transition from i to f, we must sum over all $\mathbf{k}$ and α:

$$\mathcal{P}_{if} = \sum_{\mathbf{k},\alpha} |c_f(\mathbf{k}, t)|^2$$

$$= \frac{V}{(2\pi)^3}\sum_{\alpha=1}^{2}\int \hbar^{-2}|\langle\varphi_f|H_\alpha^{\text{abs}}(\mathbf{k})|\varphi_i\rangle|^2 \frac{\sin^2[(\omega - \omega_{fi})t/2]}{[(\omega - \omega_{fi})/2]^2}\, d^3k,$$
(22.31)

where we used equation (22.2) to perform the sum over $\mathbf{k}$. If we write $d^3k = k^2\,dk\,d\Omega = c^{-3}\omega^2\,d\omega\,d\Omega$, with $H_\alpha^{\text{abs}}(\mathbf{k})$ given by equation (22.14), we get

$$\mathcal{P}_{if} = \left(\frac{e}{2\pi m_e}\right)^2 \sum_{\alpha=1}^{2}\int \frac{\mathcal{N}_\alpha(\mathbf{k})}{\hbar\omega}|\langle\varphi_f|e^{i\mathbf{k}\cdot\mathbf{x}}\mathbf{e}_\alpha(\hat{\mathbf{k}})\cdot\mathbf{p}|\varphi_i\rangle|^2$$

$$\times \frac{\sin^2[(\omega - \omega_{fi})t/2]}{[(\omega - \omega_{fi})/2]^2}\frac{\omega^2}{c^3}\, d\omega\, d\Omega.$$
(22.32)

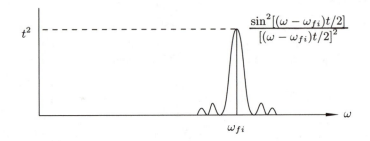

FIGURE 22.2
For time-dependent perturbation theory, we are interested in times $t \gg \omega_{fi}^{-1}$ (many oscillation periods of the emitted photon), but with t still small in comparison with the inverse of the Einstein coefficient for spontaneous emission A_{fi}^{-1} (so our iteration procedure will have formal validity). For $t \gg \omega_{fi}^{-1}$, the function $\sin^2[(\omega - \omega_{fi})t/2]/[(\omega - \omega_{fi})/2]$ has a strong peak at $\omega = \omega_{fi}$, the resonant frequency of the ultimate transition.

Notice that the dependence on V has dropped out of the expression for $\mathcal{P}_{if}$ in equation (22.32), as advertised. For $t \gg 2/\omega_{fi}$ ($\sim 10^{-15}$ s in the optical regime), the function

$$\frac{\sin^2[(\omega - \omega_{fi})t/2]}{[(\omega - \omega_{fi})/2]^2}$$

is strongly peaked at $\omega = \omega_{fi}$ (see Figure 22.2). This result has a simple physical interpretation. As time proceeds, we expect most of the absorption to occur for photons that have exactly the energy difference between two quantized energy states of the atom. There exists some "slop" because of the uncertainty principle, but the margin of this "slop" disappears in the limit $t \to \infty$. In other words, for any smoothly varying function $F(\omega)$, we may perform the integral

$$\int_0^\infty F(\omega) \frac{\sin^2[(\omega - \omega_{fi})t/2]}{[(\omega - \omega_{fi})/2]^2} \, d\omega \approx 2tF(\omega_{fi}) \int_{-\infty}^\infty \frac{\sin^2 \xi}{\xi^2} \, d\xi = 2\pi t F(\omega_{fi}),$$

in the limit of large t. Applying this result to equation (22.32), we obtain the formula

$$\mathcal{P}_{if} = t \left(\frac{e^2}{\hbar c^3 m_e^2} \right) \sum_{\alpha=1}^{2} \oint \left[\omega \mathcal{N}_\alpha(\mathbf{k}) |\langle \varphi_f | e^{i\mathbf{k}\cdot\mathbf{x}} \mathbf{e}_\alpha(\hat{\mathbf{k}}) \cdot \mathbf{p} | \varphi_i \rangle|^2 \right]_{fi} d\Omega, \quad (22.33)$$

where the subscript denotes that the quantities ω and $\mathbf{k}$ inside the brackets are to be evaluated at $\omega = \omega_{fi}$ and $\mathbf{k} = (\omega_{fi}/c)\hat{\mathbf{k}}$.

The linear increase of $\mathcal{P}_{if}$ with time t in equation (22.33) signals that our naive iteration procedure must break down for t very much in excess of ω_{fi}^{-1}. Basically, we are attempting to represent a transition from one quantum state to another as a small perturbation, which it cannot be *for the wave function* if the process can go to completion. Fortunately, the secular increase of $\mathcal{P}_{if}$ yields a constant transition rate, $d\mathcal{P}_{if}/dt$, which has a felicitous physical interpretation if applied in a statistical sense. For a transition to reach completion, it must first get going, so the initial linear increase of $\mathcal{P}_{if}$ with t, computed correctly by the current technique, constitutes the most important aspect of the current problem. (For a more rigorous approach, consult the discussion on the *interaction approximation* in Chapter 24.) To calculate the rate $d\mathcal{P}_{if}/dt$ (related to Einstein's coefficient for true absorption), we need to compute the "matrix element"

$$\langle \varphi_f | e^{i\mathbf{k}\cdot\mathbf{x}} \mathbf{e}_\alpha(\hat{\mathbf{k}}) \cdot \mathbf{p} | \varphi_i \rangle \tag{22.34}$$

associated with the radiative transition between the initial state i and the final state f. (Recall that quantum [matrix] mechanics began with Heisenberg's arrangement of tables of numbers having such a format.) We perform this chore in Chapter 23 for the problems of bound-bound and bound-free absorption. Here we finish up by recording the relationship between the preceding deliberations and what quantum-mechanical textbooks refer to as Fermi's *Golden Rule*.

FERMI'S GOLDEN RULE

Knowing that the integral over k or ω produces merely a factor of $2\pi t$, we see that our result in equation (22.31) does not depend on the specific form of the assumed interaction Hamiltonian. This observation allows us to write more generally that

$$\frac{d\mathcal{P}_{if}}{dt} = \sum_{\alpha=1}^{2} \oint w_\alpha \, d\Omega, \tag{22.35}$$

where w_α equals the constant probability rate per unit solid angle of making the (radiative) transition $i \rightarrow f$ (the absorption or emission of photons with polarization state α, frequency ω, and propagation direction $\hat{\mathbf{k}}$):

$$w_\alpha = \frac{2\pi}{\hbar} |\langle \varphi_f | H_\alpha^{\text{int}}(\mathbf{k}) | \varphi_i \rangle|^2 \rho_{\alpha\omega}(\hat{\mathbf{k}}). \tag{22.36}$$

In equation (22.36)—Fermi's Golden Rule—$H_\alpha^{\text{int}}(\mathbf{k})$ is the appropriate Fourier component of the interaction Hamiltonian, and $\rho_{\alpha\omega}(\hat{\mathbf{k}})$ is the density of photon states with energy $\hbar\omega = |E_f - E_i|$:

$$\rho_{\alpha\omega}(\hat{\mathbf{k}})d(\hbar\omega) = \frac{V}{(2\pi)^3}\frac{\omega^2}{c^3}d\omega. \tag{22.37}$$

The density $\rho_{\alpha\omega}(\hat{\mathbf{k}})$ refers to initial states for absorption processes and to final states for emission processes.

23

Radiative Absorption Cross Sections

Reference: Wu, *Quantum Mechanics*, pp. 279–288.

In this chapter we apply the formalism developed so far to the calculation of radiative absorption cross sections for simple atoms. From such cross sections, we can use the relationships derived earlier for the Einstein coefficients to obtain other transition rates of interest (review Chapters 8 and 9). At its most basic level, our job boils down to the computation of the matrix element

$$\langle \varphi_f | e^{i\mathbf{k}\cdot\mathbf{x}} \mathbf{e}_\alpha(\hat{\mathbf{k}}) \cdot \mathbf{p} | \varphi_i \rangle. \tag{23.1}$$

DIPOLE APPROXIMATION

Since atoms (or more precisely, the spatial extent where φ_i and φ_f have appreciable values) are much smaller than the wavelength of the light that they can typically emit or absorb, we find it convenient to expand the term $e^{i\mathbf{k}\cdot\mathbf{x}}$:

$$e^{i\mathbf{k}\cdot\mathbf{x}} = 1 + i\mathbf{k}\cdot\mathbf{x} + \cdots \tag{23.2}$$

The (electric) dipole approximation to the matrix element (23.1) amounts to the retention of only the first term:

$$\mathbf{e}_\alpha \cdot \langle \varphi_f | \mathbf{p} | \varphi_i \rangle. \tag{23.3}$$

This result provides the quantum-mechanical equivalent to the classical statement that in order for a charge to radiate, its momentum must change; i.e., it must undergo acceleration.

Since the wave functions φ represent eigenfunctions of H_0 and not of $\mathbf{p}$, we find it convenient to express this operator in terms of H_0 and $\mathbf{x}$. To do this, consider the commutator

$$[H_0, \mathbf{x}] \equiv H_0\mathbf{x} - \mathbf{x}H_0 = -\frac{\hbar^2}{2m_e}(\nabla^2\mathbf{x} - \mathbf{x}\nabla^2) = -\frac{\hbar^2}{m_e}\nabla = -i\frac{\hbar}{m_e}\mathbf{p},$$

as is easily verified using Cartesian tensors. Thus we have the identification

$$\mathbf{p} = i\frac{m_e}{\hbar}(H_0\mathbf{x} - \mathbf{x}H_0), \tag{23.4}$$

which leads to the result

$$\langle\varphi_f|\mathbf{p}|\varphi_i\rangle = im_e\omega_{fi}\mathbf{X}_{fi}, \tag{23.5}$$

where we have denoted $\omega_{fi} = (E_f - E_i)/\hbar$ and set

$$\mathbf{X}_{fi} \equiv \langle\varphi_f|\mathbf{x}|\varphi_i\rangle. \tag{23.6}$$

When multiplied by $-e$, the quantity in the middle of the matrix element corresponds to the *electric dipole* operator

$$\mathbf{d} \equiv -e\mathbf{x}. \tag{23.7}$$

DIPOLE TRANSITION PROBABILITY

The substitution of equation (23.5) into equation (22.33) yields, in the electric dipole approximation,

$$\frac{dP_{if}}{dt} = \frac{e^2}{hc^3}\sum_{\alpha=1}^{2}\oint\left[\mathcal{N}_\alpha(\mathbf{k})\omega^3|\mathbf{e}_\alpha(\hat{\mathbf{k}})\cdot\mathbf{X}_{fi}|^2\right]_{fi}d\Omega. \tag{23.8}$$

We are usually interested in the average absorption cross section for a random orientation of atoms. This cross section can be calculated equally well by considering a single atom in an isotropic radiation field. For the latter problem, suppose $\mathbf{k}$ makes an angle θ with respect to $\mathbf{X}_{fi}$ (see Figure 23.1). The projections of $\mathbf{X}_{fi}$ in the two polarization directions may then be written

$$\mathbf{e}_1 \cdot \mathbf{X}_{fi} = |\mathbf{X}_{fi}|\sin\theta\cos\phi, \qquad \mathbf{e}_2 \cdot \mathbf{X}_{fi} = |\mathbf{X}_{fi}|\sin\theta\sin\phi.$$

For an isotropic and unpolarized radiation field, $\mathcal{N}_1(\mathbf{k}) = \mathcal{N}_2(\mathbf{k}) \equiv \mathcal{N}(\omega)/2$, we may now integrate equation (23.8) to get

$$\frac{dP_{if}}{dt} = \frac{4\pi e^2\omega_{fi}^3}{3hc^3}\mathcal{N}(\omega_{fi})|\mathbf{X}_{fi}|^2, \tag{23.9}$$

where we have evaluated

$$\int_0^{2\pi}d\phi\int_0^{\pi}d\theta\,\sin\theta[\sin^2\theta(\cos^2\phi + \sin^2\phi)] = \frac{8\pi}{3}.$$

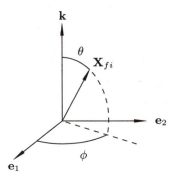

FIGURE 23.1
Geometry of angular integrations for bound-bound transitions in an atom.

BOUND-BOUND ABSORPTION CROSS SECTION

We define the atomic bound-bound absorption cross section $\sigma_{\text{bb}}(\omega)$ by the image that the transition rate $i \rightarrow f$ arises from a flux of photons $c\mathcal{N}(\omega)$, integrated over all phase-space elements, $4\pi\omega^2 \, d\omega/(2\pi)^3 c^3$, as they encounter an atom of radiative cross section σ_{bb}. Thus we require

$$\frac{d\mathcal{P}_{if}}{dt} = \int_0^\infty \sigma_{\text{bb}}(\omega) c\mathcal{N}(\omega) \frac{4\pi\omega^2 \, d\omega}{(2\pi)^3 c^3}.$$

Upon comparison with equation (23.9), this allows us to identify

$$\sigma_{\text{bb}}(\omega) = \frac{4\pi^2}{3} \left(\frac{e^2}{\hbar c}\right) |\mathbf{X}_{fi}|^2 \omega \delta(\omega - \omega_{fi}), \tag{23.10}$$

where we recognize the combination $e^2/\hbar c$ as the fine-structure constant.

Since $\omega\delta(\omega - \omega_{fi}) = \nu\delta(\nu - \nu_{fi})$, we may also write the above equation in the notation used earlier in the text where the initial and final states are now denoted by 1 and 2, respectively [see equation (10.4)]:

$$\alpha_\nu = \frac{\pi e^2}{m_e c} f_{12} \phi_{12}(\nu). \tag{23.11}$$

To make a correspondence with equation (23.10), we have taken the line profile function $\phi_{12}(\nu)$ to be a delta function $\delta(\nu - \nu_{21})$. The transition f-value we identify as

$$f_{12} = \frac{4\pi m_e}{3\hbar} \nu_{21} |\mathbf{X}_{21}|^2. \tag{23.12}$$

Notice that we may write the oscillator strength in the form

$$f_{12} = \frac{2m_e(\omega_{21}|\mathbf{X}_{21}|)^2}{3\hbar\omega_{21}},$$

which—up to a factor of order 2/3—can be thought of as the ratio of the energy of electronic oscillation to the emitted photon energy, and typically has an order-unity magnitude for allowed dipole transitions.

THOMAS-KUHN SUM RULE

Kramers and Heisenberg used the old quantum theory, some previous work by Landenburg, and the correspondence principle, to demonstrate that the sum of f-values, with a given initial state, over all possible final states (both higher and lower) *exactly* equals 1:

$$\sum_f f_{if} = 1. \tag{23.13}$$

The proof of this result using the methods of modern quantum mechanics, and its generalization to transitions involving N electrons (when the value 1 is replaced by N) is known as the *Thomas-Kuhn sum rule*. The sum rule has great utility in spectroscopy as a tool to obtain absolute f-values from the relative strengths of lines that are measured experimentally.

Proof: For simplicity we assume only bound states, and we begin with an expression for f_{if} which includes the sign contained in $(E_f - E_i)$:

$$f_{if} = \frac{2m_e}{3\hbar^2}(E_f - E_i)(\mathbf{X}_{fi}^* \cdot \mathbf{X}_{fi}).$$

With this definition, absorption f-values with $E_f > E_i$ are positive; emission f-values with $E_f < E_i$ are negative. In what follows, we simply regard f and i as indices denoting quantum states, without regard to the labels "final" and "initial." We also do not use the summation convention regarding repeated indices here.

Equation (23.5) allows us to write

$$(E_f - E_i)\mathbf{X}_{fi}^* = \frac{i\hbar}{m_e}\langle\varphi_f|\mathbf{p}|\varphi_i\rangle^* = \frac{i\hbar}{m_e}\langle\varphi_i|\mathbf{p}|\varphi_f\rangle,$$

where the last step can be obtained from $\mathbf{p} = -i\hbar\boldsymbol{\nabla}$ and an integration by parts. Thus

$$f_{if} = \frac{2i}{3\hbar}\langle\varphi_i|\mathbf{p}|\varphi_f\rangle \cdot \langle\varphi_f|\mathbf{x}|\varphi_i\rangle,$$

where we have used the definition for $\mathbf{X}_{fi}$ as the matrix element of $\mathbf{x}$. If we interchange the indices i and f and the order of the dot product, we get

$$f_{fi} = \frac{2i}{\hbar} \langle \varphi_i | \mathbf{x} | \varphi_f \rangle \cdot \langle \varphi_f | \mathbf{p} | \varphi_i \rangle.$$

Since $f_{fi} = -f_{if}$, we may form the unrestricted sum of f_{fi} as

$$\sum_f f_{if} = \frac{1}{2} \sum_f (f_{if} - f_{fi})$$

$$= \frac{i}{3\hbar} \sum_f \langle \varphi_i | \mathbf{p} \cdot | \varphi_f \rangle \langle \varphi_f | \mathbf{x} | \varphi_i \rangle - \langle \varphi_i | \mathbf{x} \cdot | \varphi_f \rangle \langle \varphi_f | \mathbf{p} | \varphi_i \rangle. \quad (23.14)$$

To simplify the last expression, we need the *closure identity*,

$$\sum_f | \varphi_f \rangle \langle \varphi_f | = I, \quad (23.15)$$

where I is the identity operator (for bound states). To see that the sum of direct products of all eigenkets and eigenbras forms the identity operator (the generalization of the delta-function representations for Fourier transforms—see Chapter 12), recall that the eigenkets of a Hermitian operator form a complete basis. Thus any ket (wave function) can be represented as a linear superposition of the eigenkets,

$$| \varphi \rangle = \sum_i c_i | \varphi_i \rangle.$$

If we left-operate with equation (23.15) on the above and use the orthonormality condition, $\langle \varphi_f | \varphi_i \rangle = \delta_{fi}$, we recover $| \varphi \rangle$, demonstrating that the left-hand side of equation (23.15) is, indeed, the identity operator (for both kets and bras). With a careful limit procedure (e.g., cubic box normalization), the same result can be extended to include the possibility of free states.

With the substitution of equation (23.15), equation (23.14) becomes

$$\sum_f f_{if} = \frac{i}{3\hbar} \langle \varphi_i | \mathbf{p} \cdot \mathbf{x} - \mathbf{x} \cdot \mathbf{p} | \varphi_i \rangle.$$

But the commutation relation for $\mathbf{p}$ and $\mathbf{x}$ implies

$$\mathbf{p} \cdot \mathbf{x} - \mathbf{x} \cdot \mathbf{p} = [p_x, x] + [p_y, y] + [p_z, z] = -3i\hbar;$$

therefore we finally obtain the desired result

$$\sum_f f_{if} = \langle \varphi_i | \varphi_i \rangle = 1,$$

which is the Thomas-Kuhn sum rule for radiative transitions involving a single electron. (Q.E.D.)

SPONTANEOUS EMISSION AND THE NATURAL LINE WIDTH

The infinitesimal line width predicted in the section before last arises from overidealization. All real lines would have a broadening by at least an amount given by the following considerations (devised by Wigner and Weisskopf). As an atom tries to absorb a resonant line photon, it simultaneously tries to decay back spontaneously to state i. To take account of this effect, define the rate of spontaneous emission A_{fi} to equal $d\mathcal{P}_{fi}/dt$ as if the initial state were f instead of i and as if the perturbing Hamiltonian came from the 1 in the term $1 + \mathcal{N}_\alpha(\mathbf{k})$ in equation (22.15) for $H_\alpha^{\mathrm{em}}(\mathbf{k})$. A calculation completely analogous to the one already carried out then yields

$$A_{fi} = \frac{e^2}{\hbar c^3} \sum_{\alpha=1}^{2} \oint \left[\omega^3 |\mathbf{e}_\alpha(\hat{\mathbf{k}}) \cdot \mathbf{X}_{fi}|^2 \right]_{fi} d\Omega, \qquad (23.16)$$

where we have made use of the fact that the absolute value of the matrix element is symmetric with respect to interchange of its indices: $|\mathbf{X}_{fi}| = |\mathbf{X}_{if}|$. (This represents a mechanistic statement of the quantum principle behind detailed balance.) Evaluating the angular part of the integration as before, we obtain

$$A_{fi} = \frac{4e^2\omega_{fi}^3}{3\hbar c^3} |\mathbf{X}_{fi}|^2, \qquad (23.17)$$

a formula first derived by Heisenberg using the correspondence principle [see equation (14.18)]. Notice that our formulae for A_{fi} and the absorption cross section are consistent with Einstein's relations because we have carried out the calculations for both to the same order of approximation. Einstein's relations for the rate coefficients A_{21}, B_{21}, and B_{12} represent more general statements, however, in that they hold to all *orders* of perturbation theory (in this case, a multipole expansion in powers of the square of the fine-structure constant, α^2).

For our original problem, denote $A_{fi} = \Gamma$ as the rate at which an atom, initially in state i and in the process of absorbing a photon of radian frequency ω_{fi}, attempts to decay back to i:

$$\left[\frac{d}{dt} \left(|c_f|^2 \right) \right]_{\text{spont. em}} = -\Gamma |c_f|^2.$$

With a spontaneous decay rate for c_f equal to $-\Gamma c_f/2$, we heuristically modify the basic equation of time-dependent perturbation theory [see equation (22.29)]:

$$\dot{c}_f = -i\hbar^{-1} \langle \varphi_f | H_\alpha^{\text{abs}} | \varphi_i \rangle e^{i(\omega_{fi} - \omega)t} - \frac{\Gamma}{2} c_f.$$

Recognizing $e^{\Gamma t/2}$ as an integrating factor for the above equation, we can solve for $c_f(t)$, subject to the initial condition $c_f(0) = 0$, to obtain

$$c_f(t) = -\hbar^{-1}\langle\varphi_f|H_\alpha^{\text{abs}}|\varphi_i\rangle \left[\frac{e^{i(\omega_{fi}-\omega)t} - e^{-\Gamma t/2}}{\omega_{fi} - \omega - i\Gamma/2}\right].$$

Multiplying the above result by its complex conjugate, we get

$$|c_f(t)|^2 = \hbar^{-2}|\langle\varphi_f|H_\alpha^{\text{abs}}|\varphi_i\rangle|^2 \left[\frac{1 + e^{-\Gamma t} - 2e^{-\Gamma t/2}\cos(\omega_{fi} - \omega)t}{(\omega_{fi} - \omega)^2 + (\Gamma/2)^2}\right].$$

Instead of diverging linearly with t in the limit $t \to \infty$, $|c_f(t)|^2$ now has a finite value:

$$|c_f(\infty)|^2 = \hbar^{-2}\frac{|\langle\varphi_f|H_\alpha^{\text{abs}}|\varphi_i\rangle|^2}{(\omega_{fi} - \omega)^2 + (\Gamma/2)^2}.$$

The above finding suggests that the bound-bound absorption cross section in equation (23.10) would more accurately read:

$$\sigma_{\text{bb}}(\omega) = \frac{4\pi^2}{3}\left(\frac{e^2}{\hbar c}\right)|\mathbf{X}_{fi}|^2\omega_{fi}\mathcal{L}(\omega),$$

where $\mathcal{L}(\omega)$ is the Lorentz profile (in ω space):

$$\mathcal{L}(\omega) \equiv \frac{1}{\pi}\left[\frac{\Gamma/2}{(\omega - \omega_{fi})^2 + (\Gamma/2)^2}\right].$$

A better generalization, which takes into account the fact that both the lower and the upper states may be "fuzzy" because of spontaneous decays out of them, gives

$$\alpha_\nu = \frac{\pi e^2}{m_e c}f_{if}\mathcal{L}(\nu), \tag{23.18}$$

where f_{if} is the f-value of the transition [see equation (23.12)] and $\mathcal{L}$ is the Lorentz profile (in ν space), with $\nu\mathcal{L}(\nu) = \omega\mathcal{L}(\omega)$:

$$\mathcal{L}(\nu) = \frac{1}{\pi}\left[\frac{\Gamma_{if}/4\pi}{(\nu - \nu_{fi})^2 + (\Gamma_{if}/4\pi)^2}\right],$$

with Γ_{if} given by

$$\Gamma_{if} = \sum_{n<f} A_{fn} + \sum_{n<i} A_{in}.$$

People also occasionally put into the formula for the width parameter, Γ_{if}, additional perturbing effects for line broadening (e.g., mechanisms such as atomic collisions that make the relevant energy states more fuzzy).

PHOTOIONIZATION

Let us now consider another prototypical quantum radiation process, the absorption by an atom of a photon energetic enough to liberate a bound electron to a free state. We use as the starting point the transition-rate formula before making the multipole expansion [see equation (22.33)]:

$$\frac{dP_{if}}{dt} = \frac{e^2}{hc^3m_e^2} \sum_{\alpha=1}^{2} \oint \left[\omega \mathcal{N}_\alpha(\mathbf{k}) |\langle \varphi_f | e^{i\mathbf{k}\cdot\mathbf{x}} \mathbf{e}_\alpha \cdot \mathbf{p} | \varphi_i \rangle|^2 \right]_{fi} d\Omega. \qquad (23.19)$$

As the simplest example of a bound-free transition, we consider an initial state i that corresponds to an electron in the K-shell ($n = 1$) of a hydrogen-like atom with effective nuclear charge Z:

$$\varphi_i = (\pi a_Z^3)^{-1/2} \exp(-|\mathbf{x}|/a_Z), \qquad (23.20)$$

where we have placed the atom at the origin of our coordinate system and where

$$a_Z = \hbar^2/Zm_e e^2 \qquad (23.21)$$

represents the analog of the Bohr radius for this problem. Notice that equation (23.20) gives φ_i as a properly normalized wave function when integrated over all space (or over $V \gg a_Z^3$).

For the final state f, we should consider, in principle, an exact free-state electron wave function that includes the presence of the nucleus; however, such a refined calculation entails more difficulty than we have space to treat here. As a pedagogical example, we consider the simplification possible by making the *Born approximation*, which takes φ_f to be the free-state solution for an electron in a *vacuum* (i.e., a plane wave). For normalization purposes, we put this electron in the same cubic box of volume $V = L^3$ that we confined our radiation fields last time:

$$\varphi_f^* = V^{-1/2} e^{-i\mathbf{k}_e \cdot \mathbf{x}}. \qquad (23.22)$$

The plane-wave assumption of equation (23.22) forms the quantum analog of the straight-line trajectory that we adopted to simplify thermal *Bremsstrahlung* calculations in Chapter 15.

On the principle that energy must be conserved in the overall photoionization process, we anticipate that the free-electron wavenumber satisfies

$$|\mathbf{k}_e| = (2m_e E_f)^{1/2}/\hbar, \qquad (23.23)$$

where

$$E_f = \hbar\omega_{fi} - Ze^2/2a_Z$$

is the final energy of the liberated electron. On substituting equations (23.20) and (23.22) into equation (23.19) and writing $\mathbf{p} = -i\hbar\nabla$, we obtain for the square of matrix element

$$\frac{\hbar^2}{\pi a_Z^3 V} \left| \int_V e^{i(\mathbf{k}-\mathbf{k_e})\cdot\mathbf{x}} \mathbf{e}_\alpha \cdot \nabla(e^{-|\mathbf{x}|/a_Z}) \, d^3x \right|^2 .$$

The ability to combine the two plane waves for the electron and the photon constitutes the motivation for not expanding in a multipole expansion and for adopting the Born approximation.

In the matrix element of the previous paragraph, integrate by parts to obtain

$$\frac{\hbar^2}{\pi a_Z^3 V} (\mathbf{k_e} \cdot \mathbf{e}_\alpha)^2 \left| \int_V \exp\left[i(\mathbf{k}-\mathbf{k_e})\cdot\mathbf{x} - |\mathbf{x}|/a_Z\right] d^3x \right|^2 ,$$

where we have used $\mathbf{k} \cdot \mathbf{e}_\alpha(\hat{\mathbf{k}}) = 0$ to simplify the coefficient in front of the integral. Define $\mathbf{q} \equiv \mathbf{k}-\mathbf{k_e}$ and notice that $\hbar\mathbf{q}$ represents the momentum not transferred from the photon to the liberated electron (i.e, the part that must be absorbed by the atom, which is not calculated in an approximation that treats the nucleus as being infinitely heavy). Adopt now a spherical polar coordinate system such that $\mathbf{q} \cdot \mathbf{x} = qr\cos\theta$, with $d^3x = 2\pi r^2 \, dr \, \sin\theta \, d\theta$. Define $\mu \equiv \cos\theta$ and evaluate the integral in the limit $V \to \infty$:

$$\int \exp\left[i\mathbf{q}\cdot\mathbf{x} - |\mathbf{x}|/a_Z\right] d^3x = 2\pi \int_0^\infty dr \, r^2 e^{-r/a_Z} \int_{-1}^{+1} e^{iqr\mu} \, d\mu$$

$$= \frac{8\pi a_Z^3}{(1 + q^2 a_Z^2)^2}.$$

The last result follows by evaluating the second integral and then integrating the first by parts twice. Collecting terms, we now obtain

$$|\langle \varphi_f | e^{i\mathbf{k}\cdot\mathbf{x}} \mathbf{e}_\alpha \cdot \mathbf{p} | \varphi_i \rangle|^2 = 64\pi \frac{\hbar^2 a_Z^3}{V} (\mathbf{k_e} \cdot \mathbf{e}_\alpha)^2 (1 + q^2 a_Z^2)^{-4}.$$

On substituting this result into equation (23.19), we obtain, upon dropping the "fi" subscript,

$$\frac{d\mathcal{P}_{if}}{dt} = 32 \frac{e^2 \hbar a_Z^3}{V c^3 m_e^2} \sum_{\alpha=1}^{2} \oint \left[\omega \mathcal{N}_\alpha(\mathbf{k})(\mathbf{k_e} \cdot \mathbf{e}_\alpha)^2\right] \left[1 + |\mathbf{k}-\mathbf{k_e}|^2 a_Z^2\right]^{-4} d\Omega.$$

For randomly oriented atoms, we may perform the integrals as if an isotropic radiation field impinged on a fixed direction $\mathbf{k_e}$ for the ejected

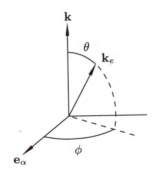

FIGURE 23.2
Geometry of angular integrations for bound-free transitions in an atom.

electron. With this geometry (see Figure 23.2), we have

$$\mathbf{k}_e \cdot \mathbf{e}_\alpha = k_e \sin \theta \cos \phi,$$

$$|\mathbf{k} - \mathbf{k}_e|^2 = k^2 + k_e^2 - 2kk_e \cos \theta,$$

where equation (23.23) implies, with $\omega = ck$,

$$k_e^2 = \frac{2m_e ck}{\hbar} - \frac{1}{a_Z^2}.$$

In the above, we have used equation (23.21) to equate the expression $m_e Z e^2 / \hbar^2 a_Z$ with $1/a_Z^2$. Collecting expressions, we may now write

$$1 + |\mathbf{k} - \mathbf{k}_e|^2 a_Z^2 = a_Z^2 \frac{2m_e ck}{\hbar} \left(1 + \frac{\hbar k}{2m_e c} - \frac{\hbar k_e}{2m_e c} \cos \theta \right) \approx a_Z^2 \frac{2m_e \omega}{\hbar},$$

for photon and electron momenta, $\hbar k$ and $\hbar k_e$, much less than $m_e c$. (This assumption must hold if we use a nonrelativistic description for the process.)

The angular integrations for dP_{if}/dt may now be performed as

$$\int_0^{2\pi} d\phi \int_0^\pi \sin^2 \theta \cos^2 \phi \sin \theta \, d\theta = 4\pi/3.$$

Thus we obtain

$$\frac{dP_{if}}{dt} = \frac{8\pi e^2 a_Z^3}{3V \hbar m_e^2 c^3} \omega^{-3} \mathcal{N}(\omega)(\hbar k_e)^2 \left(\frac{\hbar}{m_e a_Z^2} \right)^4. \qquad (23.24)$$

BOUND-FREE ABSORPTION CROSS SECTION

Let dN_f represent the number of final free-electron states between k_e and $k_e + dk_e$:

$$dN_f = \frac{V}{(2\pi)^3} 4\pi k_e^2 \, dk_e.$$

We define the bound-free absorption cross section $\sigma_{bf}(\omega)$ through

$$\frac{dP_{if}}{dt} dN_f = \sigma_{bf}(\omega) \left[c\mathcal{N}(\omega) \frac{4\pi\omega^2}{(2\pi)^3 c^3} \, d\omega \right]. \tag{23.25}$$

Write $k_e \, dk_e = (m_e/\hbar) \, d\omega$, use equations (23.24) and (23.25) to solve for $\sigma_{bf}(\omega)$, and get

$$\sigma_{bf}(\omega) = \frac{8\pi}{3} \left(\frac{e^2}{m_e c} \right) \left(\frac{\hbar}{m_e a_Z^2} \right)^4 \omega^{-5} (k_e a_Z)^3, \tag{23.26}$$

where

$$k_e^2 = \frac{2m_e}{\hbar^2} \left(\hbar\omega - \frac{Ze^2}{2a_Z} \right)$$

is positive if the incident photon energy $\hbar\omega$ exceeds the ionization potential $Ze^2/2a_Z$.

   In order for us to invoke the Born approximation self-consistently, we need $\hbar\omega$ to lie comfortably beyond the ionization edge. In this regime, $k_e \approx (2m_e\omega/\hbar)^{1/2}$, which implies, when we substitute equation (23.21) for a_Z (partially),

$$\sigma_{bf}(\omega) = \frac{8\pi}{3\sqrt{3}} \frac{Z^4 m_e e^{10}}{c\hbar^3(\hbar\omega)^3} \left[48 \frac{Ze^2/2a_Z}{\hbar\omega} \right]^{1/2}. \tag{23.27}$$

The quantity in the square bracket equals 48 times the ratio of the ionization energy to the photon energy and is dimensionless. The square root of this ratio varies only slowly with ω. We have introduced a redundant factor of $\sqrt{3}$ on the top and bottom of equation (23.27) to make comparison easier between the Born-approximation value for K-shell ionization ($n = 1$) and the standard expression for the photionization cross section from a general level n in a hydrogenic atom (see Chapter 10):

$$\alpha_\nu(n) = n^{-5} \frac{8\pi}{3\sqrt{3}} \frac{Z^4 m_e e^{10}}{c\hbar^3(h\nu)^3} g_{bf}(\nu). \tag{23.28}$$

In equation (23.28) $g_{bf}(\nu)$ is a slowly varying Gaunt factor that has a value near unity close to the ionization edge, where the Born approximation (23.27) breaks down badly. In practice, a many-electron atom will be characterized by a series of ionization edges as it loses electrons from successive

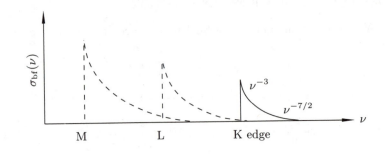

FIGURE 23.3
Schematic frequency dependence of bound-free absorption cross section for a
heavy atom as it successively undergoes M-, L-, and K-shell ionization.

shells, as depicted schematically in Figure 23.3.

Heavy elements, with many inner-shell electrons, either in the gas phase
or locked up in grains in the interstellar medium, provide substantial opacity to the propagation of X-rays softer than about 1 keV from celestial
sources. In contrast, hard X-rays with energies in the range of 10 keV or
more have no difficulty reaching us from clear across the Galaxy because
they encounter only the tail of the ν^{-3} dependence (becoming $\nu^{-3.5}$ at
very large ν) in $\alpha_\nu(1)$ due to K-shell electrons. The cutoff in the X-ray
spectral-energy distribution received from sources with both hard and soft
X-rays therefore allows astronomers one independent method for estimating the column density of matter along specific lines of sight (toward the
Galactic center, for example). Such values can be usefully compared with
estimates of column density provided by other means, such as dust extinction, molecular line surveys, and gamma-ray observations. Apart from this
astrophysical application, the bound-free absorption process by heavy elements also constitutes an important source of opacity in the interiors of
stars.

24

Selection Rules and Forbidden Transitions

References: Schiff, *Quantum Mechanics*, Chapter 10; Jackson, *Classical Electrodynamics*, Chapter 16.

In this chapter we consider the selection rules that govern radiative transitions. By convention, we call *permitted* or *allowed* those radiative transitions for which the associated dipole matrix element

$$\mathbf{X}_{fi} \equiv \langle \varphi_f | \mathbf{x} | \varphi_i \rangle$$

does not vanish. If $\mathbf{X}_{fi} = 0$, we call the transition *forbidden*. Despite this name, forbidden transitions generally have a nonzero, if much reduced, probability of occurring in comparison with allowed transitions of the same wavelength. The reduction amounts typically to a factor of α^2, where α is the fine-structure constant. Nevertheless, under terrestrial conditions, even dipole transition probabilities for spontaneous emission at optical wavelengths can compete only marginally with collisional de-excitation (both occur at rates 10^8-10^9 s^{-1}); thus the radiation of forbidden lines generally produces emission too feeble to be observed in a terrestrial laboratory. Forbidden transitions play a more important role in the astrophysics of diffuse nebular gases, where the limit on the number of downward radiative transitions may come principally from the upward rate by which inelastic collisions populate the excited states, and not by the rate at which atoms, once excited, decay to lower states. In a rarefied enough environment, a radiative decay eventually follows every upward transition (radiative or collisional), and forbidden lines can acquire greater strengths than permitted ones if the former prove easier to excite.

A SINGLE ELECTRON IN A SPHERICAL POTENTIAL

To illustrate the principle behind radiative selection rules, we consider the simple example of a single electron moving in a spherically symmetric potential created by all other charges in the system (nucleus and other electrons). The hydrogen atom satisfies this description, but so would any atom or ion which has completely closed shells except for a single valence electron (see Chapter 27). Let $-e\phi(\mathbf{x}) = V(r)$ in spherical polar coordinates (r, θ, ϕ). [Our replacement of the scalar potential by $V(r)$ is made, in part, to avoid confusion with the azimuthal angle ϕ.] With this notation, the time-independent Schrödinger equation,

$$-\frac{\hbar^2}{2m_e}\nabla^2\varphi + V(r)\varphi = E\varphi, \tag{24.1}$$

has eigensolutions of the form

$$\varphi = R(r)Y_{\ell m}(\theta, \phi), \tag{24.2}$$

where

$$Y_{\ell m}(\theta, \phi) = C_{\ell m}P_\ell^{|m|}(\cos\theta)e^{im\phi}. \tag{24.3}$$

The numbers ℓ and m, with $\ell \geq |m|$, are integers, and $P_\ell^m(\mu)$ is the associated Legendre polynomial,

$$P_\ell^m(\mu) = \frac{(-1)^{\ell+m}}{\ell!2^\ell}(1 - \mu^2)^{m/2}\frac{d^{\ell+m}}{d\mu^{\ell+m}}\left[(1 - \mu^2)^\ell\right], \tag{24.4}$$

valid for both positive and negative m as long as $|m| \leq \ell$. Note that $P_\ell^{|m|}(\mu)$ represents a polynomial in μ of degree ℓ if m is an even integer, and $(1 - \mu^2)^{1/2} = \sin\theta$ times a polynomial of degree $\ell - 1$ if m is an odd integer. Apart from a multiplicative constant, the associated Legendre polynomials with negative m equal those with positive m:

$$P_\ell^{-m}(\mu) = (-1)^m\frac{(\ell - m)!}{(\ell + m)!}P_\ell^m(\mu).$$

This fact allows us to define $Y_{\ell m}$ in terms of only $P_\ell^{|m|}$. For negative m, the coefficient $C_{\ell m}$ equals $(-1)^m$ times its value for positive m:

$$C_{\ell|m|} = \left[\frac{(\ell - |m|)!(2\ell + 1)}{(\ell + |m|)!4\pi}\right]^{1/2}. \tag{24.5}$$

This choice makes the collection of $Y_{\ell m}$ an orthonormal set:

$$\oint Y_{\ell'm'}^*(\theta, \phi)Y_{\ell m}(\theta, \phi)\,d\Omega = \delta_{\ell'\ell}\delta_{m'm},$$

while

$$Y_{\ell,-m}(\theta, \phi) = (-1)^m Y_{\ell m}^*(\theta, \phi). \tag{24.6}$$

Our definitions above conform with those of Jackson (1962).

ORBITAL ANGULAR MOMENTUM

As is well known, the angular parts of equation (24.2) have an intimate connection with the *orbital angular momentum operator*:

$$\mathbf{L} \equiv \mathbf{x} \times \mathbf{p}. \tag{24.7}$$

Of particular interest are the square of the total angular momentum, L^2, and the z-component of the angular momentum, L_z. Expressed in spherical polar coordinates, these read

$$L^2 \equiv -\hbar^2 \left(\frac{1}{\sin\theta} \frac{\partial}{\partial\theta} \sin\theta \frac{\partial}{\partial\theta} + \frac{1}{\sin^2\theta} \frac{\partial^2}{\partial\phi^2} \right), \tag{24.8}$$

$$L_z = -i\hbar \frac{\partial}{\partial\phi}. \tag{24.9}$$

Since L^2 is proportional to the angular part of the ∇^2 operator, and L_z represents a single differentiation in ϕ, they have as their eigenfunctions the spherical harmonics:

$$L^2 Y_{\ell m}(\theta, \phi) = \ell(\ell+1)\hbar^2 Y_{\ell m}(\theta, \phi), \tag{24.10}$$

$$L_z Y_{\ell m}(\theta, \phi) = m\hbar Y_{\ell m}(\theta, \phi). \tag{24.11}$$

Equations (24.10) and (24.11) show that the absolute square of the orbital angular momentum has an expectation value $\ell(\ell+1)\hbar^2$, whereas its projection onto any z axis comes with m units of $\hbar$. Except for the case $\ell = 0$, $Y_{\ell m}$ does *not* constitute an eigenfunction of the operators (see the more extended discussion in Chapter 26),

$$L_x = i\hbar \left(\sin\phi \frac{\partial}{\partial\theta} + \cot\theta \cos\phi \frac{\partial}{\partial\phi} \right), \tag{24.12}$$

$$L_y = i\hbar \left(-\cos\phi \frac{\partial}{\partial\theta} + \cot\theta \sin\phi \frac{\partial}{\partial\phi} \right). \tag{24.13}$$

Indeed, the combinations $L_x \pm iL_y$ have forms

$$L_x \pm iL_y = \hbar e^{\pm i\phi} \left(\pm \frac{\partial}{\partial\theta} + i\cot\theta \frac{\partial}{\partial\phi} \right),$$

which make them raising and lowering operators for the second index:

$$(L_x \pm iL_y)Y_{\ell m} = [(\ell \mp m)(\ell + 1 \pm m)]^{1/2} \hbar Y_{\ell,m\pm 1}. \tag{24.14}$$

As a consequence, when the z-component and the square of the angular momentum are known precisely, i.e., when m and ℓ are good eigenvalues, the expectation values of L_x and L_y equal zero, but the expectation value of the sum of their squares,

$$L_x^2 + L_y^2 = \frac{1}{2}\left[(L_x + iL_y)(L_x - iL_y) + (L_x - iL_y)(L_x + iL_y)\right],$$

equals that of $L^2 - L_z^2$, i.e., $[\ell(\ell+1) - m^2]\hbar^2$. We may regard this conclusion as arising from the uncertainty principle (because L_z commutes with L^2, but not with L_x or L_y).

ENERGY EIGENVALUES AND RADIAL WAVE FUNCTIONS

When equation (24.2) is substituted into (24.1), we get the radial wave equation:

$$\frac{1}{r^2}\frac{d}{dr}\left(r^2\frac{dR}{dr}\right) + \left\{\frac{2m_e}{\hbar^2}[E - V(r)] - \frac{\ell(\ell+1)}{r^2}\right\}R = 0. \qquad (24.15)$$

Notice that $\ell(\ell+1)\hbar^2/2m_e r^2$ represents the centrifugal potential and that equation (24.15) does not involve the quantum number m. Notice also that E must be regarded as an eigenvalue because the posed problem is homogeneous; i.e., we can freely scale R without affecting either the governing differential equation or its associated two-point boundary conditions (regularity at the origin and at infinity). In other words, the specification of the behavior of R at $r = 0$ and as $r \to \infty$ overconstrains the problem unless we can consider special choices of E. Since equation (24.15) with its associated boundary conditions is of the Sturm-Liouville type, the eigenvalues for E are purely real.

For a pure Coulomb potential, $V(r) \propto r^{-1}$, the allowed discrete values for $E < 0$ (bound states) are given by the well known Bohr formula that involves only the principal quantum number n and not ℓ. There also exists a continuum of allowed values for $E > 0$ (free states). The eigenfunctions associated with a given value of E can be found analytically in terms of confluent hypergeometric series for both bound and free states. For non-Coulomb potentials, we will generally have to rely on approximate methods or numerical integration to recover similar information.

Some general properties, however, can be deduced by performing a simple transformation. Introduce a *reduced* radial wave function $\mathcal{R}$ via

$$R(r) \equiv \frac{1}{r}\mathcal{R}(r). \qquad (24.16)$$

The factor $1/r$ removes the spherical divergence so that $\mathcal{R}$ satisfies a second-order ordinary differential equation of *normal form* (no first derivative):

$$\frac{d^2\mathcal{R}}{dr^2} + K^2(r)\mathcal{R} = 0, \qquad (24.17)$$

where the effective radial wavenumber $K(r)$ is defined through

$$K^2(r) \equiv \left\{ \frac{2m_e}{\hbar^2}[E - V(r)] - \frac{\ell(\ell+1)}{r^2} \right\}. \tag{24.18}$$

For given ℓ, we can always define a principal quantum number n for the bound states such that $n - \ell - 1$ yields the number of radial nodes (where $\mathcal{R}$ passes through zero, not counting $r = 0$ or ∞). Since the number of such nodes cannot be less than zero, we require $n \geq \ell + 1$; i.e., for given $n \geq 1$, ℓ can only take on the values: 0, 1, up to $n - 1$. (In turn, since $|m| \leq \ell$, m can assume values only from $-\ell$ to $+\ell$ in increments of 1.) The "oscillation theorem" of the theory of ordinary differential equations states that the solutions of equation (24.17) contain more oscillations as E increases algebraically in value. Thus, for fixed ℓ, increasing n orders the eigensolutions in increasing energy E. (For non-Coulomb potentials, E will generally depend on both n and ℓ in a manner that is explained qualitatively in Shu [1982, Box 3.3].) The asymptotic technique known as the WKBJ method provides excellent approximate solutions that verify the above claims for large values of n. (The solutions are often surprisingly accurate even for $n = 1$.)

Ignoring for the moment the role of electron spin (to be covered in a future chapter), we henceforth characterize a stationary quantum state of our electron by the set of three numbers (n, ℓ, m). The energy eigenvalue and the (reduced) radial wave function depend only on n and ℓ, and we will denote their values by $E_{n\ell}$ and $\mathcal{R}_{n\ell}$. For given n and ℓ, substates with different m are *degenerate*. As we will see, however, this does *not* imply that *transitions* from one energy level to another are equally likely for all choices of the degenerate m states.

RADIATIVE TRANSITIONS AND DIPOLE SELECTION RULES

Suppose we wish to consider the possibility of a radiative transition from one bound state $i \equiv (n, \ell, m)$ to another bound state $f \equiv (n', \ell', m')$. We consider the following three independent possibilities for the matrix element $\langle \varphi_f | \mathbf{e}_\alpha \cdot \mathbf{x} | \varphi_i \rangle$:

$$\langle \varphi_f | z | \varphi_i \rangle \qquad \text{and} \qquad \frac{1}{\sqrt{2}} \langle \varphi_f | x \pm iy | \varphi_i \rangle.$$

The case $\mathbf{e}_\alpha \cdot \mathbf{x} = z$ corresponds to interaction with radiation propagating in the x-y plane and linearly polarized in the z-direction, $\mathbf{e}_\alpha = \hat{\mathbf{z}}$. The case $\mathbf{e}_\alpha \cdot \mathbf{x} \propto x \pm iy$ corresponds to interaction with radiation propagating in the z-direction and circularly polarized $\mathbf{e}_\alpha \propto \hat{\mathbf{x}} \pm i\hat{\mathbf{y}}$ in a right- or left-handed sense (photons of spin ± 1). These particular components of the vector displacement $\mathbf{x}$ prove convenient to study because of their connection with rotational displacements. In particular, in spherical polar coordinates we have

$$z = r\cos\theta, \qquad x \pm iy = r\sin\theta\, e^{\pm i\phi}.$$

We define $\cos\theta \equiv \mu$ and write d^3x as equal to $r^2\,dr d\mu d\phi$. With φ given by equation (24.2), the z-component of the dipole matrix element $\langle\varphi_f|\mathbf{x}|\varphi_i\rangle$ will be proportional to the product of three terms:

$$\langle\varphi_f|z|\varphi_i\rangle \propto \left[\int_0^\infty r\mathcal{R}_{n'\ell'}\mathcal{R}_{n\ell}\,dr\right]\left[\int_{-1}^{+1}\mu P_{\ell'}^{|m'|}(\mu)P_\ell^{|m|}(\mu)\,d\mu\right]$$
$$\times \left[\int_0^{2\pi} e^{i(m-m')\phi}\,d\phi\right]. \tag{24.19}$$

The integral over r will not equal zero, because r is not an eigenvalue operator of $\mathcal{R}_{n\ell}$ (i.e., an expansion of $r\mathcal{R}_{n\ell}$ in the complete set of radial wave functions—bound and free—will contain nonvanishing coefficients in *all* of the eigenfunctions). Thus the (dipole) selection rules must originate in the considerations of the integrals over the angular parts of the wave functions, i.e., basically from considerations of *orbital angular momentum conservation* and from *parity* (see below and Chapter 27, where we give a more sophisticated rediscussion of these results from the point of view of the *Wigner-Eckart theorem*).

In order for the integral over ϕ not to vanish, we require m' to equal m. Since $\Delta m \equiv m' - m = 0$, no change occurs in the z-component of the electron's orbital angular momentum. This holds for an electric dipole oriented in the z direction because the bisymmetric radiation pattern cannot exert, on average, any z torque on the atom.

Suppose the transition does satisfy the requirement $m' = m$. What requirement does the $\mu = \cos\theta$ integral now place on the possibility of dipole radiation? Since $P_\ell^{|m|}(\mu)$ represents a polynomial with highest power μ^ℓ if m is even, and $(1-\mu^2)^{1/2}\mu^\ell$ if m is odd, it ought to be possible to write $\mu P_\ell^{|m|}(\mu)$ in terms of $P_{\ell+1}^{|m|}(\mu)$ plus polynomials of lower degree. In fact, the following recursion relation holds:

$$\mu P_\ell^{|m|}(\mu) = \frac{1}{2\ell+1}\left[(\ell-|m|)P_{\ell+1}^{|m|}(\mu) + (\ell+|m|)P_{\ell-1}^{|m|}(\mu)\right]. \tag{24.20}$$

If we substitute the above expression into the θ integral and use the orthogonality relationship for associated Legendre polynomials, we obtain the selection rule that ℓ' must equal either $\ell+1$ or $\ell-1$:

$$\langle\varphi_f|z|\varphi_i\rangle = 0 \qquad \text{unless} \qquad \Delta\ell = \pm 1 \qquad \text{and} \qquad \Delta m = 0.$$

Schiff in the cited reference points out an apparent paradox in the result $\Delta \ell = \pm 1$ for the situation described above. As we have already seen, the expectation values of L_x and L_y,

$$\langle \varphi | L_x | \varphi \rangle \quad \text{and} \quad \langle \varphi | L_y | \varphi \rangle,$$

equal zero when φ is proportional to any spherical harmonic. This condition holds for both the initial and the final states of our atom. Thus when $\Delta m = 0$ in addition, no change occurs for the expectation value of $\mathbf{L}$ along *any* of the x, y, or z projections. How then can the photon, exiting on average in the x-plane, carry off one unit of orbital angular momentum?

The resolution to this paradox comes with the realization that the expectation value of the changes of the *squares* of the orbital angular momenta in the x and y directions do not equal zero. It is the uncertainty associated with $L_x^2 + L_y^2$ that allows the photon to carry off one unit of angular momentum during transition, resulting in a change for the atom of $\Delta \ell = \pm 1$.

What happens if $\langle \varphi_f | z | \varphi_i \rangle$ vanishes because the transition being considered does not satisfy $\Delta m = 0$ and $\Delta \ell = \pm 1$? The radiative transition can still occur, via interaction with circularly polarized photons, if at least one of the other matrix elements,

$$\langle \varphi_f | x \pm iy | \varphi_i \rangle,$$

does not vanish. With the replacement $x \pm iy = r(1 - \mu^2)^{1/2} e^{\pm i\phi}$, we easily see that the associated ϕ integral implies the selection rule $m' = m \pm 1$. In other words, the emission or absorption of a photon of spin 1 by a dipole properly oriented with its axis in the x-y plane can lead to a systematic change in the z-orbital momentum of the atom. The unequal torques—like those on the arms of a windmill—will then lead to a change of the azimuthal atomic structure. (Remember that the wave functions of the initial and final states have dependences of $e^{im\phi}$ and $e^{im'\phi}$.)

Suppose the transition does satisfy the selection rule on $m' = m \pm 1$. The $\mu = \cos\theta$ integral now takes the form

$$\int_{-1}^{+1} (1 - \mu^2)^{1/2} P_\ell^{|m'|}(\mu) P_\ell^{|m|}(\mu) \, d\mu.$$

To evaluate this integral if $|m'| > |m|$, i.e., when $|m'| = |m| + 1$, we make use of the recurrence relation

$$(1 - \mu^2)^{1/2} P_\ell^{|m|}(\mu) = \frac{1}{(2\ell + 1)} \left[P_{\ell-1}^{|m'|}(\mu) - P_{\ell+1}^{|m'|}(\mu) \right].$$

For the case $|m'| < |m|$, i.e., when $|m'| = |m| - 1$, we use the analogous recurrence relation with the roles of m' and m reversed. In either case, we see that only $\ell' = \ell \pm 1$ can give a nonzero integral. Thus the matrix element

$$\langle \varphi_f | x \pm iy | \varphi_i \rangle = 0 \quad \text{unless} \quad \Delta \ell = \pm 1 \quad \text{and} \quad \Delta m = \pm 1.$$

PARITY CONSIDERATIONS AND LAPORTE'S RULE

There exists an easier derivation of the requirement that $\Delta\ell$ must equal an odd integer (in our case, $\Delta\ell = \pm 1$) for allowed dipole transitions. Electromagnetic interactions are symmetric with respect to parity reversals; i.e., the transition probability for a given process cannot depend on whether we consider it or its mirror image. The dipole operator $\mathbf{x}$ is clearly odd with respect to the parity operation $\mathbf{x} \rightarrow -\mathbf{x}$. On the other hand the spherical harmonic $Y_{\ell m}$ has even or odd parity depending on whether ℓ is even or odd. Thus, the matrix element $\langle \varphi_f | \mathbf{x} | \varphi_i \rangle$, with φ_f and φ_i proportional to spherical harmonics, will have an odd integrand that vanishes upon integration over all space unless the initial and final states have opposite parities (a result known in spectroscopy as *Laporte's rule*). This requires $\Delta\ell = \pm$ an odd integer. The maximum amount of angular momentum that can be carried away (orbital or spin) by a single photon in a dipole transition then argues that this odd integer must equal 1.

EFFECTIVE TRANSITION RATES FOR HYDROGEN

From the preceding discussion, we see that dipole selection rules produce nonzero Einstein A's in the hydrogen atom for transitions $n\ell \rightarrow n'\ell'$ only if $\ell' = \ell \pm 1$. The *effective* transition rates $A_{nn'}$ in hydrogen, where the energy levels are degenerate in ℓ (if we ignore spin-orbit coupling; see Chapter 26), were calculated in Chapter 10 assuming a statistical weight $(2\ell+1)$ for the number of independent magnetic sublevels m having the same transition rates (see Problem Set 5). Since there exist n^2 possible values of ℓ and m for given n in the upper state, we have

$$A_{nn'} = \sum_{\ell'=0}^{n'-1} \sum_{\ell=0}^{n-1} \frac{(2\ell+1)}{n^2} A_{n\ell n'\ell'}.$$

For example, the cited value for A_{21} was computed by taking 3/4 of the value of A_{2p1s}, where $A_{2p1s} = 6.2 \times 10^8$ s^{-1} is the spontaneous radiative decay rate for the transition $2p \rightarrow 1s$. (We use here the standard spectroscopic notation s, p, d, f, ... for $\ell = 0$, 1, 2, 3, ...) The $2s \rightarrow 1s$ term (with weight 1/4) contributes nothing, since $A_{2s1s} = 0$ according to the dipole selection rules. Indeed, as we will show shortly, the latter transition is *strictly forbidden*.

The procedure of using effective rates for spontaneous decay from a given n level breaks down if collisions (or some other process) fail to establish a random distribution among the ℓ sublevels of the upper state. In particular, radiative processes acting alone (e.g., a cascade from the continuum) will generally *not* populate all ℓ sublevels equally, since such radiative processes do obey selection rules. In cases where collisions do not dominate the relative population of levels, working with the $A_{n\ell n'\ell'}$'s rather than the $A_{nn'}$'s always amounts to the safer procedure. The problem does not usually arise for non-hydrogenic atoms, since the L-sublevels are then not degenerate even if we ignore spin-orbit coupling (see Chapter 27), and cited decay rates refer to specific transitions.

ELECTRIC QUADRUPOLE TRANSITIONS

What happens if the desired transition has an electric dipole moment vectorially equal to zero? For such a "forbidden" transition, we must investigate higher-order terms in the expansion

$$e^{i\mathbf{k}\cdot\mathbf{x}} = 1 + i\mathbf{k}\cdot\mathbf{x} + \ldots,$$

for substitution into the expression

$$\langle\varphi_f|e^{i\mathbf{k}\cdot\mathbf{x}}\mathbf{p}|\varphi_i\rangle \cdot \mathbf{e}_\alpha.$$

Since by assumption the term 1 vanishes upon integration, we need to consider the quantity

$$i\mathbf{k} \cdot \langle\varphi_f|\mathbf{x}\mathbf{p}|\varphi_i\rangle \cdot \mathbf{e}_\alpha.$$

Our experience in Chapter 15 suggests that we write the tensor $\mathbf{x}\mathbf{p}$ in terms of its symmetric and antisymmetric parts:

$$\mathbf{x}\mathbf{p} = \frac{1}{2}(\mathbf{x}\mathbf{p} + \mathbf{p}\mathbf{x}) + \frac{1}{2}(\mathbf{x}\mathbf{p} - \mathbf{p}\mathbf{x}).$$

Here the combinations $\mathbf{x}\mathbf{p} \pm \mathbf{p}\mathbf{x}$ represent operators. As in Chapter 22, we wish to relate these quantities to other operators for which the bra- and ket-functions $\langle\varphi_f|$ and $|\varphi_i\rangle$ may be eigenfunctions. For this purpose, we note the following identities:

$$\mathbf{x}\mathbf{p} + \mathbf{p}\mathbf{x} = i\frac{m_e}{\hbar}[H_0, \mathbf{x}\mathbf{x}], \tag{24.21}$$

$$\mathbf{k} \cdot (\mathbf{x}\mathbf{p} - \mathbf{p}\mathbf{x}) \cdot \mathbf{e}_\alpha = (\mathbf{k} \times \mathbf{e}_\alpha) \cdot (\mathbf{x} \times \mathbf{p}). \tag{24.22}$$

The proof of the first identity follows most simply by writing out the expressions in their corresponding form in Cartesian tensors. The proof of the second identity follows by noting that the right-hand side can also be expressed as

$$\mathbf{k} \cdot [\mathbf{e}_\alpha \times (\mathbf{x} \times \mathbf{p})] = \mathbf{k} \cdot (\mathbf{x}\mathbf{e}_\alpha \cdot \mathbf{p}) - (\mathbf{e}_\alpha \cdot \mathbf{x})(\mathbf{k} \cdot \mathbf{p}),$$

which equals the left-hand side of equation (24.22), except for a term representing $\mathbf{k}\cdot\mathbf{p}$ operating on $\mathbf{x}\cdot\mathbf{e}_\alpha$. The last operation yields a term proportional to $\mathbf{k} \cdot \mathbf{e}_\alpha$, which equals zero for a transverse electromagnetic wave.

From the term $\mathbf{xx}$ in equation (24.21) we may subtract $(|\mathbf{x}|^2/3)\mathbf{I}$ without affecting the result when we left-dot with $\mathbf{k}$ and right-dot with $\mathbf{e}_\alpha$ because $\mathbf{k} \cdot \mathbf{e}_\alpha = 0$. Hence

$$\mathbf{k} \cdot (\mathbf{xx}) \cdot \mathbf{e}_\alpha = -\frac{1}{3e}\mathbf{k} \cdot \mathbf{Q} \cdot \mathbf{e}_\alpha,$$

where $\mathbf{Q}$ is the *electric quadrupole* operator,

$$\mathbf{Q} \equiv -e(3\mathbf{xx} - |\mathbf{x}|^2\mathbf{I}), \tag{24.23}$$

and $-e$ is the charge of the electron. Clearly, $\mathbf{Q}$ is even under parity reversal, $\mathbf{x} \to -\mathbf{x}$; so Laporte's rule for electric quadrupole transitions states that the initial and final states must have the same parity.

The derivation of the electric quadrupole selection rules begins with the expression of $\mathbf{Q}$ in terms of spherical harmonics:

$$Q_{33} = -e(2z^2 - x^2 - y^2) = (-er^2)2(4\pi/5)^{1/2}Y_{20}, \tag{24.24}$$

$$Q_{31} \pm iQ_{23} = -e3z(x \pm iy) = (\pm er^2)3(8\pi/15)^{12}Y_{2,\pm 1}, \tag{24.25}$$

$$Q_{11} \pm 2iQ_{12} - Q_{22} = -e3(x \pm iy)^2 = (-er^2)12(2\pi/15)^{1/2}Y_{2,\pm 2}. \tag{24.26}$$

These five linearly independent relations suffice to specify the traceless symmetric tensor $\mathbf{Q}$. In particular, by arguments similar to those for permitted dipole transitions, we see that electric quadrupole radiation can connect quantum states with $\Delta\ell = 0$ and ± 2, with $\Delta m = 0$, ± 1, ± 2.

In order actually to calculate the values of the transition matrix elements for quantum multipole radiation, we need the following useful formula relating the integral of three Y's:

$$\oint Y_{\ell_3 m_3}^* Y_{\ell_2 m_2} Y_{\ell_1 m_1} d\Omega$$

$$= \left[\frac{(2\ell_1 + 1)(2\ell_2 + 1)}{4\pi(2\ell_3 + 1)}\right]^{1/2} \langle \ell_1 \ell_2 00 | \ell_1 \ell_2 \ell_3 0 \rangle \langle \ell_1 \ell_2 m_1 m_2 | \ell_1 \ell_2 \ell_3 m_3 \rangle.$$

$$\tag{24.27}$$

The symbols inside the angular brackets denote quantities called *Clebsch-Gordon coefficients*. (For tabulations, see, e.g., Condon and Shortly 1967.) Such coefficients frequently arise, explicitly or implicitly, in problems involving the addition of angular momenta (see Chapters 26 and 27 for more explicit discussions).

MAGNETIC DIPOLE TRANSITIONS

Consider now the antisymmetric part of $\mathbf{xp}$, which we have shown to be related to the angular momentum operator $\mathbf{L} = \mathbf{x} \times \mathbf{p}$. When we multiply $\mathbf{L}$ by $-e/2m_e c$, we get the orbital part of the *magnetic moment* operator:

$$\mathbf{M} \equiv -\frac{e}{2m_e c}\mathbf{L}. \qquad (24.28)$$

The selection rules for the nonvanishing of the orbital part of the magnetic dipole moment follow straightforwardly from the application of equations (24.11) and (24.14). The allowable values are $\Delta m = 0, \pm 1$, and $\Delta \ell = 0$. Moreover, since the magnetic moment operator does not involve the radial coordinate r, the remaining integral

$$\int_0^\infty \mathcal{R}_{n'\ell'}\mathcal{R}_{n\ell}\,dr,$$

for $\ell' = \ell$, will yield zero unless $n' = n$. This additional qualification makes the $\Delta \ell = 0$ case for magnetic dipole transitions (e.g., fine-structure or hyperfine transitions) somewhat different from the case for electric quadrupole transitions.

The constraint $\Delta \ell = 0$ for magnetic dipole transitions arises from the fact that the angular momentum operator has *even* parity (giving it the property of a *pseudovector*). Thus Laporte's rule demands that $\Delta \ell$ be restricted to even values. However, dipole radiation (either electric or magnetic) can carry off a maximum r.m.s. amount of angular momentum equal to 1 unit; thus $\Delta \ell = 0$ for magnetic dipole radiation, whereas $\Delta \ell = \pm 1$ for electric dipole radiation. With these examples, you should be able to generalize the results to arbitrary multipole order.

Finally, we notice that the selection rules (including parity) for one multipole transition largely preclude the satisfaction of the selection rules for another. Two well-defined quantum states may be radiatively connected by an electric dipole transition, or a magnetic dipole transition, or an electric quadrupole transition, etc., but only rarely by any two or more simultaneously. (As a counter-example, because of departures from pure LS coupling—see Chapter 27—quantum states may actually be linear combinations of the simple "pure" states. Transitions between them may then mix electric quadrupole and magnetic dipole radiation, the most famous such example being the forbidden [O III] lines originally investigated by Bowen.)

Quantum multipole radiation therefore differs from classical multipole radiation in that each member of the former represents a nearly distinct physical process (because the connected quantum states constitute distinct physical entities), rather than successive mathematical approximations of a single process (radiation by an accelerated charge). In other words, we could have derived our results above by starting, not with a systematic expansion of equation (22.4), but with a perturbational Hamiltonian,

$$H_{\mathrm{d}} = -\mathbf{E} \cdot \mathbf{d}, \qquad (24.29)$$

for electric dipole transitions; with

$$H_{\mathrm{M}} = -\mathbf{B} \cdot \mathbf{M} \qquad (24.30)$$

for magnetic dipole transitions; with

$$H_{\mathrm{Q}} = -\frac{1}{6}\frac{\partial \mathbf{E}}{\partial \mathbf{x}} : \mathbf{Q} \qquad (24.31)$$

for electric quadrupole transitions; etc. These more general forms prove especially important to adopt when we consider atomic transitions involving more than one electron, or molecular transitions involving degrees of freedom (e.g., vibrations and rotations of the nuclei) other than electronic.

In the formulae (24.29)–(24.31), $\mathbf{E}$ and $\mathbf{B}$ represent radiation electric and magnetic fields, with polarization and Fourier components given by $\mathbf{E}_\alpha(\mathbf{k}) = ika_\alpha(\mathbf{k})\mathbf{e}_\alpha$ and $\mathbf{B}_\alpha(\mathbf{k}) = ika_\alpha(\mathbf{k})(\hat{\mathbf{k}} \times \mathbf{e}_\alpha)$ (refer back to Chapter 22). When we apply these forms to electronic transitions in atoms, we readily see why, in comparison to H_{d}, H_{M} is typically of order $|\mathbf{M}|/|\mathbf{d}| \sim v/c$, whereas H_{Q} is typically of order $|\mathbf{x} \cdot \partial\mathbf{E}/\partial\mathbf{x}|/|\mathbf{E}| \sim kx$, where x equals the typical atomic dimension.

ELECTRON-SPIN MAGNETIC MOMENT

In quantum-mechanical systems, orbital angular momentum comes quantized in integer units of $\hbar$; spin angular momentum, in half-integer units of $\hbar$. Thus we intuitively expect that both will contribute comparably to atomic magnetic moments. To equation (24.28), then, one should also add the spin magnetic moment,

$$\mathbf{m} = -\frac{ge}{2m_e c}\mathbf{s}, \qquad (24.32)$$

where $\mathbf{s}$ is the spin angular momentum of the electron and g is the (infamous) electron g-factor.

Classically, g should equal 1, but experimentally, it is found to have a value close to 2. Dirac's theory of the electron (to be discussed in Chapter 25) gives a value of exactly 2, but correcting for vacuum fluctuations yields a perturbation expansion,

$$g = 2 + \left(\frac{\alpha}{\pi}\right) - 0.65696\left(\frac{\alpha}{\pi}\right)^2 + 2.98\left(\frac{\alpha}{\pi}\right)^3 + \cdots, \qquad (24.33)$$

where α is the fine-structure constant $e^2/\hbar c$ calculated with the renormal-ized charge of the electron (the value of e measured by experiments carried out at low energies that do not penetrate the layers of vacuum polarization that shield an electron's large, possibly infinite, "bare" charge).

The first term 2 in equation (24.33) comes from Dirac's theory; the second, proportional to one power of α, from Schwinger's inclusion of all the ways in which an electron can interact with one virtual photon; the third, proportional to two powers of α, from calculations (which took two independent teams more than three years to make) of all the ways in which the electron can interact with two virtual photons or a virtual electron-positron pair; the fourth, proportional to three powers of α, from computer-assisted calculations that took more than twenty years to sort out all the ways (about ten thousand Feynman diagrams) in which the electron can interact with three virtual photons or one virtual photon and an electron-positron pair. Supercomputers have been recruited to help work out the coefficient of the next term, proportional to $(\alpha/\pi)^4$.

Up to the known terms, the predicted theoretical value for g equals 2.0023193048, with an uncertainty in the last digit that arises mostly from remaining imprecision in our knowledge of α. Except for this uncertainty, the theoretical and experimental numbers are in precise agreement; quantum electrodynamics, as presently formulated, is thus the most accurate theory known to humanity. Nevertheless, the same discussion shows that under all circumstances likely to be encountered in astrophysics, corrections due to vacuum fluctuations are of little practical importance, except possibly for the earliest instants of the big bang.

STRICTLY FORBIDDEN TRANSITIONS

As our last topic for this chapter we consider the problem of *strictly forbidden transitions*—radiative transitions for which the matrix element vanishes to all perturbational order in an expansion in multipole moments (an expansion in increasing powers of α^2). To prove such a property, we begin, not with the infinite series, but with the original result that the total transition probability is proportional to the absolute square of the exact matrix element:

$$\langle \varphi_f | e^{i\mathbf{k}\cdot\mathbf{x}} \mathbf{e}_\alpha(\hat{\mathbf{k}}) \cdot \mathbf{p} | \varphi_i \rangle. \tag{24.34}$$

The cautious reader might object that such an interpretation already contains the iteration approximation of time-dependent perturbation theory (review Chapter 22); we deal with this objection directly in the section after next, where we find that a systematic expansion recovers the simple treatment as the correct answer for any process that involves the emission or absorption of a single photon.

As an example, let us consider the simplest quantum process in the universe—also, arguably, the most important—the radiative decay by the emission of a Lyman-α photon of a hydrogen atom in its first excited electronic level to the ground state. As we have already mentioned, the transition from the $2p$ state to the $1s$ state has a large associated dipole moment and an appropriately large Einstein A value. Since $\Delta\ell = 0$ and $\Delta m = 0$ for the $2s$ to $1s$ transition, however, the electric dipole moment vanishes. In fact, all multipole moments vanish (if we don't consider spin-dependent perturbations); so, by the criterion described above, the transition $2s \to 1s$ turns out to be strictly forbidden!

To see the basis for this claim, consider the situation when both the initial and the final states have *spherically symmetric* wave functions. (Recall that the angular function Y_{00} associated with any s-state is a constant.) Classically, we know that a spherically symmetric matter distribution, even if it changes dimensions but preserves its total electric charge, cannot radiate any electromagnetic waves. To see that the same holds quantum-mechanically, examine the behavior of the exact matrix element (24.34). Call the direction of the polarization vector $\mathbf{e}_\alpha(\hat{\mathbf{k}})$ the x-direction (calling it y or z would not change the basic argument); then

$$\mathbf{e}_\alpha(\hat{\mathbf{k}}) \cdot \mathbf{p} = -i\hbar \frac{\partial}{\partial x}.$$

Since radiation fields are transverse waves, $e^{i\mathbf{k}\cdot\mathbf{x}}$ varies solely with y and z; i.e., this term is an *even* function of x. But φ_f and φ_i are functions of only the combinations $x^2 + y^2 + z^2$ by the assumption of spherical symmetry; therefore they are also even functions of x. Partial differentiation with respect to x changes even functions to odd functions; hence the integrand associated with the exact matrix element (24.34) is odd in x and will vanish identically when integrated over all space. (Q.E.D.)

TWO-PHOTON PROCESS

How, then, does a hydrogen atom that finds itself in the $2s$ state ever de-excite to the ground level? One possible route involves a superelastic collision with another particle. The collision partner can carry off a different ratio of energy and angular momentum than a photon, and the transition then will not have to satisfy the stringent constraints that prevented radiative decay. A more probable alternative, requiring virtually no energy, involves an angular-momentum shifting collision that takes the hydrogen atom from $2s$ to $2p$, followed by the allowed radiative decay $2p \to 1s$. This process turns out to have a reasonably high rate for collisions with either thermal protons or thermal electrons at a temperature of 10^4 K.

An even more fascinating path, actually attained under interstellar conditions (in H$_{II}$ regions and planetary nebulae, for example), involves the spontaneous emission of *two* photons with combined frequencies (and energies) equal to that of a single Lyman-α photon:

$$\nu + \nu' = \nu_{21}. \tag{24.35}$$

The number of photons emitted per unit frequency interval has a continuous probabilistic distribution of frequencies [subject to the constraint (24.35)], symmetrically configured about the midpoint, $\nu = \nu_{21}/2$ (wavelength $=$ 2431 Å), and falling off on both sides toward $\nu = 0$ and $\nu = \nu_{21}$. The emission occurs with the help of the quadratic term H_2 [see equation (21.13)] acting as (part of) the interaction Hamiltonian (see below). The process has an Einstein A equal to 8.2 s^{-1}, which should be compared with the value $A_{2p1s} = 6.2 \times 10^8$ s^{-1} associated with the $2p \to 1s$ transition. Although small by the standards of radiative decay, the two-photon process dominates the collisional routes for hydrogen atom de-excitation in H$_{II}$ regions when the electron density $\lesssim 10^4$ cm^{-3}.

When we consider two-photon processes that involve H_2, we should also re-evaluate contributions from H_1 that enter as higher-order iterations to the time-dependent perturbation solution considered in Chapter 24. Thus motivated, we begin our discussion of the problem by first developing the notion of different *representations* for making generic calculations of this kind (quantum interactions involving two or more photons).

SCHRÖDINGER, HEISENBERG, AND INTERACTION REPRESENTATIONS

If we had a problem involving only a time-independent Hamiltonian $H = H_0$, the *Schrödinger representation* puts the time dependence of the problem entirely into the wave function ψ; so we are left to solve the evolution of an equation having the form

$$i\hbar \frac{\partial \psi}{\partial t} = H_0 \psi. \tag{24.36}$$

An equivalent way to formulate quantum mechanics, however, adopts the *Heisenberg representation*, which makes the wave function *time-independent* and puts all of the time dependence into the operators of the system (except for H_0 and operators which commute with it). To see how to do this, notice that if H_0 were a constant instead of an operator, we could directly integrate equation (24.36) in time as

$$\psi(\mathbf{x}, t) = e^{-i\hbar^{-1}tH_0}\psi(\mathbf{x}, 0), \tag{24.37}$$

where we have required ψ to satisfy the initial condition $\psi(\mathbf{x}, 0)$ at $t = 0$. In fact, equation (24.37) yields a perfectly good solution even when H_0 is an operator (in space) instead of a constant [as can be seen by substituting equation (24.37) into equation (24.36)]. By the expression $\exp(-i\hbar^{-1}tH_0)$, we mean the operator that we get by expanding the exponential:

$$e^{-i\hbar^{-1}tH_0} \equiv 1 - i\hbar^{-1}tH_0 + \cdots.$$

Notice that the inverse operator of $\exp(-i\hbar^{-1}tH_0)$ is $\exp(i\hbar^{-1}tH_0)$.

We obtain the Heisenberg representation by regarding the time-independent wave function $\varphi(\mathbf{x}) \equiv \psi(\mathbf{x}, 0)$, as obtained from $\psi(\mathbf{x}, t)$ in the Schrödinger representation by a *unitary* transformation:

$$\varphi(\mathbf{x}) = e^{i\hbar^{-1}tH_0}\psi(\mathbf{x}, t). \tag{24.38}$$

In this transformation, any operator P in the Schrödinger representation transforms as

$$\mathcal{P} = e^{i\hbar^{-1}tH_0} P e^{-i\hbar^{-1}tH_0}.$$

Notice that setting t equal to zero identifies P as $\mathcal{P}$ evaluated at time $t = 0$. In other words, in the Heisenberg representation, the time evolution of all operators satisfies the equation

$$\mathcal{P}(t) = e^{i\hbar^{-1}tH_0}\mathcal{P}(0)e^{-i\hbar^{-1}tH_0}, \tag{24.39}$$

which gives, upon differentiation once in time, the "equation of motion":

$$\frac{\partial \mathcal{P}}{\partial t} = i\hbar^{-1}(H_0\mathcal{P} - \mathcal{P}H_0) \equiv i\hbar^{-1}[H_0, \mathcal{P}]. \tag{24.40}$$

Equivalently, we could have first obtained the Heisenberg version of quantum mechanics from a Hamiltonian formulation of classical mechanics by identifying (within a factor of $i\hbar^{-1}$) Poisson brackets with commutator brackets (see Goldstein 1959). From the Heisenberg representation, we could then use a unitary transformation to obtain the Schrödinger representation.

Many representations other than those of Schrödinger and Heisenberg are also possible; in particular, the *interaction representation* yields an intermediate approach to getting the time evolution of a system when the governing Hamiltonian H *does* vary with t. Often we have the generic situation where H consists of the sum of a term H_0 that does not depend on time and another term H_1 that does. In this situation, we are motivated to write the Schrödinger equation in the form

$$i\hbar\frac{\partial \psi}{\partial t} - H_0\psi = H_1\psi, \tag{24.41}$$

and to solve it formally as if the right-hand side were a known inhomogeneous term rather than one that also involves the unknown wave function
ψ.

When H_0 does not depend on time, the left-hand side of equation (24.41)
has the integrating factor $e^{-i\hbar^{-1}tH_0}$; thus we are again motivated to introduce the transformation (24.38), except we must now allow φ to have a
dependence on t. Taking the inverse transformation now yields

$$\psi(\mathbf{x}, t) \equiv e^{-i\hbar^{-1}tH_0}\varphi(\mathbf{x}, t). \tag{24.42}$$

If we substitute this definition into equation (24.41) and left-multiply by
$\exp(i\hbar^{-1}tH_0)$, we get

$$i\hbar\frac{\partial\varphi}{\partial t} = \mathcal{H}_1\varphi, \tag{24.43}$$

where $\mathcal{H}_1$ is the (time-dependent) perturbation Hamiltonian in the interaction representation:

$$\mathcal{H}_1 \equiv e^{i\hbar^{-1}tH_0} H_1 e^{-i\hbar^{-1}tH_0}. \tag{24.44}$$

Equation (24.43) has the formal solution:

$$\varphi(t) = \varphi(0) - i\hbar^{-1}\int_0^t dt'\, \mathcal{H}_1(t')\varphi(t'), \tag{24.45}$$

where for notational simplicity we have suppressed the spatial dependences
of $\mathcal{H}_1$ and φ on $\mathbf{x}$. We may regard equation (24.45) as representing another
unitary transformation which takes the initial function $\varphi(0)$ to $\varphi(t)$:

$$\varphi(t) = U(t)\varphi(0). \tag{24.46}$$

By inspection of equation (24.45), we see that the operator $U(t)$ satisfies
the integral equation:

$$U(t) = 1 - i\hbar^{-1}\int_0^t dt'\mathcal{H}_1(t')U(t'), \tag{24.47}$$

which equals the solution of the differential equation,

$$i\hbar\frac{\partial U}{\partial t} = \mathcal{H}_1 U,$$

subject to the initial condition $U(0) = 1$.

So far we have made no approximations, and our results are as exact as
the original equation (24.41). If we substitute the formal solution (24.47)
into the integrand on the right-hand side for itself, we get

$$U(t) = 1 - i\hbar^{-1}\int_0^t dt_1\mathcal{H}_1(t_1)\left[1 - i\hbar^{-1}\int_0^{t_1} dt_2\mathcal{H}_1(t_2)U(t_2)\right].$$

Repeated iterations in this manner (a standard procedure in the solution of integral equations) yields

$$U(t) = 1 + U_1(t) + U_2(t) + \cdots, \tag{24.48}$$

which has an n-th term in the expansion of

$$U_n(t) = (-i\hbar^{-1})^n \int_0^t dt_1 \int_0^{t_1} dt_2 \ldots \int_0^{t_{n-1}} dt_n \mathcal{H}_1(t_1)\mathcal{H}_1(t_2) \ldots \mathcal{H}_1(t_n). \tag{24.49}$$

Physicists call the value of $U(t)$ as $t \to \infty$, the "S matrix," because of its central role for problems involving quantum scattering.

The transition probability from state i (with $\varphi[0] = \varphi_i$) to some state f equals

$$|\langle \varphi_f | U(t) | \varphi_i \rangle|^2 \equiv |U_{fi}(t)|^2.$$

The time-averaged rate of making the transition $i \to f$ (with $f \neq i$) is given by the expression

$$w = \frac{1}{t} |\langle \varphi_f | U_1 + U_2 + \cdots | \varphi_i \rangle|^2. \tag{24.50}$$

Notice that the U's need to be added before we square.

When $\mathcal{H}_1 \propto \mathbf{A}$, the radiation vector potential, U_n is proportional to n powers of $\mathbf{A}$ and to n powers of $\hbar^{-1}$. The latter powers get translated ultimately into powers of $\alpha = e^2/\hbar c$ if we are dealing with electromagnetic interactions. Thus the iterative procedure generally yields a series expansion whose terms rapidly diminish with increasing n. In contrast with this situation, the difficulty with carrying out calculations in quantum chromodynamics (the theory of strong interactions) arises because the analogous coupling constant is no longer small compared to unity.

Because of the additional factors of $\mathbf{A}$, iterations of higher order than linear correspond to interactions with multiple photons. As a consequence, if we wish to consider the two-photon processes contained in the term H_2, we must, for consistency, also contemplate carrying the iteration involving factors of $\mathcal{H}_1 \propto H_1$ at least to second order. Our identification of H_1 as leading to *one-photon* processes contains, therefore, an approximation. If we are willing to study events having vanishingly small probability, then an atom could, in principle, emit or absorb an *arbitrary* number of photons.

NATURE OF SUCCESSIVE APPROXIMATIONS

To linear order, we have

$$\langle \varphi_f | U_1 | \varphi_i \rangle = -i\hbar^{-1} \int_0^t \langle \varphi_f | e^{i\hbar^{-1}t_1 H_0} H_1 e^{-i\hbar^{-1}t_1 H_0} | \varphi_i \rangle \, dt_1.$$

But φ is the time-independent part of the usual Schrödinger representation; therefore,

$$\langle \varphi_f | e^{i\hbar^{-1}t_1 H_0} = e^{iE_f t_1/\hbar} \langle \varphi_f | \quad \text{and} \quad e^{-i\hbar^{-1}t_1 H_0} | \varphi_i \rangle = e^{-iE_i t_1/\hbar} | \varphi_i \rangle.$$

As a consequence, we may write

$$\langle \varphi_f | U_1 | \varphi_f \rangle = -i\hbar^{-1} \int_0^t \langle \varphi_f | H_1 | \varphi_i \rangle e^{i\omega_{fi} t_1} \, dt_1,$$

which recovers Dirac's result (22.29). (The method outlined here also originated with Dirac.)

If $\langle \varphi_f | H_1 | \varphi_i \rangle$ vanishes or is very small, we need to go to quadratic order:

$$\langle \varphi_f | U_2 | \varphi_i \rangle = -\hbar^{-2} \int_0^t dt_1 \int_0^{t_1} dt_2 \langle \varphi_f | e^{i\hbar^{-1}t_1 H_0} H_1(t_1) e^{-i\hbar^{-1}t_1 H_0} e^{i\hbar^{-1}t_2 H_0}$$

$$\times H_1(t_2) e^{-i\hbar^{-1}t_2 H_0} | \varphi_i \rangle.$$

We easily carry through the operations at the extremes (contiguous to a wave function). To decompose the operations in the middle, insert an identity operator [see equation (23.15)]:

$$I = \sum_n |\varphi_n\rangle\langle\varphi_n|$$

between the two middle exponential operators. Carrying out the indicated operations, we now have

$$\langle \varphi_f | U_2 | \varphi_i \rangle = -\hbar^{-2} \sum_n \int_0^t dt_1 \langle \varphi_f | H_1(t_1) | \varphi_n \rangle e^{i\omega_{fn} t_1}$$

$$\times \int_0^{t_1} dt_2 \langle \varphi_n | H_1(t_2) | \varphi_i \rangle e^{i\omega_{ni} t_2}. \qquad (24.51)$$

Applied to our original problem—the radiative transition $2s \rightarrow 1s$ in atomic hydrogen—equation (24.51) turns out to have the following physical interpretation. The sum over n represents a sum over *virtual* transitions to all intermediate states n for which the associated single-photon matrix elements $\langle \varphi_n | H_1 | \varphi_i \rangle$ and $\langle \varphi_f | H_1 | \varphi_n \rangle$ (taken in succession) do not vanish. These involve transitions to states n (including the continuum) that do not possess the spherical symmetry of states i and f. Although permitted by selection rules, these single-photon transitions are *virtual* because the atom does not possess enough energy to go permanently to the intermediate states. However, as long as the interval of time between the "violation" $i \rightarrow n$ and the "correction" $n \rightarrow f$ is small enough, there exists enough overlap because of the uncertainty principle to entertain temporary excursions to the state n. As t goes to infinity, all the realized outcomes, with the atom going from $2s$ to $1s$ by the emission of two photons with combined frequencies $\nu + \nu'$ equal to ν_{21}, do satisfy the conservation of energy. (For details, see L. Spitzer and J. Greenstein, 1951, *Ap. J.*, **114**, 407, and the references cited therein.)

25

Dirac's Equation and the Effects of Electron Spin

References: Dirac, *Quantum Mechanics*, Chapter XI; Sakurai, *Advanced Quantum Mechanics*, pp. 78–91, 117–119; Schiff, *Quantum Mechanics*, Chapter 12.

In this chapter we consider the problem of how to incorporate the effects of electron spin. At its most basic level, spin arises most naturally as the product of the marriage of quantum mechanics and special relativity. We begin our discussion therefore with Dirac's development of this subject.

RELATIVISTIC WAVE EQUATIONS AND PROBABILITY CONSERVATION

Consider the straightforward expression for a relativistically correct Hamiltonian, equation (21.5):

$$H = [|c\mathbf{P}|^2 + m^2 c^4]^{1/2} + q\phi \tag{25.1}$$

where $\mathbf{P}$ is the particle momentum, expressible in terms of the canonical momentum $\mathbf{p}$ and the electromagnetic vector potential $\mathbf{A}$ as

$$\mathbf{P} = \mathbf{p} - \frac{q}{c}\mathbf{A}. \tag{25.2}$$

In 1928, Dirac recognized that a fundamental difficulty arises if we try to use equation (25.1) as an operator in the relativistic generalization of Schrödinger's equation:

$$H\psi = i\hbar \frac{\partial \psi}{\partial t}. \tag{25.3}$$

To remove the difficulty caused by the square root, we could—in the absence of the electromagnetic potentials—first square both the H and the $i\hbar\partial/\partial t$ operators before applying them to ψ. The resulting second-order equation, called today the Klein-Gordon equation, reads

$$\left[\left(\nabla^2 - \frac{1}{c^2}\frac{\partial^2}{\partial t^2}\right) - \left(\frac{mc}{\hbar}\right)^2\right]\psi = 0. \tag{25.4}$$

Notice that the term in the first parenthesis equals the d'Alembertian, whereas the term in the second parenthesis equals the Compton wavenumber of a particle with mass m.

Schrödinger himself actually wrote down equation (25.4) as the straightforward relativistic generalization of his equation, but rejected it because he could not interpret ψ as a wave function. The reason is as follows. In nonrelativistic quantum mechanics, Schrödinger's equation for a free particle reads

$$i\hbar\frac{\partial\psi}{\partial t} + \frac{\hbar^2}{2m}\nabla^2\psi = 0. \tag{25.5}$$

(We could include the electromagnetic potentials in the following calculations without changing the essential conclusions, but the manipulations would be somewhat more involved.) If we left-multiply equation (25.5) by ψ^* and add the result to the complex conjugate of equation (25.5) left-multiplied by ψ, we get, after dividing by $i\hbar$,

$$\frac{\partial\rho}{\partial t} + \boldsymbol{\nabla}\cdot\mathbf{j} = 0, \tag{25.6}$$

where we have introduced the symbols

$$\rho \equiv |\psi|^2, \qquad \mathbf{j} \equiv -\frac{i\hbar}{2m}(\psi^*\boldsymbol{\nabla}\psi - \psi\boldsymbol{\nabla}\psi^*). \tag{25.7}$$

To derive equation (25.6), we have made use of the identity

$$\boldsymbol{\nabla}\cdot\mathbf{j} = -\frac{i\hbar}{2m}(\psi^*\nabla^2\psi - \psi\nabla^2\psi^*),$$

since $(\boldsymbol{\nabla}\psi^*)\cdot(\boldsymbol{\nabla}\psi) - (\boldsymbol{\nabla}\psi)\cdot(\boldsymbol{\nabla}\psi^*) = 0$. In the usual probabilistic interpretation for the wave equation (due to Born), ρ represents the probability density; and since equation (25.6) has the form of a conservation relation, $\mathbf{j}$ must then represent the probability current. The latter identification makes intuitive sense when we use $\mathbf{p} = -i\hbar\boldsymbol{\nabla}$ to write $\mathbf{j}$ as

$$\mathbf{j} = \frac{1}{2}(\psi^*\mathbf{v}\psi + \text{c.c.}),$$

where $\mathbf{v} \equiv \mathbf{p}/m$ represents the velocity operator and c.c. denotes the complex conjugation of $\psi^*\mathbf{v}\psi$ needed to make $\mathbf{j}$ real. Classically, we might have expected the identification of $\mathbf{j}$ with $\rho\mathbf{v}$ [see equation (9.8) when we deal with mass conservation rather than probability conservation]. However, in quantum mechanics, $\mathbf{v} = \mathbf{p}/m$ becomes an operator; so $\rho\mathbf{v} \neq \mathbf{v}\rho$, and $\mathbf{j}$ takes the more symmetric form given by equation (25.7).

The difficulty with equation (25.4) now becomes clear: instead of a single derivative in time, it contains two. Although we can still derive a conservation relation of the form (25.6) for equation (25.4), the corresponding expressions for ρ and $\mathbf{j}$ read

$$\rho = i\hbar \left(\psi^* \frac{\partial \psi}{\partial t} - \psi \frac{\partial \psi^*}{\partial t} \right), \qquad \mathbf{j} = -\frac{i\hbar}{2m}(\psi^* \boldsymbol{\nabla} \psi - \psi \boldsymbol{\nabla} \psi^*).$$

Unlike the nonrelativistic case, where $\rho = |\psi|^2$, the above expression for ρ is not positive definite; i.e., it cannot be interpreted as a *probability* density. Thus ψ in the Klein-Gordon equation (25.4) cannot represent a wave *function* of a single particle in the familiar and convenient sense to which we have become accustomed for the Schrödinger equation. As it turns out, quantum field theory *does* allocate a physical place for the Klein-Gordon equation, but ψ must take the interpretation of a *scalar field*. In other words, equation (25.4) constitutes a field equation for a scalar quantity $\psi(\mathbf{x}, t)$ in the same sense that equation (21.8) constitutes a field equation for the vector potential $\mathbf{A}(\mathbf{x}, t)$. Upon quantization ("second quantization"), the former yields a viable description of a collection for gauge bosons of mass m and spin 0, just as the latter does for a collection of gauge bosons of mass 0 and spin 1 (photons).

The birth of quantum field theory with Dirac probably could not have taken place without the prior appearance of nonrelativistic quantum mechanics. In hindsight, then, when we consider the extraordinary successes of nonrelativistic quantum mechanics, we must pay special tribute to Schrödinger's cleverness at arriving at a *wave* equation that uses only one differentiation in time. After all, the classical analogs of partial differential equations that have one derivative in time and two in space correspond to *diffusion* equations (see, e.g., Volume II), not to wave equations. The crucial trick to getting oscillatory behavior in space *and* time from a governing PDE that uses only one derivative in the latter is to introduce the i in front of $\partial/\partial t$—i.e., to make the equation *complex*. Then, if we were to write $\psi = \psi_R + i\psi_I$ and decompose Schrödinger's equation into its real and imaginary parts, we could eliminate ψ_I in favor of ψ_R (or vice versa) and find that the governing equation for ψ_R does indeed contain two derivatives in time (and four in space). This decomposition and elimination, of course, would mar much of the elegance of the original Schrödinger formulation, which explains why most physicists think of quantum mechanics as intrinsically requiring complex numbers for its natural expression.

DIRAC'S POSTULATE AND SPIN MATRICES

Realizing the difficulty of a probabilistic interpretation for ψ faced by the Klein-Gordon equation, Dirac looked elsewhere for a relativistically correct description for the behavior of an electron. To retain the linearity of the time derivative in the Hamiltonian formulation of quantum mechanics, equation (25.3), yet preserve the symmetry between space and time required by special relativity, Dirac proposed to write the operator H in equation (25.1) as a linear function of $\mathbf{P}$:

$$H = \mathbf{a} \cdot \mathbf{P}c + bmc^2 + q\phi, \tag{25.8}$$

where the vector $\mathbf{a}$ and scalar b are constants independent of $\mathbf{x}$ and t. [The quantities $\mathbf{a}$ and b are usually denoted as α and β, but we reserve these symbols for other purposes. Other authors use sign conventions different from those in equation (25.8); we follow the sign convention used by Dirac and Sakurai, but not by Schiff.]

If we equate equations (25.1) and (25.8), we may cancel the term $q\phi$, square both sides, and obtain the requirements

$$a_x a_x = a_y a_y = a_z a_z = 1 = bb,$$
$$a_x a_y + a_y a_x = a_y a_z + a_z a_y = a_z a_x + a_x a_z = 0$$
$$= a_x b + b a_x = a_y b + b a_y = a_z b + b a_z.$$

The above relations cannot be satisfied if a_x, a_y, a_z, and b are ordinary numbers, but they can hold if we interpret them as 4×4 matrices made up of 2×2 blocks:

$$a_x = \begin{pmatrix} \bigcirc & \sigma_x \\ \sigma_x & \bigcirc \end{pmatrix}, \tag{25.9}$$

$$a_y = \begin{pmatrix} \bigcirc & \sigma_y \\ \sigma_y & \bigcirc \end{pmatrix}, \tag{25.10}$$

$$a_z = \begin{pmatrix} \bigcirc & \sigma_z \\ \sigma_z & \bigcirc \end{pmatrix}, \tag{25.11}$$

$$b = \begin{pmatrix} I & \bigcirc \\ \bigcirc & -I \end{pmatrix}. \tag{25.12}$$

In these equations, $\bigcirc$ and I represent the 2×2 null and unit matrices,

$$\bigcirc \equiv \begin{pmatrix} 0 & 0 \\ 0 & 0 \end{pmatrix}, \qquad I \equiv \begin{pmatrix} 1 & 0 \\ 0 & 1 \end{pmatrix}, \tag{25.13}$$

and σ_x, σ_y, and σ_z form the 2×2 spin matrices that Pauli introduced (see next section) to describe spin $1/2$ in a nonrelativistic context:

$$\sigma_x = \begin{pmatrix} 0 & 1 \\ 1 & 0 \end{pmatrix}, \tag{25.14}$$

$$\sigma_y = \begin{pmatrix} 0 & -i \\ i & 0 \end{pmatrix}, \qquad (25.15)$$

$$\sigma_z = \begin{pmatrix} 1 & 0 \\ 0 & -1 \end{pmatrix}. \qquad (25.16)$$

In what follows, except to avoid confusion, we do not distinguish between I and 1, nor between $\bigcirc$ and 0. Moreover, if we add operators of apparently different matrix dimensions, say, $p_x^2 + i\sigma_z p_x p_y$, we shall implicitly understand the term p_x^2 to contain an additional factor of I.

With the above understanding, we leave as an exercise for the reader the verification that the matrices a_x, a_y, a_z, and b satisfy the desired relationships. The following formulae are useful for simplifying the proof:

$$\sigma_x \sigma_x = \sigma_y \sigma_y = \sigma_z \sigma_z = 1,$$

$$\sigma_x \sigma_y = i\sigma_z, \qquad \sigma_y \sigma_z = i\sigma_x, \qquad \sigma_z \sigma_x = i\sigma_y;$$

$$\sigma_y \sigma_x = -i\sigma_z, \qquad \sigma_z \sigma_y = -i\sigma_x, \qquad \sigma_x \sigma_z = -i\sigma_y.$$

Shorthand Cartesian-tensor notation summarizing the above are the commutation and anti-commutation relations:

$$\sigma_i \sigma_k - \sigma_k \sigma_i = 2i\epsilon_{ikm}\sigma_m, \qquad (25.17)$$

$$\sigma_i \sigma_k + \sigma_k \sigma_i = 2\delta_{ik}, \qquad (25.18)$$

where δ_{ik} is the Kronecker delta and ϵ_{ikm} is the Levi-Cevita tensor in 3-space (equaling $+1$ if ikm form an even permutation of 1, 2, 3; equaling -1 if ikm form an odd permutation of 1, 2, 3; and equaling 0 if any two or more of the ikm are the same). In equations (25.17) and (25.18), and in what follows, we use the Einstein summation convention concerning repeated indices, with the summation on Latin indices occurring over the 1, 2, 3, space directions.

CONNECTION WITH ROTATION AND ANGULAR MOMENTUM

In this section, we digress a bit to expand on the meaning of electron spin in the nonrelativistic context of two-component wave functions (see the expansion of the Dirac equation discussed later in this chapter). We begin by noting that we can write equation (25.17) in the vectorial form:

$$\boldsymbol{\sigma} \times \boldsymbol{\sigma} = 2i\boldsymbol{\sigma}.$$

The cross-product of a vector with itself need not equal zero if that vector represents an *operator*—in this case, on an internal space with only two states. Except for a missing factor $\hbar/2$, the situation has a perfect analogy with the commutation relations among the x, y, and z components of the orbital angular momentum, which we can write as

$$\mathbf{L} \times \mathbf{L} = i\hbar\mathbf{L}.$$

This suggests, as we will make more explicit below and in Chapter 27, that we should identify the quantity $\hbar\boldsymbol{\sigma}/2$ with the *spin angular momentum* $\mathbf{s}$ of the electron:

$$\mathbf{s} = \frac{\hbar}{2}\boldsymbol{\sigma} \quad \text{so that} \quad \mathbf{s} \times \mathbf{s} = i\hbar\mathbf{s}. \tag{25.19}$$

A deep geometric reason underlies the above parallelism for the operators $\mathbf{L}$ and $\mathbf{s}$. If we calculate how any scalar wave function $\psi(\mathbf{x}, t)$ changes when we displace $\mathbf{x}$ by an infinitesimal amount to $\mathbf{x} + d\mathbf{x}$, we get

$$\psi(\mathbf{x} + d\mathbf{x}, t) = \psi(\mathbf{x}, t) + d\mathbf{x} \cdot \boldsymbol{\nabla}\psi(\mathbf{x}, t).$$

If the displacement corresponds to rotation through an infinitesimal angle $d\varphi$ about a unit normal $\hat{\mathbf{n}}$, we get

$$d\mathbf{x} = (d\varphi)\hat{\mathbf{n}} \times \mathbf{x}.$$

(Recall that rotation, classically, produces a velocity $d\mathbf{x}/dt = \boldsymbol{\omega} \times \mathbf{x}$, where $\boldsymbol{\omega} = \hat{\mathbf{n}}d\varphi/dt$ is the angular velocity.) Exchanging the cross and dot products in $[(d\varphi)\hat{\mathbf{n}} \times \mathbf{x}] \cdot \boldsymbol{\nabla}$, we get

$$\psi(\mathbf{x} + d\mathbf{x}, t) = [1 + +(d\varphi)\hat{\mathbf{n}} \cdot (\mathbf{x} \times \boldsymbol{\nabla})]\psi(\mathbf{x}, t) = \left[1 + i\hbar^{-1}(d\varphi)\hat{\mathbf{n}} \cdot \mathbf{L}\right]\psi(\mathbf{x}, t),$$

where we have defined $-i\hbar\mathbf{x} \times \boldsymbol{\nabla}$ as the orbital-angular-momentum operator $\mathbf{L}$.

The above considerations lead us to identify the expression

$$1 + i\hbar^{-1}(d\varphi)\hat{\mathbf{n}} \cdot \mathbf{L}$$

as the generator of infinitesimal rotations. To obtain the operator R_L for a rotation through a finite angle φ, we perform N rotations through infinitesimal angles and take the limit $N \to \infty$, with

$$d\varphi = \varphi/N.$$

Thus

$$R_L = \lim_{N \to \infty}\left(1 + i\hbar^{-1}\frac{\varphi}{N}\hat{\mathbf{n}} \cdot \mathbf{L}\right)^N = \exp\left(i\hbar^{-1}\varphi\hat{\mathbf{n}} \cdot \mathbf{L}\right). \tag{25.20}$$

We may usefully compare the result (25.20) for the generation of advances in angle φ with the result of the previous chapter, where we demonstrated that the operator $\exp(-i\hbar^{-1}tH_0)$ generates advances in time t. In particular, consider the situation when the governing Hamiltonian H_0 is time-independent and commutes with L_z. In this case, the electronic distribution $|\psi|^2$ remains invariant with respect to translations in time t and rotations through any angle $\varphi = \phi$ about the $\hat{\mathbf{n}} = \hat{\mathbf{z}}$ axis, so the wave function ψ in the Schrödinger representation must have the eigenfunction dependences $\exp[i\hbar^{-1}(m\hbar\phi - E_0 t)]$, with E_0 and $m\hbar$ equal to constants, the eigenvalues of H_0 and L_z.

Suppose now that we consider a two-component wave function,

$$\Psi \equiv \begin{bmatrix} \psi_+(\mathbf{x}) \\ \psi_-(\mathbf{x}) \end{bmatrix}, \tag{25.21}$$

where the ratio of the scalar quantities ψ_+ to ψ_- yields, in some sense, the mixture of "spin up" to "spin down." In other words, if we define the two basis spinors

$$\alpha_+ \equiv \begin{pmatrix} 1 \\ 0 \end{pmatrix}, \qquad \alpha_- \equiv \begin{pmatrix} 0 \\ 1 \end{pmatrix}, \tag{25.22}$$

we may write Ψ as

$$\Psi = \psi_+ \alpha_+ + \psi_- \alpha_-,$$

with ψ_+ considered the "projection" of Ψ in the "up" direction, and ψ_- as the "projection" in the "down" direction.

We naturally require the two-component spinor Ψ to correspond to a normalized probability distribution, i.e.,

$$\langle \Psi | \Psi \rangle \equiv \int \Psi^\dagger \Psi \, d^3 x = \int \left(|\psi_+|^2 + |\psi_-|^2 \right) d^3 x = 1.$$

In the above, we have defined $\Psi^\dagger$ to mean the Hermitian conjugate of the spinor (25.21), obtained by converting columns to rows and complex-conjugating the result:

$$\Psi^\dagger = (\psi_+^* \quad \psi_-^*). \tag{25.23}$$

In Dirac's nomenclature, Ψ represents a ket-vector; $\Psi^\dagger$, a bra-vector.

We now ask the question: how does Ψ transform under rotations through an angle $\varphi\hat{\mathbf{n}}$? Without proof (see Merzbacher 1961, Chapter 13), we claim that the generator for rotations involving the spin-dependent part of the problem has the unitary form

$$R_s \equiv \exp\left(i\hbar^{-1}\varphi\hat{\mathbf{n}} \cdot \mathbf{s} \right), \tag{25.24}$$

a 2×2 matrix if we were to expand out the exponential. The total rotation operator $R \equiv R_s R_L$ for a nonrelativistic electron, containing both orbital and spin angular momentum, therefore reads

$$R = \exp\left(i\hbar^{-1}\varphi\hat{\mathbf{n}} \cdot \mathbf{J} \right), \tag{25.25}$$

where

$$\mathbf{J} = \mathbf{L} + \mathbf{s} \tag{25.26}$$

represents the total angular momentum. We remind the reader, who might feel queasy about adding an operator $\mathbf{L}$ (essentially $\mathbf{x} \times \nabla$) in ordinary space to an operator $\mathbf{s}$ (essentially a 2×2 matrix) in an internal spin space, that $\mathbf{L}$ in the expression (25.26) really means $\mathbf{L}I$, where I equals the unit 2×2 matrix of the last section.

To summarize, then, under the rotational operation (25.25), spinors transform as

$$\Psi' = R\Psi,$$

whereas the angular-momentum operator itself transforms as

$$\mathbf{J}' = R\mathbf{J}R^{-1},$$

with the inverse operator to R reading

$$R^{-1} = \exp\left(-i\hbar^{-1}\varphi\hat{\mathbf{n}} \cdot \mathbf{J}\right).$$

Our ability to perform such arbitrary rotations in ordinary space (or internal space, for that matter) implies that the specific representation given in the last section for the Dirac matrices and Pauli spinors do not represent unique choices, but depend on what forms we choose for the basis spinors. Nevertheless, physical measurements of angular momenta, which deal with

$$\langle\Psi'|\mathbf{J}'|\Psi'\rangle = \langle\Psi|\mathbf{J}|\Psi\rangle,$$

remain invariant under such rotations since $R^{-1}R = RR^{-1} = 1$. For example, a rotation about the z-axis can have no effect on the measurement of the z angular momentum, since the latter does not depend on the orientation of the x-y axes. In Chapter 15 of his book, Merzbacher demonstrates the general validity of the expression (25.25) when $\mathbf{J}$ represents any sum of angular momentum operators, and not just those of a single electron. To pursue these connections here, however, would take us too far afield. We return now to the discussion of the relativistic case, Dirac's equation, where we need to deal with 4×4 matrix operators.

DIRAC'S EQUATION

With equation (25.8) specifying H as a 4×4 matrix-differential operator, we need to give it something appropriate to operate on in equation (25.3); i.e., we need to generalize the wave function $\psi \to \chi$ so that χ forms a *four-component* wave function. To make sure that no misunderstanding arises, we emphasize that these four components refer to an abstract *internal* space for the spin-1/2 particle and do *not* constitute the four components of a four-vector in the sense of special relativity. We postulate, therefore,

$$\chi \equiv \begin{pmatrix} \chi_1 \\ \chi_2 \\ \chi_3 \\ \chi_4 \end{pmatrix}. \tag{25.27}$$

Written out explicitly, Dirac's equation for an electron then becomes

$$\left[c\mathbf{a} \cdot \left(-i\hbar \boldsymbol{\nabla} + \frac{e}{c}\mathbf{A} \right) + bm_e c^2 \right] \chi = \left(i\hbar \frac{\partial}{\partial t} + e\phi \right) \chi. \tag{25.28}$$

We naturally identify the product of $\chi^\dagger$ and χ with the probability density ρ:

$$\rho = \chi^\dagger \chi \equiv |\chi_1|^2 + |\chi_2|^2 + |\chi_3|^2 + |\chi_4|^2.$$

From Dirac's equation (25.28), we can straightforwardly show (see Sakurai 1967, pp. 82–83) that the continuity equation (25.6) holds if we identify

$$\mathbf{j} = c\psi^\dagger \mathbf{a}\psi.$$

Hence, in Dirac's theory, $c\mathbf{a}$ plays the role of the velocity operator. The equations of *quantum electrodynamics* result if we allow an electron to interact with its own electromagnetic field (through Maxwell's equations) by identifying the charge density and electric current associated with the electron as

$$\rho_e = -e\chi^\dagger \chi, \qquad \mathbf{j}_e = -ec\chi^\dagger \mathbf{a}\chi. \tag{25.29}$$

The study of the complete set of equations for quantum electrodynamics for particle and self-consistent field, however, lies beyond the scope of this book. In what follows, we focus on the part that governs the particle motion, Dirac's equation (25.28).

FREE PARTICLE AT REST

There are many problems that can benefit from an attack via Dirac's equation. For example, exact solutions can be derived for the the problem of the hydrogen atom. Our interest will lie mostly in taking the nonrelativistic limit of Dirac's equation. Before we do this, however, we pause to treat the simplest application of the full relativistic formulation—the problem of a free particle at rest. Naively, one might think this problem too trivial to deserve the attention of Dirac's equation; in fact, it probably constitutes its single most important application, for it led to the prediction of the positron, i.e., to the recognition of the existence of antimatter.

For a free particle, we may set ϕ and $\mathbf{A} = 0$, and Dirac's equation (25.28) becomes

$$[-i\hbar c\mathbf{a} \cdot \boldsymbol{\nabla} + bm_e c^2]\chi = i\hbar\frac{\partial\chi}{\partial t}. \qquad (25.30)$$

Since the coefficients depend on neither $\mathbf{x}$ nor t, this equation has plane-wave solutions of the type

$$\chi = \chi_0(\mathbf{k}_e)\exp[i(\mathbf{k}_e \cdot \mathbf{x} - Et/\hbar)]. \qquad (25.31)$$

For a free particle at rest, $\mathbf{k}_e = 0$, and the substitution of equation (25.31) into equation (25.30) produces the requirement

$$m_e c^2 \begin{pmatrix} I & O \\ O & -I \end{pmatrix}\begin{pmatrix} \Psi_+ \\ \Psi_- \end{pmatrix} = E\begin{pmatrix} \Psi_+ \\ \Psi_- \end{pmatrix}, \qquad (25.32)$$

where we have written b out explicitly in block 2×2 form and have decomposed χ_0 as two two-component spinors, Ψ_+ and Ψ_-.

Equation (25.32) shows that the equations governing Ψ_+ and Ψ_- completely decouple for a particle at rest: $m_e c^2\Psi_+ = E\Psi_+$ and $-m_e c^2\Psi_- = E\Psi_-$. The first gives the eigenvalue solution, $E = +m_e c^2$; the second, $E = -m_e c^2$. The solutions

$$E = \pm m_e c^2 \qquad (25.33)$$

are degenerate in that each choice of sign yields a *pair* of possible eigenvectors for the four-component wave function χ. For the positive-energy solution, $E = +m_e c^2$, we have the bottom spinor Ψ_- equal to zero, but there are two possible choices for the top spinor: $\Psi_+ = \alpha_+$ (spin *up*) and $\Psi_+ = \alpha_-$ (spin *down*), where $\alpha_\pm$ are the basis spinors defined by equation (25.22). For the negative-energy solution, $E = -m_e c^2$, we have the top spinor Ψ_+ equal to zero, but there are also two possible choices for the bottom spinor: $\Psi_- = \alpha_+$ and $\Psi_- = \alpha_-$. In other words, the independent four-component wave functions correspond to column vectors with three 0's and one 1, with the 1 occupying, successively, the first, second, third, and fourth rows.

Clearly, from the preceding discussion, the negative-energy solutions have the same formal status as the positive-energy solutions in Dirac's theory. But, what physically can be meant by states with negative *rest energies*? If such states exist and lie below the ones in which we find actual electrons, why don't observed electrons fall into the negative-energy states? Dirac supposed that the negative-energy states must already be filled with an (infinite) sea of electrons (which we somehow mistake for the vacuum!), so that the Pauli exclusion principle prevents ordinary electrons from falling into the sea (see Figure 25.1).

If, however, a "hole" is created in the sea of negative-energy states, then it could move, as neighboring electrons in the sea fall into the hole, leaving another hole behind. Such a hole would behave in all respects like

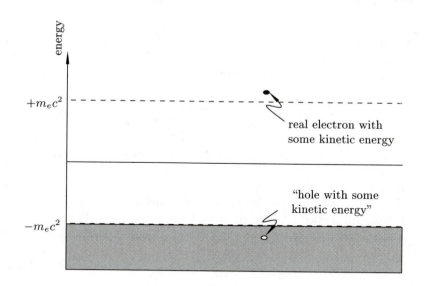

FIGURE 25.1
The positive- and negative-energy states of Dirac's theory of the electron show
a gap $2m_e c^2$, the rest-energy difference if an electron and a hole (a positron)
should annihilate.

an *anti-electron*. In practice, Dirac first tried (conservatively) to interpret
the holes as *protons*, rather than as *positrons*, but Weyl and Oppenheimer
independently pointed out the untenability of such a tactic. For example,
only a real electron not already in the sea could fall into the hole and annihilate it. The discovery by Anderson and Neddermeyer in 1932 of positrons
among the decay products of cosmic-ray showers brilliantly confirmed these
conjectures; moreover, the annihilation of electrons and positrons has become, by now, so well accepted that beginning astronomy students learn
about the concept. Ironically, however, no one any longer accepts Dirac's
original interpretation of positrons as holes in an unobserved sea of electrons. Indeed, the concept is incompatible with the logical structure of his
theory, which holds for *single particles*, not a sea of them. The rigorous
basis for the identification of antimatter has to be sought in quantum field
theory. Alternatively, we can continue to use Dirac's equation in its original
formulation, but use Feynman's interpretation of positrons as equivalent to
electrons in having positive mass-energy but going backward through time!

NONRELATIVISTIC LIMIT OF DIRAC'S EQUATION

It lies beyond the scope of this text to explore additional aspects of relativistic quantum mechanics. For now, we remark only that Dirac needs *two* two-component wave functions (spinors) because his equation contains the notion of negative energy states. If we focus on low-energy phenomena—the structure of atoms and molecules—the effects of the negative energy states on real electrons (with positive energy states) are small, and we can reduce the problem of electron spin to Pauli's description in terms of a single two-component wave function. Pauli first discovered that electrons had to have two internal states (spin "up" or "down") by considering the requirements of the periodic table; he did not connect the concept with *angular momentum*. The latter came historically only with Uhlenbeck and Goudsmit's discussion of the Stern-Gerlach experiment.

To take the nonrelativistic limit of Dirac's equation, first define the electron nonrelativistic energy and momentum operators for positive energy states:

$$\mathcal{E} \equiv i\hbar \frac{\partial}{\partial t} - m_e c^2, \tag{25.34}$$

$$\mathbf{P} \equiv -i\hbar \boldsymbol{\nabla} + \frac{e}{c}\mathbf{A}. \tag{25.35}$$

We wish to consider the limit of equation (25.28) when $\mathcal{E}$ and $\mathbf{P}$ are, in some sense, very small in comparison with $m_e c^2$ and $m_e c$. As in the previous section, we are guided by the 2×2 block nature of the governing equation to partition the four-component wave function χ into the two two-component wave functions Ψ_+ and Ψ_-:

$$\chi = \begin{pmatrix} \Psi_+ \\ \Psi_- \end{pmatrix}. \tag{25.36}$$

We may now write equation (25.28) as the two coupled equations,

$$(\mathcal{E} + e\phi)\Psi_+ - c\boldsymbol{\sigma} \cdot \mathbf{P}\Psi_- = 0, \tag{25.37}$$

$$(\mathcal{E} + 2m_e c^2 + e\phi)\Psi_- - c\boldsymbol{\sigma} \cdot \mathbf{P}\Psi_+ = 0, \tag{25.38}$$

where $\boldsymbol{\sigma} \equiv \hat{\mathbf{x}}\sigma_x + \hat{\mathbf{y}}\sigma_y + \hat{\mathbf{z}}\sigma_z$.

In the previous section, we found Ψ_+ and Ψ_- to be completely decoupled for a particle at rest. Thus we expect the coupling to be weak when $\mathcal{E} \ll m_e c^2$. We mathematically verify this expectation as follows. The presence of the large term $2m_e c^2$ in the operator in front of Ψ_- suggests that the first term in equation (25.38) can be balanced by the second only if Ψ_- is much smaller than Ψ_+ (by a factor $P/2m_e c \sim v/c$). This motivates us to solve equation (25.38) formally as

$$\Psi_- = [2m_e c^2 + (\mathcal{E} + e\phi)]^{-1}c\boldsymbol{\sigma} \cdot \mathbf{P}\Psi_+, \tag{25.39}$$

where the notation $[2m_e c^2 + (\mathcal{E} + e\phi)]^{-1}$ denotes the inverse operator to that contained within the brackets. For $\mathcal{E} + e\phi \ll 2m_e c^2$, this inverse may be found approximately as

$$\frac{1}{2m_e c^2} \left(1 - \frac{\mathcal{E} + e\phi}{2m_e c^2}\right),$$

as may be directly verified by multiplication by $[2m_e c^2 + (\mathcal{E} + e\phi)]$ and dropping terms of quadratic smallness. Substituting the approximate solution for Ψ_- into equation (25.37), we get, correct to order v^2/c^2,

$$\left[(\mathcal{E} + e\phi) - \frac{\boldsymbol{\sigma} \cdot \mathbf{P}}{2m_e} \left(1 - \frac{\mathcal{E} + e\phi}{2m_e c^2}\right) \boldsymbol{\sigma} \cdot \mathbf{P}\right] \Psi_+ = 0 \qquad (25.40)$$

To simplify this expression further, we factor out the rest-energy dependence in the remaining (two-component) wave function

$$\Psi_+(\mathbf{x}, t) \equiv \Psi(\mathbf{x}, t) e^{-im_e c^2 t/\hbar}, \qquad (25.41)$$

which upon operation by $\mathcal{E}$ equals

$$\mathcal{E}\Psi_+ = i\hbar \frac{\partial \Psi}{\partial t} e^{-im_e c^2 t/\hbar}.$$

When this expression is substituted into equation (25.40), together with the definition (25.34) for $\mathcal{E}$, we obtain, after a little manipulation,

$$\mathcal{S}\Psi = 0, \qquad (25.42)$$

where $\mathcal{S}$ is the operator

$$\mathcal{S} \equiv \mathcal{S}_0 + \mathcal{S}_2, \qquad (25.43)$$

with $\mathcal{S}_0$ being the lowest-order term,

$$\mathcal{S}_0 \equiv \left(i\hbar \frac{\partial}{\partial t} + e\phi\right) - \frac{(\boldsymbol{\sigma} \cdot \mathbf{P})^2}{2m_e}, \qquad (25.44)$$

and $\mathcal{S}_2$ being a factor v^2/c^2 smaller,

$$\mathcal{S}_2 \equiv \frac{\boldsymbol{\sigma} \cdot \mathbf{P}}{2m_e c} \left(i\hbar \frac{\partial}{\partial t} + e\phi\right) \frac{\boldsymbol{\sigma} \cdot \mathbf{P}}{2m_e c}. \qquad (25.45)$$

To manipulate $\mathcal{S}_2$ into the displayed form, we have made use of the fact that $i\hbar \partial/\partial t$ commutes with both $\boldsymbol{\sigma}$ and $\mathbf{P}$. The last statement does not hold if $\mathbf{A}$ contains time variations; however, radiative couplings through $\mathcal{S}_2$ can be estimated to be smaller than those through $\mathcal{S}_0$ by a factor $\hbar k/m_e c$, which is negligible at the low photon energies at which we wish to apply the nonrelativistic formalism.

FULLY NONRELATIVISTIC APPROXIMATION

Perturbation theory always begins with the assumption that we want to find a solution to a problem similar to one where we already have a good solution. Here our first order of business therefore consists of finding a solution to

$$S_0 \Psi = 0, \qquad (25.46)$$

with S_0 given by equation (25.44). Since $\boldsymbol{\sigma}$ operates in a different space than does $\mathbf{P}$, we may write

$$(\boldsymbol{\sigma} \cdot \mathbf{P})^2 = \mathbf{P} \cdot \boldsymbol{\sigma}\boldsymbol{\sigma} \cdot \mathbf{P}.$$

Consider the combination $\boldsymbol{\sigma}\boldsymbol{\sigma}$, whose Cartesian components we may write as

$$\sigma_i \sigma_k = \frac{1}{2}[(\sigma_i \sigma_k + \sigma_k \sigma_i) + (\sigma_i \sigma_k - \sigma_k \sigma_i)] = \delta_{ik} + i\epsilon_{ikm}\sigma_m, \qquad (25.47)$$

in accordance to equations (25.17) and (25.18). Equation (25.47) now implies that

$$\mathbf{P} \cdot \boldsymbol{\sigma}\boldsymbol{\sigma} \cdot \mathbf{P} = |\mathbf{P}|^2 + i\boldsymbol{\sigma} \cdot (\mathbf{P} \times \mathbf{P}), \qquad (25.48)$$

where $\mathbf{P} = -i\hbar\boldsymbol{\nabla} + e\mathbf{A}/c$ prevents the expression $\mathbf{P} \times \mathbf{P}$ from being trivially equal to zero. Because of this complication, the discussion that follows will not consider the most general situation possible, but will separate itself into two parts: (a) When we wish to incorporate a radiation field or a static magnetic field, we will discard terms of order v^2/c^2 in smallness. (b) When we wish to calculate accurately to v^2/c^2, we will suppose $\mathbf{A} = 0$ (i.e., we will assume no external radiation field or static magnetic field). This apparent schizophrenia causes no real difficulties when the perturbations to the dominant term remain small, because we can simply superimpose the individual terms, one by one. If the perturbations are not small, then we probably need to use the unexpanded form of the Dirac equation in any case.

RADIATIVE TRANSITIONS INVOLVING SPIN

We consider first the case when $\mathbf{A} \neq 0$. When $\mathbf{P}$ is given by $\mathbf{p} + e\mathbf{A}/c$, with $\mathbf{p} = -i\hbar\boldsymbol{\nabla}$, equation (25.48) becomes

$$\mathbf{P} \cdot \boldsymbol{\sigma}\boldsymbol{\sigma} \cdot \mathbf{P} = \left|\mathbf{p} + \frac{e}{c}\mathbf{A}\right|^2 + \frac{e\hbar}{2m_e c}\boldsymbol{\sigma} \cdot (\boldsymbol{\nabla} \times \mathbf{A}),$$

since the curl of the gradient of any (wave) function equals zero. If we now write out equation (25.46), we can put it into the Hamiltonian form:

$$(H_0 + H_1 + H_2 + H_{Bs})\Psi = i\hbar \frac{\partial \Psi}{\partial t}, \tag{25.49}$$

where $H_0 + H_1 + H_2$ are the usual operators (see Chapter 22),

$$H_0 + H_1 + H_2 = \frac{1}{2m_e} \left| \mathbf{p} + \frac{e}{c}\mathbf{A} \right|^2 - e\phi,$$

and H_{Bs} is the part of the interaction Hamiltonian related to electron spin that enters at zeroth order (in an expansion of v^2/c^2):

$$H_{Bs} \equiv \frac{e}{m_e c} \mathbf{s} \cdot \mathbf{B}. \tag{25.50}$$

In equation (25.50), we have again let $\mathbf{s} \equiv (\hbar/2)\boldsymbol{\sigma}$ denote the spin angular momentum, and we have written $\mathbf{B} = \boldsymbol{\nabla} \times \mathbf{A}$ for the external magnetic field.

To see how $\mathbf{s}$ acts mathematically as a spin-angular-momentum operator for our specific choice of basis spin kets, we note first the following easily proved relation:

$$(s_x^2 + s_y^2 + s_z^2)\alpha_\pm = \frac{3}{4}\hbar^2 \alpha_\pm; \tag{25.51}$$

i.e., the spinor eigenvectors $\alpha_\pm$ have associated eigenvalues for the square of the spin angular momentum equal to $s(s+1)\hbar^2 = (3/4)\hbar^2$, where $s = 1/2$ for a fermion like the electron. Similarly, from the basic spin-matrix definitions, we easily demonstrate

$$s_z \alpha_\pm = \pm\frac{1}{2}\hbar\alpha_\pm, \tag{25.52}$$

$$(s_x \pm is_y)\alpha_\mp = \hbar\alpha_\pm \qquad (s_x \pm is_y)\alpha_\pm = 0. \tag{25.53}$$

The term $\pm 1/2$ that appears in equation (25.52) is often denoted m_s, with $m_s = +1/2$ being the eigenvalue associated with the spin "up," α_+, and $m_s = -1/2$ with spin "down," α_-. The operators $s_x \pm is_y$ perform "spin flip" transitions, transforming α_- to α_+ and vice versa, in the process changing the spin angular momentum of the electron by one unit of $\hbar$ (which can go to a liberated photon or come from an absorbed one).

Suppose the vector potential $\mathbf{A}$ to be composed of two parts:

$$\mathbf{A} = \mathbf{A}_{rad} + \frac{1}{2}\mathbf{B}_0 \times \mathbf{x}, \tag{25.54}$$

where $\mathbf{B}_0$ is a constant vector and $\mathbf{A}_{\mathrm{rad}}$ satisfies the vacuum wave equation (i.e., what we denoted as $\mathbf{A}$ in Chapter 21). Our total $\mathbf{A}$ still satisfies the Coulomb gauge, $\boldsymbol{\nabla} \cdot \mathbf{A} = 0$, but its curl gives a static magnetic field $\mathbf{B}_0$ in addition to a radiation magnetic field $\mathbf{B}_{\mathrm{rad}}$. Write now H_{Bs} also as the sum of two terms,

$$H_{Bs} = H_{Bs}^{\mathrm{rad}} + H_{Zs}, \tag{25.55}$$

where the first contains the effect of the radiation field,

$$H_{Bs}^{\mathrm{rad}} = \frac{e}{m_e c} \mathbf{s} \cdot (\boldsymbol{\nabla} \times \mathbf{A}_{\mathrm{rad}}),$$

and the second, the effects of the static field,

$$H_{Zs} = \frac{e}{m_e c} \mathbf{s} \cdot \mathbf{B}_0.$$

Although bracketed together with H_1 [see equation (22.4)] in an expansion of v^2/c^2, H_{Bs}^{rad} does not contribute to the transition matrix element in the same *multipole* order as the largest term (electric dipole) usually present in H_1. The perturbational effect of H_{Bs}^{rad} enters only at the level of the magnetic dipole moment (see Chapter 24). In particular, the term $\mathbf{k} \times \mathbf{e}_\alpha$ in equation (24.22) arises in the same manner as the term $\boldsymbol{\nabla} \times \mathbf{A}_{\mathrm{rad}} \equiv \mathbf{B}_{\mathrm{rad}}$ in equation (25.55), because magnetic dipoles interact with only the magnetic part of electromagnetic radiation. From this point of view, a "spinning" electron constitutes the ultimate magnetic particle, and it is appropriate that the most important of all discrete transitions in the radio spectrum— the famous 21-cm line (predicted by van de Hulst and first detected by Ewen and Purcell)—arises from a magnetic dipole transition involving the spin-flip of an electron in the ground electron state of atomic hydrogen, where the two spin states ("up" and "down") have their degeneracy split by a hyperfine interaction with the spin (magnetic moment) of the nuclear proton (see Problem Set 5).

ZEEMAN EFFECT

Returning to our original line of thought, we note that the part containing the term H_{Zs} should be combined with a comparable additional term in $H_1 = (e/m_e c)\mathbf{A} \cdot \mathbf{p}$ that reads

$$H_{ZL} = \frac{e}{2m_e c}(\mathbf{B}_0 \times \mathbf{x}) \cdot \mathbf{p} = \frac{e}{2m_e c}\mathbf{B}_0 \cdot \mathbf{L},$$

where $\mathbf{L} = \mathbf{x} \times \mathbf{p}$ is the angular momentum operator. If we sum the two perturbational terms H_{ZL} and H_{Zs}, we see that the total contribution to the *Zeeman effect* reads

$$H_Z = -\mathbf{B}_0 \cdot (\mathbf{M} + \mathbf{m}), \tag{25.56}$$

where $\mathbf{M} = -(e/2m_e c)\mathbf{L}$ is the electron's orbital magnetic moment, and $\mathbf{m} = -(e/m_e c)\mathbf{s}$ is the electron's spin magnetic moment. The latter has a g-factor equal to 2 in Dirac's theory, as advertised in Chapter 24. Except for this peculiarity, the right-hand side of equation (25.56) has the form suggested by classical electrodynamics for the excess energy associated with the placement of a magnetic dipole in an external magnetic field.

Applied to astronomy, the Zeeman effect provided Hale's original demonstration that sunspots are connected with kilogauss magnetic fields. [Hale demonstrated that lines with larger so-called Landé-g factors (see Chapter 27) had larger Zeeman splittings.] The Zeeman effect on the 21-cm transition of H I and the 18-cm transition of OH yields what is currently the only reliable way to obtain the strengths of magnetic fields in interstellar clouds.

NO SPIN COUPLING

If we discard the term H_{Bs} (because we ignore the Zeeman effect and the effects of spin on magnetic dipole transitions), equation (25.49) becomes identical to the nonrelativistic Schrödinger equation with which we have been working so far, except that we must now interpret Ψ as a two-component wave function. To solve equation (25.49) in this case, we may choose Ψ as the product of a scalar spatial wave function ψ and one of the basic spinors $\alpha_{\pm}$ defined in equation (25.22),

$$\Psi = \psi(\mathbf{x}, t)\alpha_{\pm}. \tag{25.57}$$

The wave function defined this way constitutes an eigenket in spin space; the eigenbras would be row spinors instead of column spinors. Since the spin operator does not enter in the lowest-order equation (25.49) if we drop H_{Bs}, the two spin states do not mix to this order of approximation (i.e., electron spin would be conserved in radiative transitions). The inclusion of electron spin in this case adds nothing of substance to the theory developed so far for the scalar wave function $\psi(\mathbf{x}, t)$ alone.

RELATIVISTIC CORRECTION FOR THE NORMALIZATION OF THE WAVE FUNCTION

If we wish to consider effects of order v^2/c^2, we not only need to include such terms in the governing Hamiltonian, but must also account for their implications concerning the normalization of the wave function. The normalization condition for the Dirac wave function, $\langle\chi|\chi\rangle = 1$, occurs as an inner product that includes a summation over the indices of the four components as well as an integral over physical space. When expressed in terms of two (two-component) spinors, this normalization condition becomes

$$\langle\Psi_+|\Psi_+\rangle + \langle\Psi_-|\Psi_-\rangle = 1, \tag{25.58}$$

where the last inner products are formed over two-component wave functions.

Our formalism has been directed at recovering a single (two-component) equation for Ψ_+. On the other hand, the substitution of equation (25.39) suggests that an adequate approximation (to order v^2/c^2) for use in equation (25.58) is

$$\Psi_- \approx \frac{\boldsymbol{\sigma}\cdot\mathbf{P}}{2m_ec}\Psi_+,$$

so that equation (25.58) now reads

$$\langle\Psi|1 + (4m_e^2c^2)^{-1}\mathbf{P}\cdot\boldsymbol{\sigma}\boldsymbol{\sigma}\cdot\mathbf{P}|\Psi\rangle = 1, \tag{25.59}$$

where we have invoked equation (25.41) together with the facts that neither $\boldsymbol{\sigma}$ nor $\mathbf{P}$ operate on the factor $e^{-im_ec^2t/\hbar}$, and that $\boldsymbol{\sigma}$ operates in a different space than $\mathbf{P}$ and therefore commutes with it. As in the previous discussion, we will now consider relativistic corrections to the case $\mathbf{A} = 0$.

ELECTROSTATIC EXTERNAL FIELDS ONLY

When $\mathbf{P}$ can be taken to be $\mathbf{p}$, equation (25.48) becomes $\mathbf{p}\cdot\boldsymbol{\sigma}\boldsymbol{\sigma}\cdot\mathbf{p} = |\mathbf{p}|^2$, again because the curl of the gradient of any (wave) function equals zero. Since the operator

$$T^2 \equiv 1 + \frac{|\mathbf{p}|^2}{4m_e^2c^2}$$

that appears in equation (25.59) is now purely real, we easily show that

$$\Psi' \equiv T\Psi$$

forms a properly normalized two-component wave function, where, to a sufficient order of approximation, T and its inverse T^{-1} have the (operator) expressions:

$$T \equiv 1 + \frac{|\mathbf{p}|^2}{8m_e^2c^2}, \qquad T^{-1} = 1 - \frac{|\mathbf{p}|^2}{8m_e^2c^2}.$$

With $\Psi = T^{-1}\Psi'$, we can now write equation (25.42) as $\mathcal{S}'\Psi' = 0$, where $\mathcal{S}'$ represents a unitary transformation of $\mathcal{S}$: $\mathcal{S}' \equiv T\mathcal{S}T^{-1}$. To order v^2/c^2, $\mathcal{S}'$ now reads

$$\mathcal{S}' \equiv \mathcal{S}_0 + \frac{1}{8m_e^2 c^2}\left[|\mathbf{p}|^2\mathcal{S}_0 - \mathcal{S}_0|\mathbf{p}|^2\right] + \mathcal{S}_2. \qquad (25.60)$$

The operators $\mathcal{S}_0$ and $\mathcal{S}_2$ are given by equations (25.44) and (25.45) when $\mathbf{p}$ replaces $\mathbf{P}$. Applying the rule (25.47) to them, we obtain

$$\mathcal{S}_0 = D - \frac{|\mathbf{p}|^2}{2m_e},$$

$$\mathcal{S}_2 = \frac{\mathbf{p}}{2m_e c}\cdot D\frac{\mathbf{p}}{2m_e c} + i\boldsymbol{\sigma}\cdot\left(\frac{\mathbf{p}}{2m_e c}\times D\frac{\mathbf{p}}{2m_e c}\right),$$

where D is the operator

$$D \equiv i\hbar\frac{\partial}{\partial t} + e\phi.$$

Writing out $\mathcal{S}'$ explicitly, we obtain

$$\mathcal{S}' = \mathcal{S}_0 + \frac{1}{8m_e^2 c^2}\left[e|\mathbf{p}|^2\phi - e\phi|\mathbf{p}|^2 + 2\mathbf{p}\cdot D\mathbf{p} + 2i\boldsymbol{\sigma}\cdot(\mathbf{p}\times D\mathbf{p})\right].$$

With $\mathbf{p} = -i\hbar\boldsymbol{\nabla}$, we get

$$e|\mathbf{p}|^2\phi - e\phi|\mathbf{p}|^2 = -e\hbar^2(\nabla^2\phi) - 2ie\hbar(\boldsymbol{\nabla}\phi)\cdot\mathbf{p},$$

where we have adopted the notation that $\mathbf{p}$ operates on everything to its right, whereas $\boldsymbol{\nabla}$ operates only on ϕ. Similarly, if we let the operator D operate through to the right, we get

$$2\mathbf{p}\cdot D\mathbf{p} = 2ie\hbar(\boldsymbol{\nabla}\phi)\cdot\mathbf{p} + 2e\hbar^2(\nabla^2\phi) + 2|\mathbf{p}|^2 D$$

$$\approx 2ie\hbar(\boldsymbol{\nabla}\phi)\cdot\mathbf{p} + 2e\hbar^2(\nabla^2\phi) + \frac{|\mathbf{p}|^4}{m_e}.$$

The last result comes from noting that the lowest-order equation $\mathcal{S}_0\Psi' \approx 0$ allows us to approximate D operating on Ψ' as equivalent to $|\mathbf{p}|^2/2m_e$ operating on Ψ'. Finally, if we let the leftmost $\mathbf{p}$ operate through to the right, we get

$$2i\boldsymbol{\sigma}\cdot(\mathbf{p}\times D\mathbf{p}) = 2e\hbar\boldsymbol{\sigma}\cdot[(\boldsymbol{\nabla}\phi)\times\mathbf{p}].$$

Collecting terms, we may now write

$$\mathcal{S}' = \mathcal{S}_0 + \frac{1}{8m_e^2 c^2}\left\{+e\hbar^2(\nabla^2\phi) + \frac{|\mathbf{p}|^4}{m_e} + 2e\hbar\boldsymbol{\sigma}\cdot[(\boldsymbol{\nabla}\phi)\times\mathbf{p}]\right\}.$$

RELATIVISTIC CORRECTIONS AND SPIN-ORBIT COUPLING

As standard by now, we put the equation $S'\Psi' = 0$ into a Hamiltonian form,

$$H\Psi = i\hbar\frac{\partial\Psi}{\partial t}, \tag{25.61}$$

where we have dropped the prime on Ψ' for notational simplicity (i.e., Ψ now satisfies $\langle\Psi|\Psi\rangle = 1$). We find that H equals the following sum of terms:

$$H = H_S + H_K + H_D + H_{so}. \tag{25.62}$$

In the above, H_S is the standard Schrödinger (structure) operator,

$$H_S \equiv \frac{|\mathbf{p}|^2}{2m_e} - e\phi; \tag{25.63}$$

H_K is a relativistic correction for the kinetic energy,

$$H_K \equiv -\frac{|\mathbf{p}|^4}{8m_e^3 c^2}; \tag{25.64}$$

H_D is a relativistic correction for the potential energy (studied in detail by C.G. Darwin, and therefore called the Darwin term),

$$H_D \equiv -\frac{e\hbar^2}{8m_e^2 c^2}\nabla^2\phi; \tag{25.65}$$

and H_{so} is the contribution to the energy from *spin-orbit* coupling,

$$H_{so} \equiv -\frac{e\mathbf{s}\cdot[(\boldsymbol{\nabla}\phi)\times\mathbf{p}]}{2m_e^2 c^2}. \tag{25.66}$$

In the last equation, we have used $\mathbf{s} \equiv (\hbar/2)\boldsymbol{\sigma}$ to identify the spin angular momentum.

PHYSICAL INTERPRETATIONS

Physical interpretations for the relativistic corrections to the usual Schrödinger operator follow. The kinetic-energy correction H_K is easiest to understand, since an expansion of the classical formula for $p \ll m_e c$ gives

$$[c^2 p^2 + m_e^2 c^4]^{1/2} - m_e c^2 = p^2/2m_e - p^4/8m_e^3 c^2 + \cdots.$$

The potential-energy correction (the Darwin term) H_D turns out to originate in the interference between the positive and negative energy components in Dirac's four-component description, causing the electron effectively to "jitter" with a mean displacement, $|\delta \mathbf{x}| \sim \hbar/m_e c$ at a frequency $\sim 2m_e c^2/\hbar$. As a consequence, the effective potential energy $-e\langle \phi(\mathbf{x}+\delta\mathbf{x})\rangle$ that the electron samples at a mean position $\mathbf{x}$ has an uncertainty of order

$$-\frac{e}{2}\langle\delta x_i \delta x_k\rangle\frac{\partial^2 \phi}{\partial x_i \partial x_k} \sim -\frac{e}{6}\left(\frac{\hbar}{m_e c}\right)^2 \nabla^2 \phi,$$

which, apart from the 6 instead of an 8, equals the expression for H_D. In the hydrogen atom, where the charge distribution external to the electron resides completely in the nucleus, $\nabla^2 \phi$ is a delta function centered on the origin, and only electronic wave functions with ℓ equal to zero (with nonvanishing amplitude at the origin) would be affected by the Darwin term H_D.

From the point of view of a classical description, spin-orbit coupling arises for the following reason. An observer stationary with respect to the atomic nucleus measures an electric field $\mathbf{E} = -\nabla\phi$; an electron which orbits at velocity $\mathbf{v}$ will also experience, by Lorentz transformation [consult, e.g., equation (16.14)], an effective magnetic field $\mathbf{B} = \mathbf{E} \times \mathbf{v}/c = (-\nabla\phi) \times (\mathbf{p}/m_e c)$. Naively (by analogy with the spin Zeeman effect, for example), we expect this magnetic field interacting with a spin magnetic moment $\mathbf{m}$ to have an associated energy equal to $-\mathbf{m}\cdot\mathbf{B}$. The last expression actually equals twice the value H_{so}, if we identify the spin magnetic moment $\mathbf{m} = -(ge/2m_e c)\mathbf{s}$ to contain a g factor equal to 2 (see the discussion of Chapter 24). The discrepancy arises because our classical expectations failed to account for the effects of the electron's being in a noninertial frame of reference. Correction for this effect, called *Thomas precession*, results in a reduction by a factor of 2. *As we have seen from the above derivation, Dirac's theory gets every factor of 2 correct automatically, without additional equivocation.* Of course, making two mistakes—forgetting that $g = 2$ and forgetting the Thomas precession factor of $1/2$—could also have recovered the correct formula for the spin-orbit coupling. This is not a recommended procedure.

THE TRUE IMPORTANCE OF SPIN IN ATOMIC STRUCTURE

The terms H_K, H_D, and H_{so} formally all arise at the same perturbational order; they are all order v^2/c^2 corrections. Nevertheless, spectroscopists generally attach more importance to the spin-orbit term, because it often contributes to a splitting of states that would otherwise be *degenerate*, leading to so-called *fine structure* of atomic energy levels. Fine-structure transitions constitute an important source of lines for far-infrared astronomy. Nevertheless, had the effects of electron spin been confined to only small terms in the Hamiltonian, it would be far less prominent in physics than it actually is.

We do not allude here to the fact that spin-orbit coupling can acquire substantial magnitude in heavy atoms. No, the true importance of electron spin resides in *quantum statistics*—in its central role for the Pauli exclusion principle—and not on its direct presence in the system Hamiltonian. The need to account for this indirect presence in the allowable wave functions for a multi-electron atom explains a large part of the systematics of atomic structure and spectroscopy (as well as chemistry, quantum electronics, etc.). It forms the topic for study in Chapter 27.

26

Single-Electron Atoms

References: Leighton, *Principles of Modern Physics*, Chapter 5; Wu, *Quantum Mechanics*, Chapter 8.

With this chapter we begin the consideration of the structure of atoms and molecules. Except for the hydrogen atom, our discussion will proceed on a fairly qualitative level. For atoms, our strategy consists of proceeding from (1) the hydrogen atom, where we work out the details of spin-orbit coupling in the context of a single-electron atom, to (2) the helium atom, where we confront the problem of the requirements of the Pauli principle when we have just one more electron, to (3) brief comments about the rest of the periodic table, where the technical difficulties of the problem generally require laboratory work guided by the qualitative ideas of the existing theoretical framework. Before we start, however, we make a few remarks about the systematic problem confronting us, in order to set the mathematical scene for much of what follows.

SPECIFIC EIGENFUNCTION REPRESENTATIONS

We wish to solve the Schrödinger equation,

$$H\Psi = i\hbar \frac{\partial \Psi}{\partial t},$$

in the case when the Hamiltonian H does not contain any time-dependent perturbations (i.e., when we are interested in the problem of *structure* rather than in radiative transitions). In this case, we may look for stationary states having the form

$$\Psi = \Phi e^{-iEt/\hbar},$$

where Φ carries the spatial and spin dependences of Ψ. The function Φ satisfies the time-independent Schrödinger equation:

$$H\Phi = E\Phi. \tag{26.1}$$

We speak of Φ as an eigenfunction of H, and E as its eigenvalue. Suppose, now, that H has enough symmetry in it to allow some operator X to commute with it. It then becomes possible to choose an eigenfunction basis such that eigenfunctions Φ of H are also eigenfunctions of X. To remind ourselves of the truth of this result, left-multiply equation (26.1) by X to obtain $XH\Phi = EX\Phi$. But $XH\Phi = HX\Phi$, if H and X commute; thus $H(X\Phi) = E(X\Phi)$, and $X\Phi$ must be an eigenfunction of H. If E corresponds to a nondegenerate eigenvalue, $X\Phi$ must be linearly proportional to Φ itself, say, $\xi\Phi$, where ξ is a constant; i.e., $X\Phi = \xi\Phi$, demonstrating that Φ is an eigenfunction of X with the associated eigenvalue ξ. If E corresponds to a degenerate eigenvalue, a Gram-Schmidt procedure still allows us to construct an eigenfunction basis such that our statement remains true. (Q.E.D.)

Moreover, if another operator Y commutes with both H and X, it becomes possible to pick the basis such that Φ is simultaneously an eigenfunction of all three. On the other hand, if Y commutes with H, but not X, then we may choose a basis in which Φ is an eigenfunction of H and Y, but not X; or of H and X, but not Y. As a simple example of this situation, note by direct calculation that the square of the orbital-angular-momentum operator L^2 and the z-component of the angular momentum L_z, given by equations (24.8) and (24.9) for a single electron, commute with each other and with H_0 if the potential $V(r)$ in the Schrödinger operator

$$H_0 = \frac{1}{2m_e r^2} \left(-\hbar^2 \frac{\partial}{\partial r} r^2 \frac{\partial}{\partial r} + L^2 \right) + V(r)$$

possesses spherical symmetry. (For maximum accuracy, we should really replace m_e by the reduced mass of the electron $\mu_e \equiv m_e M/[M + m_e]$, where M is the mass of the rest of the atom.)

With H_0 as the Hamiltonian, separation of variables allowed us in Chapter 24 to find (scalar) wave functions φ that are simultaneously eigenfunctions of H_0, L^2, and L_z. However, written in Cartesian coordinates, with $L^2 \equiv L_x^2 + L_y^2 + L_z^2$, it should be clear that if L^2 commutes with L_z, it should also commute with L_x and L_y; i.e., the following statements must all be true:

$$[L^2, L_x] = [L^2, L_y] = [L^2, L_z] = 0. \tag{26.2}$$

Indeed, the Cartesian expressions for $\mathbf{L} \equiv \mathbf{x} \times \mathbf{p}$, with $\mathbf{p} = -i\hbar\boldsymbol{\nabla}$, provide an easy way to verify equation (26.2) directly.

If such symmetry exists, why should we single out the z direction for special attention? Because, although H_0 and L^2 commute with L_x and L_y, their direct counterpart L_z does not. Indeed, the last three form a mutually non-commuting triplet:

$$[L_x, L_y] = i\hbar L_z, \qquad [L_y, L_z] = i\hbar L_x, \qquad [L_z, L_x] = i\hbar L_y. \tag{26.3}$$

Thus, to have a definite eigenfunction basis, we have to choose among L_x, L_y, and L_z. The choice of a standard spherical-polar coordinate system and L_z then results in the spherical harmonics $Y_{\ell m}$ as the angular part of this eigenfunction basis.

Suppose, now, we wish to include the effects of spin, still within the context of a single-electron atom:

$$\mathbf{S} \equiv \frac{\hbar}{2}\boldsymbol{\sigma}.$$

(We have changed notation here from $\mathbf{s}$ to $\mathbf{S}$ for the sake of notational parallelism in what follows.) To the extent that H is well approximated by a spin-independent Hamiltonian H_0 (i.e., if we can drop the term H_{so}, which is of order v^2/c^2 in smallness), the operators S^2, S_z, S_x, and S_y also all commute with H. However, S_x, S_y, S_z form a non-commuting triplet:

$$[S_x, S_y] = i\hbar S_z, \qquad [S_y, S_z] = i\hbar S_x, \qquad [S_z, S_x] = i\hbar S_y; \qquad (26.4)$$

and this explains our choice in singling out S_z for special treatment once we have already done so for L_z.

SPIN-ORBIT INTERACTION AND THE TOTAL ANGULAR MOMENTUM

When we allow for nonzero spin-orbit interaction, we get

$$H = H_0 + H_{\mathrm{so}},$$

where H_{so} here [see equation (25.66)] reads

$$H_{\mathrm{so}} = \frac{1}{2m_{\mathrm{e}}^2 c^2}\left(\frac{1}{r}\frac{dV}{dr}\right)\mathbf{S}\cdot\mathbf{L}. \qquad (26.5)$$

To derive equation (26.5), we have used $-e\boldsymbol{\nabla}\phi = \hat{\mathbf{r}}dV/dr = (r^{-1}dV/dr)\mathbf{r}$ together with $\mathbf{L} = \mathbf{r}\times\mathbf{p}$. In the presence of H_{so}, neither $\mathbf{L}$ nor $\mathbf{S}$ commutes with H anymore,

$$[H_{\mathrm{so}}, \mathbf{L}] \neq 0 \qquad \text{and} \qquad [H_{\mathrm{so}}, \mathbf{S}] \neq 0,$$

since the different components of $\mathbf{L}$ and $\mathbf{S}$ do not commute with each other. In other words, spin-orbit interaction introduces a noncentral force field of electromagnetic origin, which prevents either the orbital or the spin angular momentum from being separately conserved in a vectorial manner. Their magnitudes *are* conserved, since L^2 and S^2 commute with H_{so}. Moreover, since the forces are entirely internal, and we have ignored the motion of the nucleus, we anticipate that the electron's total vector angular momentum [see equation (25.26)],

$$\mathbf{J} \equiv \mathbf{L} + \mathbf{S}, \qquad (26.6)$$

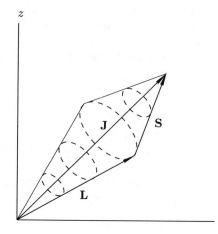

FIGURE 26.1
Vector model for the superposition of orbital and spin angular momenta when spin-orbit interaction causes a slow precession of **L** and **S** about the fixed total angular momentum **J** = **L** + **S**.

ought to have stationary eigenvalues. Using a classical image of the "vector" model for the addition of angular momentum (see Figure 26.1), we may picture **L** and **S** combining to form a constant vector **J** = **L** + **S**. The spin-orbit interaction causes a precession of both **L** and **S** about the fixed direction associated with **J**. This precession preserves the constancy of L^2 and S^2, i.e., the "lengths" of **L** and **S** will remain fixed, but the projections L_z and S_z onto any axis z not in the **J** direction will vary with time. Nevertheless, the precession occurs slowly in comparison to the characteristic orbital speed, and this slowness allows the perturbation procedure for spin-orbit interaction that we will pursue later in this chapter.

We wish to see quantum mechanically that **J** commutes with H and therefore possesses stationary eigenvalues in common with it. We begin by noting that **J** commutes with H_0 since **L** and **S** individually do. Moreover, since **L** involves differentiation only with respect to the angular variables (θ, ϕ), and since **S** refers only to inner states of the electron, **J** clearly commutes with the coefficient of **S** · **L** that depends only on r in H_{so}. Consequently, we wish to examine the commutator

$$[\mathbf{S} \cdot \mathbf{L}, \mathbf{J}]$$

to see if it equals zero. For this purpose, we need choose only one component, say, $J_z = L_z + S_z$, since if it commutes with **S** · **L**, so will J_x and

J_y by the symmetry behind equations (26.3) and (26.4). Because $L_z + S_z$ trivially commutes with the term $S_z L_z$ in

$$\mathbf{S} \cdot \mathbf{L} = S_x L_x + S_y L_y + S_z L_z,$$

we then need only concern ourselves with

$$[S_x L_x + S_y L_y, L_z + S_z] = S_x[L_x, L_z] + S_y[L_y, L_z] + [S_x, S_z]L_x + [S_y, S_z]L_y.$$

Applying the commutation relations (26.3) and (26.4), we obtain the identification,

$$[\mathbf{S} \cdot \mathbf{L}, J_z] = i\hbar \left(-S_x L_y + S_y L_x - S_y L_x + S_x L_y \right) = 0.$$

Consequently,

$$[H_{so}, \mathbf{J}] = 0,$$

and spin-orbit interaction does not affect the conservation of total vector angular momentum, orbit plus spin. (Q.E.D.) As an aside, we note that the conservation of $\mathbf{J} = \mathbf{L} + \mathbf{S}$ in the presence of a noncentral perturbational force field provides the ultimate justification for the identification of $\mathbf{S} = \hbar\boldsymbol{\sigma}/2$ as the *spin-angular-momentum operator* for the electron, fully coequal to the orbital-angular-momentum operator $\mathbf{L}$.

Having proven this important result, we still need to construct the eigenkets of the operators $H = H_0 + H_{so}$, L^2, S^2, J^2, and J_z. In what follows, we perform this construction by using as a basis the eigenkets of the unperturbed problem when we ignore spin-orbit interactions. We then obtain the eigenkets of the complete problem, including the perturbation H_{so}, as a linear superposition of the relevant eigenkets of the unperturbed problem.

ELECTRON ORBITAL REPRESENTATION

We can identify the two-component spinor:

$$\Phi = \frac{1}{r} R_{n\ell}(r) Y_{\ell m_\ell} \alpha_\pm \equiv |n, \ell, m_\ell, m_s\rangle \qquad \text{where} \qquad m_s = \pm 1/2, \quad (26.7)$$

as the basic eigenket for the zeroth-order equation

$$H_0 \Phi = E_0 \Phi. \tag{26.8}$$

This is called the (single-electron) *orbital representation* (or the $m_\ell m_s$ representation). It is possible to include the perturbational term H_K in H_0 without destroying the validity of ℓ, m_ℓ, and m_s as good eigenvalues. (We formerly denoted m_ℓ by m.) Hence our discussion even at this stage constitutes a slight generalization over the usual hydrogen-atom treatment, and we must allow for the ("zeroth-order") eigenvalue E_0 to depend on both n and ℓ:

$$E_0 = E_{n\ell}. \tag{26.9}$$

Notice that in accordance with the discussion of Chapter 24, the energy eigenvalue $E_{n\ell}$ is degenerate with respect to m_ℓ and m_s.

TIME-INDEPENDENT PERTURBATION THEORY FOR DEGENERATE STATES

Inclusion of the effects of spin-orbit coupling will remove the degeneracy; i.e., the solution of the equation,

$$(H_0 + H_{\text{so}})\Phi = E\Phi, \tag{26.10}$$

does not have E_0 as an eigenvalue nor $|n, \ell, m_\ell, m_s\rangle$ as an eigenket. However, since H_0 and L^2 commute with $\mathbf{S} \cdot \mathbf{L}$, we anticipate that n and ℓ will remain valid eigenvalues. (In practice, because H_0 does not commute with the coefficient $r^{-1}dV/dr$, the presence of H_{so} will shift the unperturbed energy $E_{n\ell}$ by a slight amount, and the unperturbed radial function $R_{n\ell}$ cannot be strictly an eigenfunction of the *perturbed* Hamiltonian $H = H_0 + H_{\text{so}}$. However, corrections due to this effect enter in second order in the perturbation, and will be ignored here.) Thus we look for a solution of equation (26.10) which consists of an expansion involving the degenerate eigenkets common to a given n and ℓ:

$$\Phi = \sum_{m'_\ell m'_s} C(m'_\ell, m'_s)|n, \ell, m'_\ell, m'_s\rangle.$$

If we substitute the above into equation (26.10), we get

$$\sum_{m'_\ell m'_s} (E_{n\ell} - E + H_{\text{so}})|n, \ell, m'_\ell, m'_s\rangle C(m'_\ell, m'_s) = 0.$$

Left-multiply by $\langle n, \ell, m_\ell, m_s|$ to obtain

$$\sum_{m'_\ell m'_s} (-E_{\text{so}}\delta_{m_\ell m'_\ell}\delta_{m_s m'_s} + \langle n, \ell, m_\ell, m_s|H_{\text{so}}|n, \ell, m'_\ell, m'_s\rangle)C(m'_\ell, m'_s) = 0, \tag{26.11}$$

where we have defined the change of the energy of the eigenstate as

$$E_{\text{so}} \equiv E - E_0 = E - E_{n\ell}.$$

Since m_s and m'_s assume just two values, $-1/2$ and $+1/2$, we may regard equation (26.11) as a 2×2 block-matrix equation of form:

$$\begin{pmatrix} A_{++} & A_{+-} \\ A_{-+} & A_{--} \end{pmatrix} \begin{pmatrix} C_+ \\ C_- \end{pmatrix} = 0, \tag{26.12}$$

where A_{++}, A_{+-}, A_{-+}, A_{--} are the $(2\ell + 1) \times (2\ell + 1)$ matrices

$$A_{++} \equiv \begin{pmatrix} H_{\ell,\ell}^{++} - E_{\text{so}} & H_{\ell,\ell-1}^{++} & \cdots & H_{\ell,-\ell}^{++} \\ H_{\ell-1,\ell}^{++} & H_{\ell-1,\ell-1}^{++} - E_{\text{so}} & \cdots & H_{\ell-1,-\ell}^{++} \\ \cdots & \cdots & \cdots & \cdots \\ H_{-\ell,\ell}^{++} & H_{-\ell,\ell-1}^{++} & \cdots & H_{-\ell,-\ell}^{++} - E_{\text{so}} \end{pmatrix},$$

$$A_{+-} \equiv \begin{pmatrix} H_{\ell,\ell}^{+-} & H_{\ell,\ell-1}^{+-} & \cdots & H_{\ell,-\ell}^{+-} \\ H_{\ell-1,\ell}^{+-} & H_{\ell-1,\ell-1}^{+-} & \cdots & H_{\ell-1,-\ell}^{+-} \\ \cdots & \cdots & \cdots & \cdots \\ H_{-\ell,\ell}^{+-} & H_{-\ell,\ell-1}^{+-} & \cdots & H_{-\ell,-\ell}^{+-} \end{pmatrix},$$

$$A_{-+} \equiv \begin{pmatrix} H_{\ell,\ell}^{-+} & H_{\ell,\ell-1}^{-+} & \cdots & H_{\ell,-\ell}^{-+} \\ H_{\ell-1,\ell}^{-+} & H_{\ell-1,\ell-1}^{-+} & \cdots & H_{\ell-1,-\ell}^{-+} \\ \cdots & \cdots & \cdots & \cdots \\ H_{-\ell,\ell}^{-+} & H_{-\ell,\ell-1}^{-+} & \cdots & H_{-\ell,-\ell}^{-+} \end{pmatrix},$$

$$A_{--} \equiv \begin{pmatrix} H_{\ell,\ell}^{--} - E_{\text{so}} & H_{\ell,\ell-1}^{--} & \cdots & H_{\ell,-\ell}^{--} \\ H_{\ell-1,\ell}^{--} & H_{\ell-1,\ell-1}^{--} - E_{\text{so}} & \cdots & H_{\ell-1,-\ell}^{--} \\ \cdots & \cdots & \cdots & \cdots \\ H_{-\ell,\ell}^{--} & H_{-\ell,\ell-1}^{--} & \cdots & H_{-\ell,-\ell}^{--} - E_{\text{so}} \end{pmatrix}.$$

In the above, the symbol $H_{m_\ell m_\ell'}^{\pm\pm}$ represents the matrix element for all four possible sign choices in the superscript

$$H_{m_\ell m_\ell'}^{\pm\pm} \equiv \langle n, \ell, m_\ell, \pm 1/2 | H_{\text{so}} | n, \ell, m_\ell', \pm 1/2 \rangle,$$

with both m_ℓ and m_ℓ' running from $-\ell$ to $+\ell$ in increments of 1. The symbols C_+ and C_- represent the $(2\ell+1) \times 1$ column vectors whose elements are, respectively, $C(m_\ell', +1/2)$ and $C(m_\ell', -1/2)$:

$$C_+ \equiv \begin{bmatrix} C(\ell, +1/2) \\ C(\ell - 1, +1/2) \\ \cdots \\ C(-\ell, +1/2) \end{bmatrix},$$

$$C_- \equiv \begin{bmatrix} C(\ell, -1/2) \\ C(\ell - 1, -1/2) \\ \cdots \\ C(-\ell, -1/2) \end{bmatrix}.$$

In order for equation (26.12) to have any nontrivial solutions, we require the determinant of the coefficient matrix to equal zero:

$$\begin{vmatrix} A_{++} & A_{+-} \\ A_{-+} & A_{--} \end{vmatrix} = 0.$$

This condition yields a *characteristic equation* for E_{so}, in the form of a polynomial of order $2(2\ell + 1)$, whose roots (eigenvalues) represent the changes in the energies brought about by spin-orbit coupling in the corresponding number $2(2\ell + 1)$ of formerly degenerate quantum states. Since H_{so} is a Hermitian operator (only such terms are acceptable in a valid Hamiltonian formulation), the matrix formed from the elements $H^{\pm\pm}_{m_\ell, m'_\ell}$ is a Hermitian matrix (opposite off-diagonal elements are the complex conjugates of each other; diagonal elements are real), guaranteeing, therefore, that real values will emerge from the characteristic equation for the associated eigenvalues E_{so}. The mathematical problem therefore reduces to finding the diagonal form of the matrix associated with the perturbation Hamiltonian H_{so}. (This is a general feature of the time-independent perturbation theory involving degenerate states.) Although it would be possible, in principle, to work out the problem by brute force, the solution is greatly facilitated by using physical arguments to deduce the form of the relevant eigenkets.

THE jm REPRESENTATION

To begin, we compute the value of the matrix elements. With H_{so} given by (26.5), we obtain

$$H^{\pm\pm}_{m_\ell, m'_\ell} = \left[\frac{1}{2m_e^2 c^2} \int_0^\infty \frac{1}{r} \frac{dV}{dr} R_{n\ell}^2(r)\, dr \right] \langle Y_{\ell m_\ell} \alpha(m_s) | \mathbf{S} \cdot \mathbf{L} | Y_{\ell, m'_\ell} \alpha(m'_s) \rangle,$$

where we have introduced the symbol $\alpha(m_s)$ to mean $\alpha_\pm$ depending on whether $m_s = \pm 1/2$, with a similar definition for $\alpha(m'_s)$. By the symbol $\mathbf{S} \cdot \mathbf{L}$, we mean the operator

$$\mathbf{S} \cdot \mathbf{L} \equiv S_x L_x + S_y L_y + S_z L_z.$$

The piece $S_z L_z$ requires no additional manipulation, since $Y_{\ell m_\ell} \alpha_\pm$ represents an eigenket for it. For the rest we substitute the identity:

$$S_x L_x + S_y L_y = \frac{1}{2} \left[(S_x + iS_y)(L_x - iL_y) + (S_x - iS_y)(L_x + iL_y) \right],$$

with $S_x \pm iS_y$ and $L_x \pm iL_y$ representing raising and lowering operators for the eigenvalues m_s and m_ℓ, respectively (consult Chapters 24 and 25). Because the raising of m_s is paired with the lowering of m_ℓ and vice versa, we find that

$$\langle Y_{\ell m_\ell} \alpha(m_s) | S_x L_x + S_y L_y | Y_{\ell, m'_\ell} \alpha(m'_s) \rangle$$

equals zero unless

$$m'_s + m'_\ell = m_s + m_\ell.$$

On the other hand,

$$\langle Y_{\ell m_\ell} \alpha(m_s) | S_z L_z | Y_{\ell,m'_\ell} \alpha(m'_s) \rangle$$

equals zero unless

$$m'_\ell = m_\ell, \qquad \text{and} \qquad m'_s = m_s.$$

In all cases, the matrix element $H^{\pm\pm}_{m_\ell m'_\ell}$ vanishes unless the spin and orbital angular momentum couple in such a way that

$$m'_s + m'_\ell = m_s + m_\ell \equiv m. \tag{26.13}$$

The result (26.13) follows from our earlier proof that the total-angular-momentum operator,

$$\mathbf{J} \equiv \mathbf{S} + \mathbf{L},$$

and in particular its z-component, commutes with H_{so}. Thus the sum $m = m_s + m_\ell$ forms a good quantum number (eigenvalue) even if m_s and m_ℓ do not individually (because $\mathbf{S}$ and $\mathbf{L}$ do not commute with H_{so}, although S^2 and L^2 do). Since J^2 commutes with H as well as J_z, we must be able to obtain a diagonal representation in which

$$J^2 \Phi = j(j+1)\hbar^2 \Phi,$$

with the number j being a good quantum number. From $J^2 = S^2 + L^2 + 2\mathbf{S} \cdot \mathbf{L}$, we see that the maximum value of $j(j+1)$ equals $s(s+1) + \ell(\ell+1) + 2\ell s$, whereas the minimum value equals $s(s+1) + \ell(\ell+1) - 2\ell s$. Hence j runs from $\ell - s$ to $\ell + s$ in increments of 1 (just two possibilities for $s = 1/2$; for $\ell = 0$, j has only one value, $j = s = 1/2$).

To summarize, for the particular form of $H = H_0 + H_{\text{so}}$ being considered, we have a situation in which J^2, L^2, S^2, and J_z all commute with each other and with H. Consequently, our posed eigenvalue matrix problem must contain as a possible solution a diagonal representation—the jm representation—in which

$$|n, \ell, j, m\rangle$$

is an eigenket, with m running from $-j$ to j. We do not bother to list s along with ℓ since the former can have a value of only $1/2$ for a single electron atom. Since there are $2j + 1$ possible m states for each of two possible j states ($\ell - 1/2$ and $\ell + 1/2$), we see that the total number of elementary jm states equals $2\ell + 2\ell + 2 = 2(2\ell + 1)$ for given ℓ, which equals (as it must) the number under the $m_\ell m_s$ representation. The $2(2\ell + 1)$ eigenkets of the jm representation, however, give the appropriate description of conserved quantities in a central force field, *including* the effects of spin-orbit coupling. We proceed in the following section to construct the eigenkets of the jm representation as a linear combination of the eigenkets of the $m_\ell m_s$ representation.

EXPLICIT SOLUTION

Physically, we anticipate that an eigenket $|n, \ell, j, m\rangle$ with given n and ℓ in the jm representation must form from a linear combination of the only two eigenkets in the $m_\ell m_s$ representation, with the same values of n and ℓ, whose m_ℓ and m_s can sum to m:

$$m_\ell = m - \frac{1}{2}, \quad m_s = \frac{1}{2} \quad \text{and} \quad m_\ell = m + \frac{1}{2}, \quad m_s = -\frac{1}{2}.$$

The brute-force way of doing things, equation (26.13), bears out this expectation. Equation (26.13) implies that only the diagonal elements of A_{++} and A_{--} are nonzero, and A_{+-} and A_{-+} each have only one immediate off-diagonal of nonzero elements; i.e. (except at the top and bottom), H_{so} couples one element from C_+ to one element from C_-. In other words, the eigenket $|n, \ell, j, m\rangle$ in the jm representation must equal a linear combination of $|n, \ell, m_\ell = m-1/2, m_s = +1/2\rangle$ and $|n, \ell, m_\ell = m+1/2, m_s = -1/2\rangle$ in the $m_\ell m_s$ representation if $J_z = L_z + S_z$ operating on $|jm\rangle$ is to recover $m\hbar|jm\rangle$. (Q.E.D.)

In what follows, we will suppress display of the common indices n and ℓ. From the ratio of the connected matrix elements and from the requirement that $\langle jm|jm\rangle = 1$, we can show for $j = \ell + 1/2$:

$$|jm\rangle = \left(\frac{\ell + m + 1/2}{2\ell + 1}\right)^{1/2} |m - 1/2, +1/2\rangle$$

$$- \left(\frac{\ell - m + 1/2}{2\ell + 1}\right)^{1/2} |m + 1/2, -1/2\rangle.$$

Similarly, we can show for $j = \ell - 1/2$:

$$|jm\rangle = \left(\frac{\ell - m + 1/2}{2\ell + 1}\right)^{1/2} |m - 1/2, +1/2\rangle$$

$$+ \left(\frac{\ell + m + 1/2}{2\ell + 1}\right)^{1/2} |m + 1/2, -1/2\rangle.$$

The quantities in front of the basic kets of the $m_\ell m_s$ representation are called Clebsch-Gordon coefficients (to be discussed in greater detail in the next chapter). They are sometimes given the shorthand notation $\langle m_\ell m_s|jm\rangle$ or $\langle jm|m_\ell m_s\rangle$, depending on the direction of the transformation being considered. How many quantum numbers are contained inside the angular brackets depends on the number of angular momenta being combined (two—orbital and spin—in this example).

The substitution of the solution $\Phi = |jm\rangle$ into equation (26.10) and left-multiplication by $\langle jm|$ now yields the eigenvalue E:

$$E = \langle jm|H_0|jm\rangle + \langle jm|H_{\text{so}}|jm\rangle.$$

Since $|jm\rangle$ corresponds to a normalized linear superposition of degenerate eigenfunctions of H_0, each with unperturbed energy E_0, the first term must equal E_0 (as can be verified by direct substitution of the Clebsch-Gordon solutions). The second term represents E_{so}, which reads (if we put back the $n\ell$ dependence),

$$E_{so} = \left[\frac{1}{2m_e^2 c^2} \int_0^\infty \frac{1}{r} \frac{dV}{dr} R_{n\ell}^2 \, dr \right] \langle jm | \mathbf{S} \cdot \mathbf{L} | jm \rangle.$$

But $\mathbf{S} \cdot \mathbf{L} = (J^2 - L^2 - S^2)/2$ is diagonal in the jm representation; therefore

$$\langle jm | \mathbf{S} \cdot \mathbf{L} | jm \rangle = \frac{1}{2}[j(j+1) - \ell(\ell+1) - s(s+1)]\hbar^2.$$

With $s = 1/2$, the expression in the square bracket equals ℓ for $j = \ell + 1/2$, but it equals $-(\ell+1)$ for $j = \ell - 1/2$.

For a hydrogenic atom (where $V = -Ze^2/r$, with Z being the atomic number of the nucleus),

$$E_{so} = |E_n| \frac{Z^2\alpha^2}{n\ell(\ell+1/2)(\ell+1)} \begin{cases} \ell/2, & \text{if} \quad j = \ell + 1/2, \\ -(\ell+1)/2, & \text{if} \quad j = \ell - 1/2. \end{cases}$$

In the above, α is the fine-structure constant; n equals the principal quantum number; and E_n yields the associated energy (negative) given by the Bohr formula for a hydrogenic atom of nuclear charge Z. The corresponding energy perturbation associated with the relativistic correction to the kinetic energy E_K works out to be

$$E_K = -|E_n| \frac{Z^2\alpha^2}{n^2} \left(\frac{n}{\ell+1/2} - \frac{3}{4} \right).$$

If we add, we get for the sum:

$$E_K + E_{so} = -|E_n| \left(\frac{Z^2\alpha^2}{n^2} \right) \left(\frac{n}{j+1/2} - \frac{3}{4} \right), \quad \text{for } j = \tfrac{1}{2}, \tfrac{3}{2}, \ldots, n - \tfrac{1}{2}.$$

$$(26.14)$$

Amazingly, this turns out to be identical in form to Sommerfeld's relativistic corrections, based on the old quantum theory, for Bohr's energy levels of hydrogenic atoms. Sommerfeld, of course, had no notion of electron spin, nor did his integer k have the same interpretation as $j+1/2$. The agreement arises from sheer coincidence!

Notice that the levels remain degenerate with respect to m. The m sublevels can be split via the Zeeman effect in the presence of a magnetic field; thus m is sometimes called the *magnetic quantum number*. The electric-dipole selection rules now become

$$\Delta m = 0, \pm 1; \qquad \Delta \ell = \pm 1, \qquad \Delta j = 0, \pm 1.$$

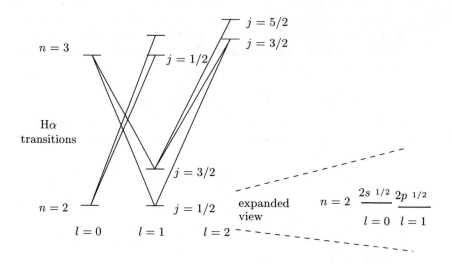

FIGURE 26.2

Energy splittings for the levels involved in the Hα transition. The blow up of the level splittings for $2s_{1/2}$ and $2p_{1/2}$ shows the Lamb shift that is not predicted by standard Dirac theory.

The usual qualification, "except for $j = 0$ to $j = 0$," does not apply to a single-electron atom, where j must have half-integer values.

As an example, we apply these considerations to the emission and absorption of the Balmer-alpha (Hα) line of atomic hydrogen. The allowed transitions are schematically linked by solid lines in Figure 26.2.

THE LAMB SHIFT

According to the above theory based on an expansion of Dirac's equation, the energy levels associated with the $2s_{1/2}$ ($n = 2$, $\ell = 0$, $j = 1/2$) and the $2p_{1/2}$ ($n = 2$, $\ell = 1$, $j = 1/2$) states should have identical values for atomic hydrogen because they have the same n and j. This result holds also when we use the full Dirac theory, and not just its nonrelativistic expansion. For a long time, spectroscopists thought that the experimental evidence agreed with the theoretical expectations, although slight discrepancies occasionally appeared. Shortly after World War II (after Allied physicists returned from the Manhattan project to their universities), Lamb and Retherford performed a very careful experiment that showed the $2s_{1/2}$ level actually lay a little above the $2p_{1/2}$ level. The separation has come to be called the

Lamb shift, and it amounts, in frequency units, to about 1057.9 MHz. This discrepancy is too large to be explained by known effects, such as hyperfine interaction with the nucleus. At nearly the same time, Kusch found that the *g* factor for the electron seemed slightly, but distinctly, larger than 2 (review Chapter 24). Theorists in the 1930s had already started to worry about various divergences that appeared in their equations when they tried to carry out the program of "second quantization" (the quantization of matter and radiation fields to allow the possibility of creation and annihilation). Now discrepancies began to show up also in experiments. The findings caused much debate, particularly at a famous Shelter Island meeting of 1947, and from these discussions emerged Bethe's (nonrelativistic) theory for the Lamb shift and Schwinger's calculation of the first-order correction for the electron *g*-factor.

The breakthrough involved the so-called renormalization technique for computing radiative corrections to the electron's rest mass that arise because of the interaction of the electron with virtual photons that it itself continually emits and reabsorbs. According to this picture, states (such as 2*s*) that carry the cloud of virtual photons, which always surround the electron, close to the nucleus allows the proton there to sample an electron mass that differs from the empirical value m_e measured at laboratory distances, and that in fact, the 2*p* state samples. A similar effect affects charge renormalization. The ability for vacuum fluctuations to create virtual electron-positron pairs means that the vacuum becomes (infinitely) polarized near a real electron, with the virtual positrons attracted in net toward the real electron, and the virtual electrons repelled. This polarization provides a divergent amount of shielding of the "bare charge" of the electron unless we introduce a cutoff in energy considered for the vacuum fluctuations. Renormalization theory then takes the "bare charge" to have just the right compensating divergent value to provide the finite value $-e$ measured by conventional laboratory experiments. The dubious process of subtracting two infinities to obtain a finite answer becomes more plausible because we can demonstrate that the answers do not depend on the value chosen for the cutoff energy, as long as it is sufficiently large. High-energy collisions that allow another charged particle to penetrate some of the shielding would then sample a larger effective electronic charge than do experiments at lower energies, and the strength of the "real" electromagnetic interaction would rise. At sufficiently high energies, the different forces of nature—strong, electromagnetic, weak, and gravitation—may all have the same "intrinsic" strength, and the forces would presumably unify in one super scheme. A completely satisfactory theory has yet to emerge from this line of speculation; nevertheless, it is interesting that the historical development that led ultimately to the present grand endeavor to unify all the forces of nature, at energies well beyond the reach of even the most ambitious accelerator projects yet envisioned, arose not from such high-energy experiments, but from exquisite spectroscopic effects measured for the tiniest of level splittings (Figure 26.2).

27

Multi-Electron Atoms

References: Leighton, *Principles of Modern Physics*, Chapters 7, 8; Wu, *Quantum Mechanics*, Chapter 9; Merzbacher, *Quantum Mechanics*, pp. 509–526.

In this chapter we consider the structure and spectroscopy of multi-electron atoms. In particular, we confront the complications presented by the need to construct a wave function that accounts (a) for the indistinguishability of the electrons in the atom, and (b) for the fact that electrons are fermions and satisfy Pauli's exclusion principle. To fix ideas, we begin with the specific example of the helium atom.

THE TWO-ELECTRON ATOM

If we ignore all terms of order v^2/c^2 including spin-orbit coupling for the time being, the Hamiltonian for two electrons in an atom reads

$$H = \frac{|\mathbf{p}_1|^2}{2m_\text{e}} + \frac{|\mathbf{p}_2|^2}{2m_\text{e}} - \frac{Ze^2}{r_1} - \frac{Ze^2}{r_2} + \frac{e^2}{r_{12}}, \qquad (27.1)$$

where the r_{12} is the separation distance between the two electrons:

$$r_{12} = |\mathbf{r}_1 - \mathbf{r}_2|. \qquad (27.2)$$

For quantum mechanics, we interpret $\mathbf{p}_1 = -i\hbar\boldsymbol{\nabla}_1$ and $\mathbf{p}_2 = -i\hbar\boldsymbol{\nabla}_2$, where $\boldsymbol{\nabla}_1$ and $\boldsymbol{\nabla}_2$ denote vector differentiation with respect to the independent coordinate variables $\mathbf{r}_1$ and $\mathbf{r}_2$ centered on the atomic nucleus, which is located at the origin.

If r_{12} is comparable in order of magnitude to r_1 and r_2, then the mutual repulsion between the two electrons can amount to 25 percent of their binding to the nucleus for $Z = 2$ (Helium). For some purposes, 25 percent can be considered small (allowing e^2/r_{12} to be treated as a perturbation, in which the first correction term to the energy is of order 25 percent, the next is $\sim [25\text{ percent}]^2$, etc.). For other purposes, it can be considered large (e.g., we should take account of its effects before considering terms like spin-orbit coupling that enter as $Z^2\alpha^2$). The same line of reasoning demonstrates why the two-electron system H^- (with $Z = 1$) constitutes a much more delicate quantum-mechanical calculation than helium, since a 50 percent term does not admit a rapidly convergent perturbational treatment. By the same token, we see that a crossover will occur at large enough Z between the importance of mutual electrostatic repulsion of the electrons and their spin-orbit couplings. From experience, people have learned that this crossover occurs roughly at $Z \sim 40$, beyond which we should go from a LS coupling scheme to a jj coupling scheme (see below).

Schrödinger's equation for the electron system,

$$H\Psi = i\hbar\frac{\partial\Psi}{\partial t}, \tag{27.3}$$

has stationary states of the form,

$$\Psi = U(\mathbf{r}_1, \mathbf{r}_2)\Sigma(1, 2)e^{-iEt/\hbar}, \tag{27.4}$$

where $U(\mathbf{r}_1, \mathbf{r}_2)$ carries the information concerning the spatial distribution, and $\Sigma(1, 2)$ describes the spin states. We begin our formal deliberations by enumerating the possible configurations for the latter.

POSSIBLE CONFIGURATIONS FOR SPIN STATES

Suppose we know that one of the electrons is in spin state C, the other in spin state D. We are tempted to write $\Sigma(1, 2) = \alpha_C(1)\alpha_D(2)$, but since the two electrons are indistinguishable, we must actually write

$$\Sigma(1, 2) \propto \alpha_C(1)\alpha_D(2) \pm \alpha_D(1)\alpha_C(2), \tag{27.5}$$

where the choice of the plus sign yields a symmetric spin ket, $\Sigma(2, 1) = \Sigma(1, 2)$, whereas the choice of the minus sign yields an antisymmetric spin ket, $\Sigma(2, 1) = -\Sigma(1, 2)$. (Recall that the probability of occurrence of a state is proportional to $\langle\Psi|\Psi\rangle$.)

Spinors of different arguments refer to different internal spaces. The postulate of quantum statistics states that *allowable wave functions of identical particles take either a completely symmetric or a completely antisymmetric form. They must either remain the same or change sign upon the interchange of the coordinates (spin or spatial) of any two particles.* In other words, the behaviors of all known systems of identical particles satisfy this postulate. If they didn't, it would in theory be possible to distinguish between the constituent particles. (See, however, Schiff 1955, pp. 225–226 for a discussion of the effective distinguishability of identical particles whose wave functions do not overlap.)

For two electrons, each of which can only have two possible spin states, $\alpha_\pm$, we have the following enumeration of possible spin-state configurations. There exist three possible symmetric combinations:

both electrons having spin "up,"

$$\Sigma_{+1}^{\text{sym}} = \alpha_+(1)\alpha_+(2), \tag{27.6}$$

one electron having spin "up," the other "down,"

$$\Sigma_0^{\text{sym}} = \frac{1}{\sqrt{2}}[\alpha_+(1)\alpha_-(2) + \alpha_-(1)\alpha_+(2)], \tag{27.7}$$

or both electrons having spin "down,"

$$\Sigma_{-1}^{\text{sym}} = \alpha_-(1)\alpha_-(2). \tag{27.8}$$

In the above, the subscript on Σ refers to the numerical value of $m_{s1} + m_{s2}$ for the configuration.

In contrast, there exists only one possible antisymmetric combination: one electron having spin "up," the other "down,"

$$\Sigma_0^{\text{anti}} = \frac{1}{\sqrt{2}}[\alpha_+(1)\alpha_-(2) - \alpha_-(1)\alpha_+(2)]. \tag{27.9}$$

We leave it as an exercise for the reader (see Problem Set 6) to show that

$$S_z\Sigma_{\pm1}^{\text{sym}} == \pm\hbar\Sigma_{\pm1}^{\text{sym}}, \qquad S_z\Sigma_0^{\text{sym}} = 0 = S_z\Sigma_0^{\text{anti}},$$

$$S^2\Sigma_{\pm1}^{\text{sym}} = 2\hbar^2\Sigma_{\pm1}^{\text{sym}}, \qquad S^2\Sigma_0^{\text{sym}} = 2\hbar^2\Sigma_0^{\text{sym}}, \qquad S^2\Sigma_0^{\text{anti}} = 0.$$

In other words, $S^2\Sigma = S(S+1)\hbar^2\Sigma$, with $S = 1$ for the symmetric spin states and $S = 0$ for the antisymmetric spin states, whereas $S_z\Sigma = M_S\hbar\Sigma$, where M_S can run from $-S$ to $+S$ in increments of 1. For Σ_0^{sym}, where $S = 1$, we still speak of the pair as having "aligned" spins, even though m_s may be $+1/2$ for one particle and $-1/2$ for the other particle, since $M_S = 0$ means only that they are not "aligned" in the z direction. (We defer to the standard notation which uses the same symbols for [some of] the operators and their associated integer [or half-integer] eigenvalues [see below]. Usually, the context of the usage will prevent confusion.)

Apart from the eigenvalue S, what distinguishes symmetric spin configurations from antisymmetric ones? To answer that question, we must consider the allowable *spatial* wave functions U that can be paired with the spin kets.

SYMMETRY OR ANTISYMMETRY OF TOTAL WAVE FUNCTION

According to the postulate of quantum statistics stated in the previous section, the total wave function $\Psi \propto U\Sigma$ of any system of identical particles must be either symmetric or antisymmetric. For electrons, we have only the *antisymmetric* choice, i.e., we must form the allowable Ψ for a system of two electrons either from

$$U^{\text{anti}}(1,2)\Sigma^{\text{sym}}(1,2), \tag{27.10}$$

or from

$$U^{\text{sym}}(1,2)\Sigma^{\text{anti}}(1,2), \tag{27.11}$$

because electrons are fermions and must, therefore, satisfy *Pauli's exclusion principle*. Suppose two electrons have identical spatial states; such a situation has a nontrivial description only in terms of a symmetric U, since $U^{\text{anti}}(1,2) = -U^{\text{anti}}(2,1) = 0$ if 1 and 2 have identical states. Equation (27.11) then requires that we choose Σ to be antisymmetric; i.e., Σ must be given by equation (27.9) in accordance with Pauli's exclusion principle, which requires two electrons to have opposed spins if they have identical spatial quantum numbers.

Conversely, suppose two electrons to have identical spin states, given by either of equations (27.6) or (27.8). Equation (27.10) then demands that the spatial wave function U must be antisymmetric, so that the probability of the two electrons having the same spatial quantum state vanishes if they both have spin "up" or both spin "down." Finally, consider a pair of electrons having one spin "up" and one spin "down." They may or may not have the same spatial quantum state; thus we must allow for both possibilities, equations (27.7) and (27.9).

More generally, for an N-electron atom, whenever we can ignore spin-orbit coupling, the governing Hamiltonian formally allows us to look for eigensolutions where the spinor part of the wave function factors out from the spatial part:

$$\Psi(1,2,\ldots,N) = U(1,2,\ldots,N)\Sigma(1,2,\ldots N)e^{-iEt/\hbar}.$$

When $N > 2$, however, the symmetry of allowable spinors Σ will remain incomplete, and we will generally need to consider sums of the product $U\Sigma$'s. For example, consider the case of three electrons, for which there must exist $2 \times 2 \times 2 = 8$ independent combinations of spinors. Of these, we can choose four spinors totally symmetric with respect to all three arguments:

$$\Sigma^{\text{sym }123}_{+3/2} = \alpha_+(1)\alpha_+(2)\alpha_+(3),$$

$$\Sigma^{\text{sym }123}_{+1/2} = \frac{1}{\sqrt{3}}[\alpha_+(1)\alpha_+(2)\alpha_-(3) + \alpha_+(1)\alpha_-(2)\alpha_+(3) + \alpha_-(1)\alpha_+(2)\alpha_+(3)],$$

$$\Sigma^{\text{sym}\,123}_{-1/2} = \frac{1}{\sqrt{3}}[\alpha_+(1)\alpha_-(2)\alpha_-(3) + \alpha_-(1)\alpha_+(2)\alpha_-(3) + \alpha_-(1)\alpha_-(2)\alpha_+(3)],$$

$$\Sigma^{\text{sym}\,123}_{-3/2} = \alpha_-(1)\alpha_-(2)\alpha_-(3);$$

two spinors symmetric in the arguments 2, 3:

$$\Sigma^{\text{sym}\,23}_{+1/2} = \frac{1}{\sqrt{6}}\{\alpha_+(1)[\alpha_+(2)\alpha_-(3) + \alpha_-(2)\alpha_+(3)] - 2\alpha_-(1)\alpha_+(2)\alpha_+(3)\},$$

$$\Sigma^{\text{sym}\,23}_{-1/2} = \frac{1}{\sqrt{6}}\{\alpha_-(1)[\alpha_-(2)\alpha_+(3) + \alpha_+(2)\alpha_-(3)] - 2\alpha_+(1)\alpha_-(2)\alpha_-(3)\};$$

and two spinors antisymmetric in the arguments 2, 3:

$$\Sigma^{\text{anti}\,23}_{+1/2} = \frac{1}{\sqrt{2}}\alpha_+(1)[\alpha_+(2)\alpha_-(3) - \alpha_-(2)\alpha_+(3)],$$

$$\Sigma^{\text{anti}\,23}_{-1/2} = \frac{1}{\sqrt{2}}\alpha_-(1)[\alpha_+(2)\alpha_-(3) - \alpha_-(2)\alpha_+(3)].$$

The last four have no particular symmetries with respect to the pairs 1, 2 or 3, 1. But since our spinor space has only eight dimensions (which we may think of as a $2 \times 2 \times 2$ array), any other spinor can be formed as a linear combination of the eight listed above (whose coefficients have been chosen to make them an orthonormal set; see Problem Set 5). For example, consider a normalized spinor symmetric in, say, the argument 1, 2, and having $+1/2$ as the eigenvalue for S_z:

$$\Sigma^{\text{sym}\,12}_{+1/2} = \frac{1}{\sqrt{6}}\{[\alpha_+(1)\alpha_-(2) + \alpha_-(1)\alpha_+(2)]\alpha_+(3) - 2\alpha_+(1)\alpha_+(2)\alpha_-(3)\}.$$

We can easily compute that $\Sigma^{\text{sym}\,12}_{+1/2}$ is orthogonal to $\Sigma^{\text{sym}\,123}_{+1/2}$, and thus contains no contribution from it. It must be obtainable, therefore, as a linear combination of $\Sigma^{\text{sym}\,23}_{+1/2}$ and $\Sigma^{\text{anti}\,23}_{+1/2}$, which we can verify as

$$\Sigma^{\text{sym}\,12}_{+1/2} = -\frac{1}{2}\Sigma^{\text{sym}\,23}_{+1/2} - \frac{\sqrt{3}}{2}\Sigma^{\text{anti}\,23}_{+1/2}.$$

Note that although we may construct spinors antisymmetric with respect to the interchange of 2 and 3, or 3 and 1, or 1 and 2, there exists no (nonzero) spinor completely antisymmetric with respect to 1, 2, 3. For example, if we try to make $\Sigma^{\text{anti}\,23}_{+1/2}$ also antisymmetric with respect to the interchange of 1 with respect to 2 or 3, we need to add inside the bracket a term $-\alpha_-(1)\alpha_+(2)\alpha_+(3)$ to balance the first term, whereas we need to add another term $\alpha_-(1)\alpha_+(2)\alpha_+(3)$ to balance the second. However, these two added terms cancel, so we've made no progress toward the desired goal. Our inability to construct a completely antisymmetric spinor for more than two electrons rests fundamentally with the fact that single-fermion spinors possess only two states corresponding to $m_s = \pm 1/2$.

To conclude this section, we mention for completeness that, in contrast to the situation for fermions, a system of identical *bosons* must have a wave function that is completely *symmetric*; so as many photons as we like can occupy a given quantum state. This allows many photons to cooperate collectively and to give the macroscopic illusion that they behave, in the classical limit, as if they satisfied Maxwell's field equations. In contrast, Pauli's exclusion principle (or, alternatively, the rule concerning the anti-symmetry of their governing wave functions) prevents the wave properties of the fermions that make up ordinary matter—the electron, the proton, and the neutron—from reinforcing each other, thereby leading to the classical illusion that they behave more like particles than like waves. As a consequence, the wave-particle duality of fermions generally can manifest itself—one particle at a time—only on a microscopic level. Exception to this rule can occur if fermions can "pair up" at low temperatures. Pairs of fermions with half-integer values of angular momenta can behave, as an aggregate, like a boson (with an integer quantity of angular momentum). Spectacular changes in the macroscopic properties of matter—e.g., super-fluidity and superconductivity—may occur under these circumstances.

SPIN-SPIN CORRELATION

To illustrate the importance of the concepts discussed in the previous section, we consider a simple example. Consider a pair of electrons whose spin state consists of either an antisymmetric spin ket Σ^{anti} (i.e., the $S = 0$ state), or a symmetric one Σ^{sym} (i.e., one of the the three possible $S = 1$ states). Suppose, moreover, we know that the two electrons occupy single-particle spatial states A and B. The spatial wave function must be symmetric if $\Sigma = \Sigma^{\text{anti}}$, and it must be antisymmetric if $\Sigma = \Sigma^{\text{sym}}$,

$$U(1,2) = \frac{1}{\sqrt{2}} \left[u_A(\mathbf{r}_1) u_B(\mathbf{r}_2) \pm u_B(\mathbf{r}_1) u_A(\mathbf{r}_2) \right], \qquad (27.12)$$

where the upper sign applies to the case of $S = 0$, the latter to $S = 1$.

Consider, now, the expected mean-square spatial separations for the two electrons:

$$\langle r_{12}^2 \rangle \equiv \langle U | (\mathbf{r}_1 - \mathbf{r}_2) \cdot (\mathbf{r}_1 - \mathbf{r}_2) | U \rangle.$$

If we substitute in U as given by equation (27.12), we get

$$\langle r_{12}^2 \rangle = \frac{1}{2} \langle u_A(1)u_B(2) \pm u_B(1)u_A(2) |$$
$$\times r_1^2 + r_2^2 - 2\mathbf{r}_1 \cdot \mathbf{r}_2 | u_A(1)u_B(2) \pm u_B(1)u_A(2) \rangle.$$

Proper normalization requires $\langle u_A(1)|u_A(1)\rangle = 1$, with a similar requirement if we replace both A's by B's, or if we replace both 1's by 2's. On the other hand, the orthogonality of independent eigenstates implies $\langle u_A(1)|u_B(1)\rangle = 0$, with a similar result if we replace both 1's by 2's. With these identities, we obtain, after a little algebra,

$$\langle r_{12}^2 \rangle = \langle A|r^2|A|\rangle + \langle B|r^2|B|\rangle - 2\langle A|\mathbf{r}|A\rangle \cdot \langle B|\mathbf{r}|B\rangle \mp \langle A|\mathbf{r}|B\rangle \cdot \langle B|\mathbf{r}|A\rangle,$$

where we have written $\langle |u_A(1)|\mathbf{r}_1|u_A(1)\rangle \equiv \langle A|\mathbf{r}|A\rangle \equiv \langle u_A(2)|\mathbf{r}_2|u_A(2)\rangle$, since the subscripts 1 and 2 are superfluous for dummy variables that get integrated inside the angular brackets. The notations for the other quantities have a similar interpretation. Even more suggestively, we may write the above relationship as

$$\langle |\mathbf{r}_1 - \mathbf{r}_2|^2 \rangle = \left[\langle r^2 \rangle_A + \langle r^2 \rangle_B - 2\langle \mathbf{r} \rangle_A \cdot \langle \mathbf{r} \rangle_B \right] \mp |\langle A|\mathbf{r}|B\rangle|^2. \qquad (27.13)$$

The terms in the square bracket above represent the classical expectation value for $|\mathbf{r}_1 - \mathbf{r}_2|^2$ if we are told that one electron occupies spatial state A and the other B. The extra term $\mp|\langle A|\mathbf{r}|B\rangle|^2$ gives the quantum correction because of *spin-spin correlation* (upper sign for $S = 0$, lower for $S = 1$). From equation (27.13), we see that a pair of electrons (or any other pair of identical fermions) with misaligned spins ($S = 0$) has a mean-square separation *smaller* than the classical expectation value, but a pair with aligned spins ($S = 1$) has a *larger* mean-square separation. In some sense, Pauli's exclusion principle causes two electrons with aligned spins (total $S = 1$) to "repel" each other, and two electrons with misaligned spins (total $S = 0$) to "attract." The tendency toward "attraction" always holds for bosons (either elementary particles with whole integer spins, or compound particles made of fermions with half-integer spins that have combined in pairs with opposing spins), which must have symmetric spatial wave functions; this tendency for bosons to attract other bosons into the same quantum state underlies the phenomenon of stimulated emission.

Notice, however, that the tendency for like bosons to aggregate and for like fermions with aligned spins to disperse lies not in forces contained as terms in the governing Hamiltonian, but *in the quantum statistics of allowable wave functions*. The tendency for electrons with aligned spins to stay apart, and electrons with misaligned spins to come closer together, has profound consequences for the spectroscopy of multi-electron atoms, a problem whose intricacies we now return to explore.

LS COUPLING

When we substitute equation (27.4) into equation (27.3), we get the time-independent Schrödinger equation:

$$HU = EU.$$

If equation (27.1) gives H, we have the complication presented by the term $e^2/|\mathbf{r}_1 - \mathbf{r}_2|$, which represents the mutual electrostatic repulsion of the two orbital electrons. This mutual repulsion destroys the central-force-field nature of the corresponding one-electron problem; consequently, the individual orbital and spin angular momenta of the electrons are no longer conserved. Consider, however, the operators

$$\mathbf{L} \equiv \mathbf{L}_1 + \mathbf{L}_2, \tag{27.14}$$

$$\mathbf{S} \equiv \mathbf{S}_1 + \mathbf{S}_2, \tag{27.15}$$

$$\mathbf{J} = \mathbf{L} + \mathbf{S}. \tag{27.16}$$

In the approximation (27.1), where we drop spin-orbit coupling, we claim that L^2, S^2, J^2, L_z, S_z, and J_z all commute with H.

Our claim holds obviously for S^2 and S_z, since spin does not enter anywhere in the Hamiltonian when we drop the spin-orbit-coupling terms. It holds intuitively for $\mathbf{J}$ and J^2, because total angular momentum of the electron system must be conserved if we ignore radiative perturbations; it also holds intuitively for $\mathbf{L}$ and L^2, since (classically) the *total* orbital angular momentum cannot be affected by opposite and equal torques exerted by the mutually repelling electrons.

To demonstrate the conservation, say, of L_z quantum-mechanically, notice that the troublesome term has a dependence on spatial coordinates equal to

$$|\mathbf{r}_1 - \mathbf{r}_2|^{-1} = (r_1^2 + r_2^2 - 2r_1 r_2 \cos\gamma)^{-1/2},$$

where γ is the angle between $\hat{\mathbf{r}}_1$ and $\hat{\mathbf{r}}_2$:

$$\cos\gamma = \hat{\mathbf{r}}_1 \cdot \hat{\mathbf{r}}_2 = \cos\theta_1 \cos\theta_2 + \sin\theta_1 \sin\theta_2 [\cos\phi_1 \cos\phi_2 + \sin\phi_1 \sin\phi_2],$$

if we adopt the usual spherical polar coordinates. A familiar trigonometric identity gives the quantity in the square bracket as

$$\cos(\phi_1 - \phi_2).$$

Because ϕ_1 and ϕ_2 enter only in the combination of their difference, if we operate on $|\mathbf{r}_1 - \mathbf{r}_2|^{-1}$ by

$$L_z = -i\hbar \left(\frac{\partial}{\partial\phi_1} + \frac{\partial}{\partial\phi_2} \right),$$

we get no net contribution. Thus L_z commutes with H, as does S_z; and so must $J_z = L_z + S_z$. (Q.E.D.) The nonuniqueness of the z direction then shows that the other components L_x, L_y, S_x, and S_y also commute with H; although this fact provides no practical computational help, since these components do not commute with each other.

To summarize, J^2, L^2, S^2, J_z, L_z, and S_z all commute with each other as well as with H. Hence we could have an eigenket representation in which these operators in matrix form would all have only diagonal elements, with associated eigenvalues $J(J+1)$, $L(L+1)$, $S(S+1)$, M_J, M_L, M_S, and E. The energy E will turn out to be degenerate with respect to M_L, M_S, and M_J, because there is nothing special physically about the z direction. The energy E will also be independent of J, because the Hamiltonian does not depend on how orbit and spin angular momenta add if we ignore the term H_{so}. (Conversely, we anticipate that the inclusion of spin-orbit coupling will split the degeneracy in J.) Although the Hamiltonian without H_{so} does not depend on spin, we cannot conclude that E will be degenerate in S, because of the indirect effects of spin-spin correlations discussed in the previous section. From that discussion, we anticipate that states corresponding to $S = 1$ will lie lower in energy than states with $S = 0$, because aligned spins tend to keep electrons farther apart than when they have misaligned spins. The increased separation reduces the positive contribution to the total energy from mutual electrostatic repulsion. If the energy E depends on S, it will generally also vary with L. (Basically, the individual electrons do not move in the pure Coulomb potential of the nucleus.) States with large values of L correspond classically to the two electrons revolving more in the same direction than states with small values of L. Thus the electrons (or their "clouds") keep apart better for large L than small L, reducing again the positive contributions of the mutual electrostatic repulsion. The importance of these two considerations lends the present coupling scheme for angular momentum its name: *LS coupling* (as discussed originally in the old quantum theory [vector model of angular momentum] by Russell and Saunders, 1925, *Ap. J.*, **61**, 38). The various splittings—ordered first in S (due to spin-spin correlation), then in L (due to the "sense" of orbital revolution), and finally in J (due to spin-orbit coupling if we include the effects of H_{so})—look schematically as depicted in Figure 27.1 (for less than half-filled subshells).

The above results were first discerned from empirical spectroscopic examination and are known as *Hund's rules*:

1. Higher S $\Rightarrow$ lower energy.

2. Higher L $\Rightarrow$ lower energy.

3. Higher J $\Rightarrow$ higher energy if less than half-filled, lower energy if more than half-filled.

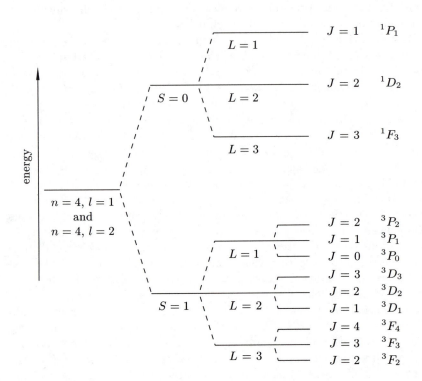

FIGURE 27.1
Example of energy splittings in the LS coupling scheme.

The spectroscopic term stands for the quantities $^{2S+1}L_J$, where the symbol S, P, D, F, ... denote $L = 0, 1, 2, 3, ...$. The electric-dipole selection rules for LS coupling are

$$\Delta S = 0, \qquad \Delta L = 0, \pm 1, \qquad \Delta J = 0, \pm 1 \qquad (\text{except } J = 0 \text{ to } J = 0).$$

To lowest (dipole) order, spin flips cannot occur; so $\Delta S = 0$. The maximum orbital angular momentum that can be carried by a photon in a dipole transition equals 1 unit, so $\Delta L = 0$ or ± 1 ($\Delta L = 0$ is no longer eliminated by parity considerations when we have two electrons), which then implies $\Delta J = 0$, or ± 1.

ADDITION OF ANGULAR MOMENTA

In the last section, we took for granted that two angular momentum operators (orbital or spin), $\mathbf{J}_1$ and $\mathbf{J}_2$, which act on two different vector spaces (either because one refers to orbital coordinates and the other to spin coordinates, or because the two refer to different particles), could be added

$$\mathbf{J} = \mathbf{J}_1 + \mathbf{J}_2,$$

with obvious consequences for the eigenvalues in which the product-space representation is diagonal. Except for the case $\mathbf{J} = \mathbf{L} + \mathbf{S}$ for the sum of orbital and spin angular momenta of a single electron covered in Chapter 26, however, we have not dwelled on how to obtain such a diagonal representation in practice. In this section, we remedy this shortcoming by generalizing the discussion of Chapter 26.

Suppose $|j_1 m_1\rangle$ and $|j_2 m_2\rangle$ represent eigenkets of J_1^2, J_{1z} and J_2^2, J_{2z}, with eigenvalues $j_1(j_1 + 1)$, m_1 and $j_2(j_2 + 1)$, m_2. Let the product of these kets be denoted by

$$|j_1 m_1\rangle |j_2 m_2\rangle \equiv |j_1 j_2 m_1 m_2\rangle.$$

The totality of such product kets forms a basis in the product space on which $\mathbf{J} = \mathbf{J}_1 + \mathbf{J}_2$ operates. We wish to use this basis to construct the eigenkets of J^2 and J_z, the totality of which can form a new basis set.

We begin by noting that J_1^2 and J_2^2 commute with J^2 and J_z, so that our new basis can simultaneously be eigenvectors of J_1^2, J_2^2, J^2, and J_z. We denote the eigenket corresponding to the eigenvalues $j_1(j_1 + 1)$, $j_2(j_2 + 1)$, $j(j + 1)$, and m by $|j_1 j_2 j m\rangle$. We wish to obtain the latter as a linear combination of the kets of the old basis:

$$|j_1 j_2 j m\rangle = \sum_{m_1 m_2} |j_1 j_2 m_1 m_2\rangle \langle j_1 j_2 m_1 m_2 | j_1 j_2 j m\rangle. \tag{27.17}$$

We have encountered the transformation coefficients $\langle j_1 j_2 m_1 m_2 | j_1 j_2 j m\rangle$ before as Clebsch-Gordon coefficients. Our task here reduces to obtaining the relevant formulae by which we may calculate these coefficients.

If we apply $J_z = J_{1z} + J_{2z}$ to equation (27.17), we get the requirement,

$$\langle j_1 j_2 m_1 m_2 | j_1 j_2 j m\rangle = 0 \qquad \text{unless} \qquad m_1 + m_2 = m. \tag{27.18}$$

For given m, this requirement simplifies, if we wish, the unrestricted double summation in equation (27.17) to a single sum. If we apply the raising or lowering operator, $J_x \pm i J_y = (J_{1x} \pm i J_{1y}) + (J_{2x} \pm i J_{2y})$, to equation (27.17), rewrite $m_1 \pm 1$ as m_1' for the double sum of kets operated upon by $(J_{1x} \pm i J_{1y})$, rewrite $m_2 \pm 1$ as m_2' for the double sum of kets operated upon by $(J_{2x} \pm i J_{2y})$, notice that equation (27.18) yields no contribution in the two sums except for $m_1' + m_2 = m \pm 1$ and $m_1 + m_2' = m \pm 1$, and suppress for brevity the dependence on j_1, j_2, we obtain the recursion relation [see equation (24.14)]

$$c_\pm(j, m)\langle m_1 m_2 | j, m \pm 1 \rangle$$
$$= c_\mp(j_1, m_1)\langle m_1 \mp 1, m_2 | jm \rangle + c_\mp(j_2, m_2)\langle m_1, m_2 \mp 1 | jm \rangle, \quad (27.19)$$

where we have introduced the notation

$$c_\pm(j, m) \equiv [(j \mp m)(j + 1 \pm m)]^{1/2}.$$

Having a recursion relation, we may obtain all the Clebsch-Gordon coefficients in terms of one of them, e.g., the one with the maximal values for the z-projection indices:

$$\langle j_1, j - j_2 | jj \rangle \equiv \langle j_1 j_2 j_1, j - j_2 | j_1 j_2 jj \rangle.$$

By convention, the value of this last coefficient is set to be real and positive, with the value determined by the condition that the eigenkets $|jm\rangle \equiv |j_1 j_2 jm\rangle$ form an orthonormal set:

$$\sum_{m_1 m_2} \langle j_1 j_2 m_1 m_2 | j_1 j_2 jm \rangle \langle j_1 j_2 m_1 m_2 | j_1 j_2 j'm' \rangle = \delta_{jj'} \delta_{mm'}. \quad (27.20)$$

The derivation of the Clebsch-Gordon formulae completes our discussion of this section. Once we know how to add two angular momenta operators, we know, in principle, how to superimpose any number of them.

MULTI-ELECTRON ATOMS

Schematically, we may think of the total Hamiltonian in an N-electron atom, including spin-orbit coupling, as being composed of three pieces:

$$H = H_{\text{cf}} + H_{\text{ee}} + H_{\text{so}}. \quad (27.21)$$

In equation (27.21), H_{cf} gives the central-field approximation of N independent electrons moving in a spherically symmetric potential:

$$H_{\text{cf}} = \sum_{a=1}^{N} H_a, \quad (27.22)$$

with H_a being the effective one-electron Hamiltonian

$$H_a \equiv -\frac{\hbar^2}{2m_{\text{e}}} \nabla_a^2 + V_a(r_a), \quad (27.23)$$

where the effective potential energy $V_a(r_a)$ equals the nuclear contribution plus a spherical approximation for a smeared-out electron-cloud distribution:

$$V_a(r_a) = -\frac{Ze^2}{r_a} + W_a(r_a). \tag{27.24}$$

In equation (27.21), H_{ee} gives the electron-electron repulsion after subtracting out their net spherically symmetric effect,

$$H_{\text{ee}} = \frac{e^2}{2} \sum_{a \neq b}^{N} \frac{1}{r_{ab}} - \sum_{a=1}^{N} W_a(r_a), \tag{27.25}$$

and H_{so} arises from spin-orbit interaction (see Chapters 25 and 26):

$$H_{\text{so}} = \sum_{a=1}^{N} \frac{1}{2m_{\text{e}}^2 c^2} \left(\frac{1}{r_a} \frac{dV_a}{dr_a} \right) \mathbf{S}_a \cdot \mathbf{L}_a. \tag{27.26}$$

In equations (27.22) and (27.25), we have added and subtracted the same term,

$$\sum_{a=1}^{N} W_a(r_a),$$

which therefore can be chosen as we please, either for convenience or for maximum computational accuracy. The *Hartree-Fock Procedure* begins by using a *Slater determinant* to construct an antisymmetric total wave function $\Psi = \Phi e^{-iEt/\hbar}$. Let $\phi \equiv u\alpha$ represent a single-particle spin-orbital, where u gives the (yet unknown) spatial dependence and α equals either α_+ or α_-. Suppose our N electrons occupy the set of spin-orbitals: $\{\phi_a\}_{a=1}^{N}$. Consider the total wave function Φ obtained by forming the following determinant:

$$\Phi(1, 2, \ldots, N) = \frac{1}{\sqrt{N!}} \begin{vmatrix} \phi_1(1) & \phi_1(2) & \cdots & \phi_1(N) \\ \phi_2(1) & \phi_2(2) & \cdots & \phi_2(N) \\ \vdots & \vdots & \vdots & \vdots \\ \phi_N(1) & \phi_N(2) & \cdots & \phi_N(N) \end{vmatrix}.$$

Since a determinant changes sign whenever any two columns are interchanged, Φ is clearly antisymmetric when any two electrons exchange places. Moreover, should any two rows be identical (i.e., should any two electrons occupy the same spin-orbital), the wave function Φ vanishes, as required by Pauli's exclusion principle.

As a simple example, the two electrons in the ground state of a helium atom both occupy the $1s$ orbital ($n = 1$ and $\ell = 0$). Let u denote the spatial wave function for the $1s$ orbital, and construct the elementary spin-orbitals as $\phi_1 = u\alpha_+$ and $\phi_2 = u\alpha_-$. The corresponding Slater determinant reads

$$\Phi(1,2) = \frac{1}{\sqrt{2}} \begin{vmatrix} u(1)\alpha_+(1) & u(2)\alpha_+(2) \\ u(1)\alpha_-(1) & u(2)\alpha_-(2) \end{vmatrix} = u(1)u(2)\Sigma_0^{\text{anti}},$$

where Σ_0^{anti} is given by equation (27.9). A choice for the two-electron wave function Φ which separates out the individual spatial dependences as $u(\mathbf{r}_1)u(\mathbf{r}_2)$, however, clearly cannot satisfy a Schrödinger equation, $H\Phi = E\Phi$, where the Hamiltonian H depends on the combination $r_{12} = |\mathbf{r}_1 - \mathbf{r}_2|$ [see equation (27.1)]. Intuitively, we might guess that the best we could manage would be to make $u(\mathbf{r})$ satisfy

$$\left[-\frac{\hbar^2}{2m_e}\nabla^2 - \frac{Ze^2}{r} + W(\mathbf{r}) \right] u(\mathbf{r}) = \epsilon u(\mathbf{r}), \qquad (27.27)$$

where ϵ is the effective energy associated with the single-particle state and $W(\mathbf{r})$ is the value of electronic repulsive energy averaged over the distribution of the other (identical) electron:

$$W(\mathbf{r}) = \int \frac{e^2}{|\mathbf{r} - \mathbf{r}'|} u^*(\mathbf{r}')u(\mathbf{r}')\, d^3r', \qquad (27.28)$$

with $W(\mathbf{r}) = W(r)$ being spherically symmetric if $u(\mathbf{r}') = u(r')$ is an s orbital. Problem Set 6 shows that equation (27.27), a special case of the general Hartree-Fock procedure, does indeed represent the *best* that we can do within the framework of the central-field approximation. To do better, we must be prepared to take a *sum* of Slater determinants.

In the general case, the quantity $W_a(r_a)$ that enters in the single-particle Schrödinger equation should equal, for self-consistency, the expectation value of e^2/r_{ab}, summed and integrated over the distribution $|\Phi(1, 2, \ldots, N)|^2$ of the coordinates of every electron except the a-th one. Since we don't know the latter distribution until after we've solved the single-particle problem for each a, the solution generally needs to proceed by iteration. (Approximate eigenvalues may be obtained, however, by variational techniques; see Chapter 28.)

LIMITATIONS OF THE HARTREE-FOCK METHOD

In any case, the Hartree-Fock method yields solutions adequate to describe the ground state configuration of N-electron atoms (e.g., it can explain the arrangement of the periodic table). The solution is not very accurate, because, being dependent only on the scalar values of $r_1, r_2, \ldots, r_N$, it does

FIGURE 27.2
Relative electron positions not distinguished by the Hartree-Fock
approximation.

not distinguish between different vector separations of the type illustrated
in Figure 27.2, for which the actual spin-spin correlation would give different weights to what the interacting electrons contribute to the electrostatic
energy.

For calculations of excited states accurate enough for spectroscopic purposes, then, we need to account for the effects of H_{ee}. For relatively light
atoms, where $H_{cf} \gg H_{ee} \gg H_{so}$, this can be carried out, in principle, via
the LS coupling scheme, which accounts for the perturbation of H_{ee}. After
the electronic electrostatic splittings have been included, we can, in principle, perform another perturbation calculation to take care of the spin-orbit
coupling term H_{so}. The latter calculation removes the degeneracy with respect to the quantum number J, yielding fine-structure energy levels that
are proportional to (see Chapter 26)

$$\frac{1}{2}[J(J+1) - L(L+1) - S(S+1)].$$

The fine-structure energy levels remain degenerate with respect to the
$2J + 1$ sublevels, $M_J = -J, \ldots, +J$. The latter can be split through the
application of an external magnetic field (Zeeman effect).

The LS coupling scheme proves adequate for most astronomical applications, where we usually deal with relatively light atoms. For heavier
atoms, $Z > 40$, the LS coupling scheme breaks down, because H_{so} becomes
comparable to or larger than H_{ee}. In the approximation that $H_{so} \gg H_{ee}$,
it is possible to introduce another angular-momentum coupling scheme,
which has its basis in our study of hydrogenic atoms in Chapter 26. The
idea of jj coupling is that spin-orbit coupling by itself preserves j^2 and j_z
(where $\mathbf{j} = \mathbf{l} + \mathbf{s}$) of the individual electrons. The individual $\mathbf{j}$'s may then
be added to give a total $\mathbf{J}$, whose absolute square and z component are
preserved even in the presence of electronic electrostatic interactions, H_{ee}.

SPHERICAL SYMMETRY OF FILLED SUBSHELLS

The Hartree-Fock procedure works as well as it does because the filled electronic shells act as if they were completely spherically symmetric, so that we usually need to account for nonspherical effects for only the valence electrons. For a filled shell, the angular distribution of the $2\ell + 1$ pairs of electrons (with opposed spins), would equal, if the electrons were distinguishable and labeled by their m-values,

$$\sum_{m=-\ell}^{\ell} Y_{\ell m}^*(\theta_m, \phi_m) Y_{\ell,m}(\theta_m, \phi_m).$$

Since the electrons are not distinguishable, each electron label in the spatial wave function for a completely filled shell must contain all $2\ell+1$ terms in the spherical-harmonic decomposition, and the angular distribution of electric charge in the closed shell behaves as

$$\sum_{m=-\ell}^{\ell} Y_{\ell m}^*(\theta, \phi) Y_{\ell m}(\theta, \phi) = \frac{2\ell + 1}{4\pi},$$

a constant independent of θ and ϕ, if we use the "sum rule" derivable from the addition theorem of spherical harmonics [see Jackson 1962, equation (3.69)]. As a consequence, if all of the m sublevels of any ℓ level are filled with electrons, these electrons have a mean spatial distribution that is spherically symmetric. (Q.E.D.)

The above result has an interesting corollary. An ℓ subshell more than half-filled has a spectroscopic behavior similar to that of a subshell that has the complementary number of missing electrons (holes). According to this equivalence between electrons and holes, a subshell missing one electron can be assigned the same spectroscopic term as a subshell containing one electron. We can reason heuristically as follows. For the subshell missing an electron, add and subtract the missing electron. The added electron fills the subshell, so that it acts as if it were spherically symmetric. The subtracted electron behaves like a positron (as a "hole"), which can reside in a filled electronic subshell because it is not an electron. The spectroscopic properties of the positron in the "new" atom are identical to those of a single electron in an atom of the same effective charge. Thus we have demonstrated our corollary for the case of one hole. We can now extend the argument to two holes, etc. This corollary provides the theoretical basis for Hund's third rule.

WIGNER-ECKART THEOREM

The above property simplifies computations considerably; nevertheless, direct methods to calculate matrix elements for complex atoms with many valence electrons (or holes) would require much laborious work if we were to start from scratch each time in forming the sums and products of various spherical harmonics. Fortunately, a device from group theory exists to relieve much of the tedium.

The crucial idea in what follows is contained in the so-called *Wigner-Eckart theorem*. The theorem intends to separate the purely geometric part of the matrix element from the physical part. As an example, consider the computation of the dipole matrix element $\langle f|\mathbf{x}|i\rangle$ for a single electron atom. In Chapter 25 we showed that if we considered the three quantities

$$T_1^0 \equiv z, \qquad T_1^{\pm 1} \equiv \mp \frac{1}{\sqrt{2}}(x \pm iy), \qquad (27.29)$$

we could reduce the calculation to angular integrals (geometric part) times radial integrals (physical part). [The coefficients are chosen so that, apart from a factor of $(4\pi/3)^{1/2}$, we get rY_{10} and $rY_{1,\pm 1}$.] The Wigner-Eckart theorem states that this example constitutes a special case of a more general result, which we invoke here without formal proof.

Consider a tensor operator of rank j of the rotation group (see below) consisting of $2j + 1$ elements, T_j^m, with $m = -j$ to $+j$ in increments of 1. We wish to calculate the matrix element formed by T_j^m with the states $\langle N_1 j_1 m_1|$ and $|N_2 j_2 m_2\rangle$, where j_1, m_1 and j_2, m_2 constitute angular-momentum quantum numbers, and N_1 and N_2 represent the totality of other quantum numbers needed to specify the states 1 and 2. The Wigner-Eckart theorem states that this matrix element can always be factored to read

$$\langle N_1 j_1 m_1|T_j^m|N_2 j_2 m_2\rangle = \langle j_2 j m_2 m|j_2 j j_1 m_1\rangle\langle N_1 j_1||T_j||N_2 j_2\rangle, \quad (27.30)$$

where $\langle j_2 j m_2 m|j_2 j j_1 m_1\rangle$ is a Clebsch-Gordon coefficient and $\langle N_1 j_1||T_j||N_2 j_2\rangle$ is the *reduced matrix element*, with $|T_j|$ being the *irreducible spherical tensor operator* whose square is the norm of T_j^m:

$$|T_j|^2 = \sum_{m=-j}^{j} |T_j^m|^2. \qquad (27.31)$$

The reduced matrix element depends only on $|T_j|$ and the quantum numbers N_1, N_2, j_1, j_2, and j. It does not depend on m_1, m_2, and m, which give the orientation of the system in space; so this is an eminently reasonable result.

For applications, the trick boils down to the issues of how to recognize what quantities qualify as "tensor operators of the rotation group" and what constitutes an "irreducible representation." This requires an excursion into group theory, which we are not prepared to do. Let us merely note that the usual quantities needed for the computation of electric dipole moments, magnetic dipole moments, electric quadrupole moments, etc., can all be put into this form.

APPLICATION TO MULTIPOLE RADIATION

To fix ideas, let us consider some examples. If we take as the tensor of rank 1 the case represented by equation (27.29), and we let $n\ell jm$ denote the complete set of quantum numbers for a one-electron atom, we obtain, upon application of the Wigner-Eckart theorem,

$$\langle n_1\ell_1 j_1 m_1 | T_1^m | n_2\ell_2 j_2 m_2 \rangle = \langle j_2 1 m_2 m | j_2 1 j_1 m_1 \rangle \langle n_1\ell_1 \| T_1 \| n_2\ell_2 \rangle,$$

where $|T_1| = r$ and

$$\langle n_1\ell_1 \| T_1 \| n_2\ell_2 \rangle = \int_0^\infty r \mathcal{R}_{n_1\ell_1} \mathcal{R}_{n_2\ell_2}\, dr,$$

with $\mathcal{R}_{n\ell}$ being the reduced radial wave function for a hydrogenic atom [see equation (24.19)]. The Clebsch-Gordon coefficient $\langle j_2 1 m_2 m | j_2 1 j_1 m_1 \rangle$ involves the addition of angular momenta of one unit to j_2 to obtain j_1; it vanishes unless $j_2 - 1 \geq j_1 \geq j_2 + 1$. This implies the selection rule $\Delta j \equiv j_1 - j_2 = 0, \pm 1$. Moreover, since m has only three possible values in our example, 0, ± 1, we have $\Delta m \equiv m_1 - m_2 = 0, \pm 1$. These are the usual selection rules for electric dipole transitions in a single-electron atom. The argument is essentially the same for electric dipole transitions in multi-electron atoms and molecules (see Chapter 30), except that accurate values for the reduced matrix element must usually be obtained by laboratory experiments rather than by explicit calculation.

For magnetic dipole transitions, we define the irreducible spherical tensor operator of rank 1 as

$$T_1^0 = J_z, \qquad T_1^{\pm 1} = \mp \frac{1}{\sqrt{2}}(J_x \pm iJ_y). \tag{27.32}$$

For electric quadrupole transitions, we define the irreducible spherical tensor operator of rank 2 as [see equations (24.24) to (24.26)]

$$T_2^0 = Q_{zz}, \quad T_2^{\pm 1} = \mp\sqrt{\frac{2}{3}}(Q_{zx} \pm iQ_{yz}), \quad T_2^{\pm 2} = \frac{1}{\sqrt{6}}(Q_{xx} \pm 2iQ_{xy} - Q_{yy}). \tag{27.33}$$

(We choose the coefficients so that $|T_j|^2 \propto r^4$.) Application of the Wigner-Eckart theorem now yields the standard selection rules for magnetic dipole and electric quadrupole transitions.

LINEAR ZEEMAN EFFECT

Consider next the linear Zeeman effect in an atom that satisfies LS coupling. If we adopt Dirac's value 2 for the g factor of the electron, the perturbation Hamiltonian is given by equation (25.56):

$$H_Z = \frac{e}{2m_e c} \mathbf{B}_0 \cdot (\mathbf{L} + 2\mathbf{S}). \tag{27.34}$$

The linear treatment given below applies if $H_Z \ll H_{so}$ (which, in turn, $\ll H_{ee}$ if LS coupling holds). In order of magnitude for electronic levels in typical atoms, this condition requires that $|\mathbf{B}_0| \ll \alpha e/a_0^2 \sim 10^5$ Gauss. Except for strongly magnetized white dwarfs and neutron stars, this range of external field strengths covers those likely to be found in most astrophysical environments.

When $H_Z \ll H_{so}$, we may regard L, S, and J as good quantum numbers that define the unperturbed state (including spin-orbit coupling) for the Zeeman calculation. Let Φ be the time-independent part of the wave function that describes the unperturbed state whose spectroscopic term is $^{2S+1}L_J$. According to time-independent perturbation theory, the following formula gives the energy-level splittings of the $2J+1$ sublevels corresponding to different M_J:

$$E_Z = \langle \Phi | H_Z | \Phi \rangle. \tag{27.35}$$

We give below two methods for computing this energy splitting: a heuristic treatment that makes clear the underlying physics, and a more rigorous treatment that exploits the Wigner-Eckart theorem.

Our heuristic treatment makes use of the vector model of quantum angular momenta. In this model, the external magnetic field induces a precession of the vector magnetic moment $\mathbf{M} + \mathbf{m} \propto \mathbf{L} + 2\mathbf{S}$ about $\mathbf{B}_0$ that occurs slowly in comparison to the rate that $\mathbf{L}$ and $\mathbf{S}$ precess about $\mathbf{J}$ because of the much stronger LS coupling. Hence we may obtain the projection of $\mathbf{L} + 2\mathbf{S}$ onto $\mathbf{B}_0$, which we define for definiteness to lie in the z direction, by first projecting $\mathbf{L} + 2\mathbf{S}$ onto the $\mathbf{J}$ direction. This yields, in some sense, the "mean" vector value of $\mathbf{L} + 2\mathbf{S}$ as

$$[(\mathbf{L} + 2\mathbf{S}) \cdot \mathbf{J}]\mathbf{J}/J(J+1)\hbar^2,$$

when we "average" over the rapid precessional motion associated with LS coupling. The factor $J(J+1)\hbar^2$ in the denominator represents the normalization needed to make the two $\mathbf{J}$'s in the numerator "unit vectors." We now project this mean value, instead of the rapidly changing $\mathbf{L} + 2\mathbf{S}$ itself, onto the $\mathbf{B}_0$ direction, obtaining

$$[(\mathbf{L} + 2\mathbf{S}) \cdot \mathbf{J}]M_J/J(J+1)\hbar, \tag{27.36}$$

where $M_J \hbar$ is the eigenvalue associated with the operator J_z. Our heuristic treatment then suggests that we replace equation (27.34) by

$$H_Z = \frac{e}{2m_e c\hbar} B_0 \frac{M_J}{J(J+1)} (\mathbf{L} + 2\mathbf{S}) \cdot \mathbf{J}. \qquad (27.37)$$

If we write $\mathbf{L} + 2\mathbf{S}$ as $\mathbf{J} + \mathbf{S}$, use the usual conversion of operators, $\mathbf{S} \cdot \mathbf{J} = (J^2 + S^2 - L^2)/2$, and substitute the expression (27.37) into equation (27.36), we get

$$E_Z = M_J g_L \mu_0 B_0, \qquad (27.38)\cdot$$

where $\mu_0 \equiv e\hbar/2m_e c$ is the *Bohr magneton* and g_L is the Landé-g factor given by the remarkable formula found *empirically* by Landé in 1923:

$$g_L = 1 + \frac{J(J+1) + S(S+1) - L(L+1)}{2J(J+1)}. \qquad (27.39)$$

We recognize the term in g_L beyond the 1 as the excess that arises because the electron g factor equals 2 rather than 1, giving $\mathbf{M} + \mathbf{m} \propto \mathbf{L} + 2\mathbf{S} = \mathbf{J} + \mathbf{S}$ rather than the "naive" expectation $\mathbf{M} + \mathbf{m} \propto \mathbf{J}$. If the naive expectation had held up, g_L would have equaled 1 for all states; so when an allowed electric dipole transition occurred between an upper and a lower state, with $\Delta M_J = -1, 0, +1$, Zeeman splitting would have produced three lines where, in the absence of the magnetic field, we would have had just one (since the different M_J levels are degenerate in both the upper and the lower levels). This expectation constitutes the "normal" Zeeman effect. It corresponds with the classical result that a magnetic field B_0 splits the radiation of an oscillating electron with natural frequency ω_0 into three separate (polarized) components: ω_0, and $\omega_0 \pm eB_0/2m_e c$. Only in special cases, however, do spectroscopists see this "normal" pattern of three lines. More generally, we have the "anomalous" Zeeman effect, with a pattern of splittings different from the classical expectation, because the true factor g_L *does* possess the additional terms, and the quantum numbers J, L, and S usually change in the transition in such a way that the g_L factors differ for the upper and the lower states.

We now give a more rigorous derivation of the same basic result. To compute the energy splitting (27.35) purely quantum-mechanically, we first denote Φ in the LS coupling scheme by $|NJM_J\rangle$, where we allow N to stand for quantum numbers that do not figure in our manipulations (like L and S):

$$E_Z = \langle NJM_J | H_Z | NJM_J \rangle.$$

The Wigner-Eckart theorem implies that the matrix elements with different M_J for given J differ only by Clebsch-Gordon coefficients. Thus we may concentrate on one of them, the most convenient one usually being $M_J = J$.

Defining $\mathbf{B}_0$ to lie along the z direction again, we then need to calculate the matrix element

$$\mathcal{M} = \frac{e}{2m_e c}\langle NJJ|L_z + 2S_z|NJJ\rangle,$$

which equals, by the Wigner-Eckart theorem,

$$\mathcal{M} = \frac{e}{2m_e c}\langle J1J0|J1JJ\rangle\langle NJ||\mathbf{L} + 2\mathbf{S}||NJ\rangle.$$

The Clebsch-Gordon coefficient that appears above has the value

$$\langle J1J0|J1JJ\rangle = \left(\frac{J}{J+1}\right)^{1/2}.$$

To evaluate the reduced matrix element of any vector $\mathbf{V}$, we invoke the projection identity, also derivable from the Wigner-Eckart theorem,

$$\langle NJ||\mathbf{V}||NJ\rangle = \hbar^{-1}[J(J+1)]^{-1/2}\langle NJM_J|\mathbf{V}\cdot\mathbf{J}||NJM_J\rangle.$$

This theorem replaces our heuristic argument, based on the vector model for angular momenta, that projected $\mathbf{L}+2\mathbf{S}$ first onto $\mathbf{J}$ and then onto $\mathbf{B}_0$. In any case, if we apply the projection theorem to $\mathbf{V} \equiv \mathbf{L} + 2\mathbf{S} = \mathbf{J} + \mathbf{S}$, we get, with the usual manipulation to eliminate $\mathbf{S}\cdot\mathbf{J}$,

$$\langle NJ||\mathbf{L} + 2\mathbf{S}||NJ\rangle = \hbar J(J+1)g_L,$$

where g_L equals the expression (27.39). Thus, collecting terms, we have

$$\mathcal{M} = J\mu_0 g_L. \tag{27.40}$$

The Zeeman energy H_Z associated with a particular magnetic sublevel M_J equals $B_0\mathcal{M}$ times the ratio of the relevant Clebsch-Gordon coefficient to $\langle J1J0|J1JJ\rangle$. If we deal with a general value of M_J instead of $M_J = J$, the latter ratio equals M_J/J; thus we obtain the splitting

$$E_Z = M_J g_L \mu_0 B_0,$$

which reproduces equation (27.38) (Q.E.D.).

28

Diatomic Molecules

Reference: McQuarrie, *Quantum Chemistry*, pp. 469–472.

In this chapter we study the structure of diatomic molecules. We begin with an order of magnitude demonstration that a clean separation exists for the energy levels that characterize the electronic, vibrational, and rotational transitions in such molecules. We then argue that this forms a basis for the *Born-Oppenheimer approximation* that allows a separation of the quantum-mechanical wave function along these lines.

ORDER OF MAGNITUDE DISCUSSION

Molecular structure constitutes a more complicated topic than atomic structure, since even a diatomic molecule has many more degrees of freedom than an atom. Because of the two centers of nuclear attraction, the molecular orbitals occupied by the electrons possess fewer symmetries than atomic orbitals do (Figure 28.1). In addition, if we think classically, the two nuclei can vibrate along their line of centers, as well as rotate about the two axes perpendicular to this line. (Because of their small dimensions, the nuclei cannot be excited under normal circumstances to rotate about the axis defined by the line of centers itself.) If all these motions coupled, we would face a much more complex problem than we actually do. Fortunately, much decoupling occurs, largely because of a small parameter in the problem: the ratio of the mass of an electron m_e to the mass of a typical nucleus M.

The fact that $m_e \ll M$ implies that the electrons in a molecule possess relatively little inertia compared to the nuclei. As a consequence, when subjected to the basic electromagnetic forces of the problem, the electrons move quickly in comparison with the nuclei. For any instantaneous separation distance R of the two nuclei, the electrons adopt a quasi-stationary orbital configuration (see Figure 28.1) that then defines their electrostatic

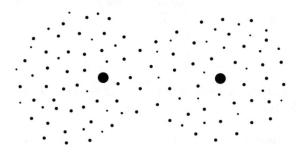

FIGURE 28.1
Schematic electronic orbital for the ground state of a diatomic molecule.

contribution to the further motion of the two nuclei. The resultant decoupling forms the basis of the *Born-Oppenheimer* approximation that we shall discuss shortly. For now, however, we merely pursue the order-of-magnitude estimates allowed by the concept of the existence of an electrostatic potential energy $E_{el}(R)$ for the system at any internuclear separation R.

We anticipate the result of Chapter 29 that a stable chemical bond can form between two atoms that approach within a distance of each other comparable to the Bohr radius $a_0 = \hbar^2/m_e e^2$. The internuclear potential-energy curve might look as depicted schematically in Figure 28.2. The

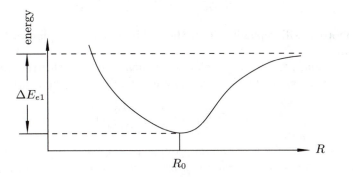

FIGURE 28.2
Schematic internuclear potential of a diatomic molecule.

location of the potential minimum corresponds typically to a_0, whereas the depth of the potential minimum has the same order of magnitude as the electronic energy levels in atoms:

$$E_{\text{el}} \sim e^2/a_0 = \hbar^2/m_{\text{e}}a_0^2. \tag{28.1}$$

Radiative transitions associated with the different electronic states of a molecule will then have radian frequencies ω_{el} such that $\hbar\omega_{\text{el}} \sim E_{\text{el}}$, and will lie typically in the visible or ultraviolet.

If the two nuclei are displaced from the equilibrium separation R_0 by a distance comparable to a_0, they will vibrate about the equilibrium position with a frequency ω_{vib} such that the vibrational energy contained in the motion and displacements of the two nuclei (of typical mass M) will be comparable to the depth of the electronic potential well: $M\omega_{\text{vib}}^2 a_0^2 \sim E_{\text{el}}$. The emitted photons from the associated process will have energies $E_{\text{vib}} = \hbar\omega_{\text{vib}}$ given by

$$E_{\text{vib}} = \hbar\omega_{\text{vib}} \sim \frac{\hbar^2}{m_{\text{e}}^{1/2}M^{1/2}a_0^2} \sim \left(\frac{m_{\text{e}}}{M}\right)^{1/2} E_{\text{el}}. \tag{28.2}$$

For $m_{\text{e}}/M \sim 10^{-3}$ or 10^{-4}, such photons will typically lie in the near- or mid-infrared.

The molecule may also rotate about any one of two axes not along the line of centers joining the two nuclei. The moment of inertia for such rotations will be of order Ma_0^2, and the angular momentum will be $\sim Ma_0^2\omega_{\text{rot}}$ if the rotation occurs at a characteristic frequency ω_{rot}. Quantum-mechanically, such an angular momentum will typically come quantized in units of $\hbar$: $Ma_0^2\omega_{\text{rot}} \sim \hbar$, yielding for an emitted photon from rotational transitions

$$E_{\text{rot}} = \hbar\omega_{\text{rot}} \sim \frac{\hbar^2}{Ma_0^2} \sim \left(\frac{m_{\text{e}}}{M}\right) E_{\text{el}}. \tag{28.3}$$

Such photons will typically lie in the millimeter-wave region of the radio spectrum.

To summarize, because the ratio of masses of the electron to atomic nuclei always constitutes a small number, the ratios of the energy levels,

$$E_{\text{rot}} : E_{\text{vib}} : E_{\text{el}} \sim \left(\frac{m_{\text{e}}}{M}\right) : \left(\frac{m_{\text{e}}}{M}\right)^{1/2} : 1,$$

form a neat hierarchy that should provide a clean basis for a perturbational attack. The *Born-Oppenheimer approximation* formulated below represents just such an approach. Notice, in particular, that the small parameter for the molecular-structure problem, m_{e}/M, still greatly exceeds the small parameter for the radiation problem, Ea_0^2/e, with E equal to the mean electric vector of the radiation field (see Chapter 22). As a consequence,

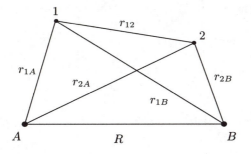

FIGURE 28.3
Coordinate system for a diatomic molecule with two valence electrons.

we can calculate the small splittings of level spacings due to vibrations and rotations independently of the radiative perturbations that might produce transitions between the different levels.

A DIATOMIC MOLECULE WITH TWO VALENCE ELECTRONS

Consider a diatomic molecule with two nuclei, A and B, and two (valence) electrons, 1 and 2. Set up a coordinate system in which the various inter-particle separations are denoted by the symbols in Figure 28.3. If we ignore small terms of order v^2/c^2 such as spin-orbit coupling, we may write the system Hamiltonian as

$$H = H_{AB} + H_{\text{el}}, \tag{28.4}$$

where H_{AB} represents the contribution from the kinetic energies of the two nuclei,

$$H_{AB} \equiv -\frac{\hbar^2}{2M_A}\nabla_A^2 - \frac{\hbar^2}{2M_B}\nabla_B^2, \tag{28.5}$$

and H_{el} represents the Hamiltonian for the electronic structure,

$$H_{\text{el}} \equiv -\frac{\hbar^2}{2m_{\text{e}}}(\nabla_1^2 + \nabla_2^2) + V, \tag{28.6}$$

with V representing the potential energy due to all the electrostatic interactions in the system:

$$V \equiv -\frac{Z_A e^2}{r_{1A}} - \frac{Z_B e^2}{r_{1B}} - \frac{Z_A e^2}{r_{2A}} - \frac{Z_B e^2}{r_{2B}} + \frac{Z_A Z_B e^2}{R} + \frac{e^2}{r_{12}}. \tag{28.7}$$

In equations (28.5) and (28.6), the symbols ∇_A, ∇_B, ∇_1, ∇_2 refer to partial differentiation with respect to the variables $\mathbf{x}_A$, $\mathbf{x}_B$, $\mathbf{x}_1$, and $\mathbf{x}_2$, whereas $R = |\mathbf{x}_A - \mathbf{x}_B|$, $r_{12} = |\mathbf{x}_1 - \mathbf{x}_2|$, etc.

Ignoring the effects of spins for the time being (we will put them in at the end), we write the Schrödinger equation for the scalar part of the system wave function as

$$H\psi = i\hbar \frac{\partial \psi}{\partial t}.$$

Since H does not depend explicitly on time, we may look for stationary solutions of the usual form:

$$\psi = \varphi(\mathbf{x}_A, \mathbf{x}_B, \mathbf{x}_1, \mathbf{x}_2)e^{-iEt/\hbar}, \tag{28.8}$$

with the spatial part φ satisfying the time-independent Schrödinger equation:

$$H\varphi = E\varphi.$$

THE BORN-OPPENHEIMER APPROXIMATION

Because of the denominators M_A and M_B, the operator H_{AB} is a factor m_e/M smaller than the corresponding (kinetic energy) terms in H_{el}. This suggested to Born and Oppenheimer a useful approximation procedure that begins by factoring φ (approximately) as a product of the configurations of the electrons and the two nuclei:

$$\varphi \approx \varphi_{el}(\mathbf{x}_1, \mathbf{x}_2, R)\varphi_{AB}(\mathbf{x}_A, \mathbf{x}_B). \tag{28.9}$$

Notice that we have left the factoring deliberately incomplete, in that we allow the electronic configuration to depend (as it must) on the internuclear (scalar) separation distance R.

With φ given by equation (28.9), we note that H_{el} operates only on φ_{el}, but H_{AB} operates on both φ_{AB} and φ_{el} (through the latter's dependence on $R = |\mathbf{x}_A - \mathbf{x}_B|$). Thus we have

$$\varphi_{AB}H_{el}\varphi_{el} + H_{AB}\varphi_{el}\varphi_{AB} = E\varphi_{el}\varphi_{AB}. \tag{28.10}$$

Since H_{AB} is a second-order differential operator, we may use Leibnitz's rule to write

$$H_{AB}\varphi_{el}\varphi_{AB} = \varphi_{el}H_{AB}\varphi_{AB} - \frac{\hbar^2}{M_A}\nabla_A\varphi_{el} \cdot \nabla_A\varphi_{AB}$$

$$- \frac{\hbar^2}{M_B}\nabla_B\varphi_{el} \cdot \nabla_B\varphi_{AB} + \varphi_{AB}H_{AB}\varphi_{el}. \tag{28.11}$$

The last term is smaller than the first term in equation (28.10) by a factor of m_e/M. On the other hand, $\nabla_A\varphi_{el} = (\mathbf{R}/R)\partial\varphi_{el}/\partial R = -\nabla_B\varphi_{el}$, where $\mathbf{R} \equiv \mathbf{x}_A - \mathbf{x}_B$. Thus the two middle terms in equation (28.11) read

$$\hbar^2\frac{\partial\varphi_{el}}{\partial R}\frac{\mathbf{R}}{R} \cdot \left(\frac{1}{M_B}\nabla_B - \frac{1}{M_A}\nabla_A\right)\varphi_{AB}.$$

The quantity in the parenthesis is proportional to the relative-velocity operator for the two nuclei, $\mathbf{v}_B - \mathbf{v}_A$, which should equal zero on average for a bound molecule. If we throw away every term on the right-hand side of equation (28.11) except for the first, we have the approximation

$$H_{AB}\varphi_{el}\varphi_{AB} \approx \varphi_{el}H_{AB}\varphi_{AB}.$$

In this case, equation (28.10) now becomes

$$\varphi_{AB}H_{el}\varphi_{el} + \varphi_{el}H_{AB}\varphi_{AB} = E\varphi_{el}\varphi_{AB}. \tag{28.12}$$

Divide equation (28.12) by $\varphi_{el}\varphi_{AB}$, and transpose terms to write

$$\frac{1}{\varphi_{el}}H_{el}\varphi_{el} = -\frac{1}{\varphi_{AB}}H_{AB}\varphi_{AB} + E.$$

The right-hand side depends only on the coordinates of the two nuclei, $\mathbf{x}_A$ and $\mathbf{x}_B$, whereas, apart for a parametric dependence on R, the left-hand side depends only on the coordinates of the two electrons, $\mathbf{x}_1$ and $\mathbf{x}_2$. For these two different functional dependences to be consistent, by the usual separation-of-variables argument, both sides had better equal a constant, or more accurately, some function of the parameter R. We call this "constant" $E_{el}(R)$. Thus we get the pair of equations:

$$H_{el}\varphi_{el} = E_{el}(R)\varphi_{el}, \tag{28.13}$$

$$H_{AB}\varphi_{AB} + E_{el}(R)\varphi_{AB} = E\varphi_{AB}. \tag{28.14}$$

The derivation of this pair of equations completes the desired factorization of the problem into electronic degrees of freedom and those associated with the motion of the two nuclei.

TRANSLATIONAL AND INTERNAL DEGREES OF FREEDOM

We further decompose the Schrödinger equation (28.14) for the motion of the two nuclei into a translation of the molecule as whole plus (as we will see) internal vibrations and rotations. To perform this decomposition, we introduce the center-of-mass coordinate,

$$\mathbf{X}_{CM} \equiv \frac{M_A\mathbf{x}_A + M_B\mathbf{x}_B}{M_A + M_B}, \tag{28.15}$$

the relative displacement,

$$\mathbf{R} \equiv \mathbf{x}_A - \mathbf{x}_B, \tag{28.16}$$

the total mass,

$$M \equiv M_A + M_B, \tag{28.17}$$

and the reduced mass,

$$\mu \equiv \frac{M_A M_B}{M_A + M_B}. \tag{28.18}$$

It then forms a standard exercise to show that H_{AB} transforms to the center-of-mass kinetic energy (operator) plus the relative kinetic energy (operator):

$$H_{AB} = -\frac{\hbar^2}{2M} \nabla^2_{CM} - \frac{\hbar^2}{2\mu} \nabla^2_R, \tag{28.19}$$

where ∇_{CM} and ∇_R denote partial differentiation with respect to $\mathbf{X}_{\mathrm{CM}}$ and $\mathbf{R}$.

Proof: Consider the transformation of variables $(\mathbf{x}_A, \mathbf{x}_B) \rightarrow (\mathbf{X}_{\mathrm{CM}}, \mathbf{R})$. The x-component of the operators ∇_A and ∇_B transform as

$$\frac{\partial}{\partial x_A} = \frac{\partial X_{\mathrm{CM}}}{\partial x_A} \frac{\partial}{\partial X_{\mathrm{CM}}} + \frac{\partial R_x}{\partial x_A} \frac{\partial}{\partial R_x} = \frac{M_A}{M} \frac{\partial}{\partial X_{\mathrm{CM}}} + \frac{\partial}{\partial R_x},$$

$$\frac{\partial}{\partial x_B} = \frac{\partial X_{\mathrm{CM}}}{\partial x_B} \frac{\partial}{\partial X_{\mathrm{CM}}} + \frac{\partial R_x}{\partial x_B} \frac{\partial}{\partial R_x} = \frac{M_B}{M} \frac{\partial}{\partial X_{\mathrm{CM}}} - \frac{\partial}{\partial R_x}.$$

Hence we have

$$\frac{1}{M_A} \frac{\partial^2}{\partial x_A^2} = \frac{M_A}{M^2} \frac{\partial^2}{\partial X_{\mathrm{CM}}^2} + \frac{1}{M_A} \frac{\partial^2}{\partial R_x^2} + \frac{2}{M} \frac{\partial^2}{\partial X_{\mathrm{CM}} \partial R_x},$$

$$\frac{1}{M_B} \frac{\partial^2}{\partial x_B^2} = \frac{M_B}{M^2} \frac{\partial^2}{\partial X_{\mathrm{CM}}^2} + \frac{1}{M_B} \frac{\partial^2}{\partial R_x^2} - \frac{2}{M} \frac{\partial^2}{\partial X_{\mathrm{CM}} \partial R_x}.$$

If we sum the two above expressions, we get

$$\frac{1}{M_A} \frac{\partial^2}{\partial x_A^2} + \frac{1}{M_B} \frac{\partial^2}{\partial x_B^2} = \frac{1}{M} \frac{\partial^2}{\partial X_{\mathrm{CM}}^2} + \frac{1}{\mu} \frac{\partial^2}{\partial R_x^2},$$

where we made use of equation (28.17) and defined μ so that

$$\frac{1}{\mu} = \frac{1}{M_A} + \frac{1}{M_B},$$

which we recognize as equivalent to equation (28.18). We can write down similar expressions for the y and z components of the transformation, yielding upon summation

$$\frac{1}{M_A} \nabla^2_A + \frac{1}{M_B} \nabla^2_B = \frac{1}{M} \nabla^2_{\mathrm{CM}} + \frac{1}{\mu} \nabla^2_R,$$

which yields equation (28.19) after we multiply by $-\hbar^2/2$. (Q.E.D.)

Obtaining $H_{AB} + E_{el}(R)$ as the sum of operators that depend on $\mathbf{X}_{CM}$ and $\mathbf{R}$ separately allows us to consider eigenfunctions φ_{AB} of equation (28.14) that equal a product of the translational degrees of freedom and the internal degrees of freedom:

$$\varphi_{AB} = \varphi_{trans}(\mathbf{X}_{CM})\varphi_{int}(\mathbf{R}). \qquad (28.20)$$

With this simplification, we may manipulate equation (28.14) into the form:

$$\frac{1}{\varphi_{trans}}\left(-\frac{\hbar^2}{2M}\nabla^2_{CM}\varphi_{trans}\right) = \frac{1}{\varphi_{int}}\left(\frac{\hbar^2}{2\mu}\nabla^2_R\varphi_{int}\right) - E_{el}(R) + E.$$

The left-hand side depends only on $\mathbf{X}_{CM}$, whereas the right-hand side depends only on $\mathbf{R}$; thus the standard argument states that both sides must equal a constant, which we denote as E_{trans}. We have now succeeded in the desired separation:

$$-\frac{\hbar^2}{2M}\nabla^2_{CM}\varphi_{trans} = E_{trans}\varphi_{trans}, \qquad (28.21)$$

$$-\frac{\hbar^2}{2\mu}\nabla^2_R\varphi_{int} + E_{el}(R)\varphi_{int} = E_{int}\varphi_{int}, \qquad (28.22)$$

where we have defined E_{int} as $E - E_{trans}$, i.e.,

$$E = E_{int} + E_{trans}. \qquad (28.23)$$

Notice that equation (28.21) represents the Schrödinger equation for a free particle with translational energy E_{trans} and mass M equal to that of the molecule as a whole. Equation (28.22) gives, on the other hand, the Schrödinger equation for an effective particle of (reduced) mass μ with energy E_{int} placed in an effective potential $E_{el}(R)$.

VIBRATIONS AND ROTATIONS

Insofar as the internal degrees of freedom of the nuclei go, the last section reduced the relative motion of nuclei A relative to B to the dynamics of a single particle of reduced mass μ in a central force field. We introduce spherical coordinates (R, θ, ϕ) to describe the vector position $\mathbf{R}$ of this hypothetical particle (Figure 28.4). In the coordinates (R, θ, ϕ), the Laplacian operator ∇^2_R can be written as

$$\nabla^2_R = \frac{1}{R^2}\left[\frac{\partial}{\partial R}\left(R^2\frac{\partial}{\partial R}\right) - \frac{L^2}{\hbar^2}\right],$$

where $\mathbf{L}$ is the angular momentum operator

$$\mathbf{L} \equiv \mathbf{R} \times (-i\hbar\mathbf{\nabla}_R).$$

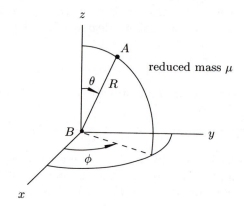

FIGURE 28.4
Spherical polar coordinates for the equivalent one-body problem describing the relative motion of nucleus A to nucleus B in a diatomic molecule.

Because $E_{\text{el}}(R)$ depends on only the magnitude of $\mathbf{R}$—as we will explicitly justify shortly—equation (28.22) may be separated into a radial part and a spherical harmonic

$$\varphi_{\text{int}}(\mathbf{R}) = \frac{1}{R} Z_{\text{vib}}(R) Y_{Jm}(\theta, \phi), \tag{28.24}$$

where the quantum numbers J and m in molecules play the same roles as ℓ and m in atoms:

$$L^2 Y_{Jm} = J(J+1)\hbar^2 Y_{Jm}, \qquad L_z Y_{Jm} = m\hbar Y_{Jm}.$$

In equation (28.24), we have introduced the standard factor of $1/R$ to eliminate the spherical divergence. The substitution of equation (28.24) into equation (28.22) now yields

$$-\frac{\hbar^2}{2\mu}\frac{d^2 Z_{\text{vib}}}{dR^2} + E_{\text{el}}(R) Z_{\text{vib}} = \left[E_{\text{int}} - \frac{J(J+1)\hbar^2}{2\mu R^2} \right] Z_{\text{vib}}. \tag{28.25}$$

Equation (28.25) is the Schrödinger equation for an anharmonic oscillator.

Notice that $Y_{Jm}(\theta, \phi)$ describes two independent degrees of freedom for the directional orientation of a diatomic molecule. These two degrees of freedom correspond (in number, but not in detail) to the classical image of the molecule rotating independently about the two axes perpendicular to the line of centers. After taking out these two internal degrees of freedom, the nuclei can still vibrate along the direction of the line of centers.

HARMONIC-OSCILLATOR APPROXIMATION

To obtain the familiar harmonic-oscillator approximation for a vibrating diatomic molecule, assume that $E_{\text{el}}(R)$ has the shape shown in Figure 28.1. Suppose also that the nuclear separation does not depart far from the equilibrium position R_0, where $E'_{\text{el}}(R) = 0$. A Taylor-series expansion about R_0 then yields

$$E_{\text{el}}(R) = E_{\text{el}}(R_0) + \frac{\mu}{2}\omega_0^2(R - R_0)^2 + \cdots,$$

where we have defined the natural vibration frequency ω_0 by

$$\mu\omega_0^2 \equiv E''_{\text{el}}(R_0). \tag{28.26}$$

If we denote $x \equiv R - R_0$, we obtain, upon keeping only the lowest-order terms,

$$-\frac{\hbar^2}{2\mu}\frac{d^2 Z_{\text{vib}}}{dx^2} + \frac{\mu}{2}\omega_0^2 x^2 Z_{\text{vib}} = E_{\text{vib}} Z_{\text{vib}}, \tag{28.27}$$

where we have defined

$$E_{\text{vib}} \equiv E_{\text{int}} - E_{\text{el}}(R_0) - \frac{J(J+1)\hbar^2}{2\mu R_0^2}.$$

Rewriting the last expression, we obtain the suggestive equation that the internal energy consists of the sum of electronic, vibrational, and rotational contributions:

$$E_{\text{int}} = E_{\text{el}}(R_0) + E_{\text{vib}} + E_{\text{rot}}, \tag{28.28}$$

where we have identified the quantized values of the rotational energy of the molecule as

$$E_{\text{rot}} = J(J+1)\frac{\hbar^2}{2\mu R_0^2} \equiv J(J+1)B. \tag{28.29}$$

In the above, E_{rot} takes the mnemonic form: the square of the angular momentum divided by twice the moment of inertia. Laboratory spectroscopy usually fixes the experimental value of $B \equiv \hbar^2/2\mu R_0^2$, the rotational constant of the molecule.

Equation (28.27) is the Schrödinger equation for the harmonic oscillator and has well-known eigenfunction solutions in terms of Hermite polynomials times Gaussians. The associated eigenvalues equal

$$E_{\text{vib}} = \left(v + \frac{1}{2}\right)\hbar\omega_0, \qquad v = 0, 1, 2, \ldots \tag{28.30}$$

Notice that the ground-state vibrational level equals $\hbar\omega_0/2$ because of zero-point oscillations.

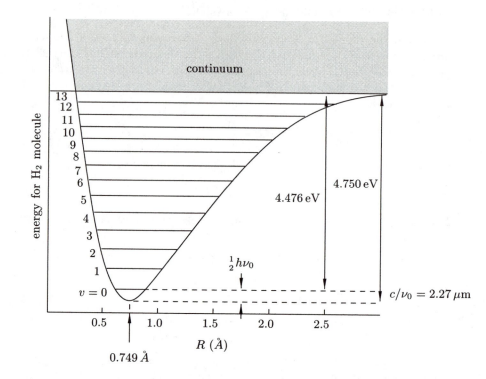

FIGURE 28.5
The vibrational ladder associated with the ground electronic state of the hydrogen molecule H_2.

For large values of v, the Taylor-series expansion for $E_{el}(R)$ about R_0 breaks down, and we need to consider anharmonic corrections (see Problem Set 6 and Figure 28.5). A related concern affects the model of the molecule as a rigid rotator. The approximation of a nonvarying rotation constant B may fail at large J if centrifugal forces stretch the molecular separation significantly beyond the equilibrium value R_0. Corrections can be applied to each of the effects, but discussions of these topics belong to more specialized texts. For now, we return to study the electronic structure in greater detail. This turns out physically to involve the issue of the nature of the chemical bond, our topic for Chapter 29.

29

The Nature of the Chemical Bond

Reference: McQuarrie, *Quantum Chemistry*, pp. 469–472.

In this chapter we develop the notion of the electronic structure of diatomic molecules in the context of the theory of Heitler and London for the covalent bond in molecular hydrogen. In particular, we wish to focus on the issue of how the sharing of electrons by atoms can lead to attractive chemical forces. Although electron-spin considerations stay in the background of our formal calculations, we will find at the end that their statistics largely determine the stability of the covalent chemical bond.

VARIATIONAL PRINCIPLE TO OBTAIN GROUND-STATE EIGENVALUES

As preparation for our assault, we first discuss a powerful approximate method for computing the ground state of any quantum-mechanical system. Consider the time-independent Schrödinger equation:

$$H\Phi = E\Phi. \tag{29.1}$$

If we knew Φ, we could compute E as

$$E = \frac{\langle \Phi | H | \Phi \rangle}{\langle \Phi | \Phi \rangle}, \tag{29.2}$$

where we have allowed the possibility that Φ might not be normalized. Equation (29.2) would yield E as the ground-state energy level E_0 if Φ were the ground-state eigenfunction Φ_0. What value could E have if Φ were *any function*? We might be tempted to reply, *any value*, but that would be wrong. In fact, no matter what function is chosen for Φ, E cannot be (algebraically) less than E_0.

This result, which we prove below, provides an immediate idea for estimating E_0, *without the need to solve equation (29.1) for Φ*. The (variational) technique involves, in essence, merely making a series of guesses for

333

Φ, calculating the necessary integrals from equation (29.2) to compute the associated E, and choosing the lowest algebraic value for E to emerge from the exercise as an estimate for the ground-state energy E_0. The method even has the virtue that we know *a priori* the sense of the error being made, since the actual E_0 can have only a smaller value yet.

Proof: To prove that the expression (29.2) exceeds E_0, we note that if we *did* know the complete set of orthonormal eigenfunctions Φ_j of equation (29.1), we could, in principle, expand any wave function for the system Φ in terms of this set:

$$\Phi = \sum_j c_j \Phi_j.$$

Substitute the above expression into the right-hand side of equation (29.2) and use the fact that $H\Phi_j = E_j \Phi_j$ plus $\langle \Phi_i | \Phi_j \rangle = \delta_{ij}$ to write

$$E = \frac{\sum_j E_j |c_j|^2}{\sum_j |c_j|^2}.$$

The right-hand side is a weighted average of all of the eigenvalues E_j. Since the weights $|c_j|^2$ are positive definite, the weighted average cannot be less than the smallest E_j, which equals E_0 by assumption. Thus $E \geq E_0$. (Q.E.D.)

PRACTICAL USAGE OF THE VARIATIONAL PRINCIPLE

Suppose we have an arbitrary sequence of n wave kets $|j\rangle$, and we construct the possible guess

$$\Phi = \sum_{j=1}^{n} c_j |j\rangle, \tag{29.3}$$

where we vary the set of coefficients $\{c_j\}_{j=1}^{n}$ to minimize E in equation (29.2). This procedure then requires us to minimize the expression

$$E = \frac{\sum_{i,j=1}^{n} c_i^* H_{ij} c_j}{\sum_{j=1}^{n} c_i^* N_{ij} c_j}, \tag{29.4}$$

where H_{ij} and N_{ij} are defined as

$$H_{ij} \equiv \langle i|H|j \rangle, \tag{29.5}$$
$$N_{ij} \equiv \langle i|j \rangle. \tag{29.6}$$

The normalization elements N_{ij} will not be Kronecker deltas unless we have a set of orthonormal kets $|j\rangle$ (not necessary for the procedure to work). Because H is Hermitian, we have $H_{ji} = H_{ij}^*$, and $N_{ji} = N_{ij}^*$ from the basic definition of the inner product $\langle i|j \rangle$.

To find the extremum of equation (29.4), we require $\partial E/\partial c_k^*$ to equal zero for each allowable value of k. A little calculus yields the resulting requirement as the simultaneous set of linear homogeneous equations:

$$\sum_{j=1}^{n} (H_{kj} - EN_{kj})c_j = 0 \qquad \text{for} \qquad k = 1, 2, \ldots, n. \qquad (29.7)$$

A similar calculation demonstrates that $\partial E/\partial c_k = 0$ would simply generate the complex conjugate of the above requirement. In order for equation (29.7) to contain nontrivial solutions for c_k, the determinant of the coefficient matrix must equal zero:

$$\det(H_{kj} - EN_{kj}) = 0. \qquad (29.8)$$

The determinantal equation (29.8) yields n possible values for E, of which the lowest will constitute our guess for E_0. The associated matrix is Hermitian, and the energy values E derived from the characteristic equation are real. (The values larger than the lowest, when ordered algebraically, yield guesses for the successively higher [excited] states, but the truncation errors associated with choosing a finite value for n will generally make the approximations for the latter poorer and poorer.)

APPLICATION TO THE HYDROGEN MOLECULE

As a relatively simple example, we apply the considerations of the previous section to the hydrogen molecule, H_2. The calculations repeated below were carried out by Heitler and London in 1927, and constitute the first successful elucidation of the nature of the covalent bond. For simplicity, we shall ignore, as before, any explicit role at first for electron spin. We therefore compute everything on the basis of a scalar wave function $\varphi_{el}(\mathbf{x}_1, \mathbf{x}_2, R)$. Only after we are through do we reconstruct the total (electronic) wave function, $\Phi_{el} = \varphi_{el}\Sigma$, with Σ being one of the (sums of) products of two two-component spinors discussed in Chapter 27 for the two-electron atom. This simplified procedure proves possible only because we have restricted ourselves to a problem containing only two electrons. In general, for three or more (valence) electrons, we could not perform a simple *a priori* splitting in the way that we have assumed here.

With this general preamble, we begin our calculations. We wish to obtain the eigenvalue $E_{el}(R)$ in equation (28.13) for the ground electronic state of H_2 when the instantaneous separation of the two nuclei equals R. (Recall that the small inertia of the orbital electrons allows us to assume in the Born-Oppenheimer approximation that the electronic configuration instantaneously relaxes to the quasistationary state appropriate for the nuclear locations, which may themselves be undergoing rotations and vibrations.) We adopt atomic units in which we use the Bohr radius $a_0 = \hbar^2/m_e e^2 = 0.529\,\text{Å}$ as the unit of length, and $e^2/a_0 = 27.2\,\text{eV}$ as the unit of energy. For the H_2 molecule, the electronic Hamiltonian H_{el} then reads

$$H_{\text{el}} = -\frac{1}{2}(\nabla_1^2 + \nabla_2^2) - \frac{1}{r_{1A}} - \frac{1}{r_{1B}} - \frac{1}{r_{2A}} - \frac{1}{r_{2B}} + \frac{1}{r_{12}} + \frac{1}{R}. \qquad (29.9)$$

Taking Heitler and London's lead, we guess φ_{el}, for variational purposes, to have the relatively simple form

$$\varphi_{\text{el}} = c_1 u(r_{1A})u(r_{2B}) + c_2 u(r_{1B})u(r_{2A}), \qquad (29.10)$$

where u represents the wave function for a single *isolated hydrogen atom* in its ground electronic state $n = 1$, $\ell = 0$, and $m_\ell = 0$. In atomic units, u has the form

$$u = \frac{1}{\sqrt{\pi}}e^{-r}. \qquad (29.11)$$

By the symbols $u(r_{1A})$, $u(r_{2B})$, etc., we mean the function u when r_{1A}, r_{2B}, etc., are substituted for r. In other words, equation (29.10) represents two hydrogen atoms (with the electrons 1 or 2 surrounding nuclei A or B taken in various combinations) simply jammed together as a zeroth-order approximation for the hydrogen molecule. The approximation is exact for infinite internuclear separation R; as we will see below, it proves qualitatively adequate—for the purpose of calculating the ground-state energy—for the actual hydrogen molecule at a physically realistic separation $\sim R_0$. (This is a general property of variational methods; they yield much better eigenvalues than we naively have a right to expect for relatively crude guesses at the eigenfunctions.)

Our discussions of the previous chapters lead us to anticipate that the physically relevant eigenfunctions correspond to the symmetric and antisymmetric choices $c_2 = c_1$ and $c_2 = -c_1$, with $c_1 = 1/\sqrt{2}$ if we want normalized wave functions. The detailed analysis below confirms this expectation; however, we do not make these choices at the outset, because we wish to demonstrate that the extremization procedure itself, with c_1 and c_2 regarded as variables, automatically leads to the "correct" solution (within the context of the Heitler-London *Ansatz*). The underlying physics of the problem itself dictates here that the distinct quantum states separate themselves out with respect to the allowable symmetries of the spatial wave function.

With φ given by equation (29.10), equation (29.8) becomes

$$\begin{vmatrix} H_{11} - EN_{11} & H_{12} - EN_{12} \\ H_{21} - EN_{21} & H_{22} - EN_{22} \end{vmatrix} = 0, \qquad (29.12)$$

where the product of single-atom wave functions allows us to write

$$N_{11} = \langle u(r_{1A})|u(r_{1A})\rangle\langle u(r_{2B})|u(r_{2B})\rangle,$$
$$N_{22} = \langle u(r_{1B})|u(r_{1B})\rangle\langle u(r_{2A})|u(r_{2A})\rangle,$$
$$N_{12} = \langle u(r_{1A})|u(r_{1B})\rangle\langle u(r_{2B})|u(r_{2A})\rangle,$$
$$N_{21} = \langle u(r_{1B})|u(r_{1A})\rangle\langle u(r_{2A})|u(r_{2B})\rangle,$$
$$H_{11} = \langle u(r_{1A})u(r_{2B})|H_{el}|u(r_{1A})u(r_{2B})\rangle,$$
$$H_{22} = \langle u(r_{1B})u(r_{2A})|H_{el}|u(r_{1B})u(r_{2A})\rangle,$$
$$H_{12} = \langle u(r_{1A})u(r_{2B})|H_{el}|u(r_{1B})u(r_{2A})\rangle,$$
$$H_{21} = \langle u(r_{1B})u(r_{2A})|H_{el}|u(r_{1A})u(r_{2B})\rangle.$$

Because u is a normalized wave function and is real, the above expressions (in atomic units) are given by

$$N_{11} = 1 = N_{22},$$

$$N_{12} = \Theta^2 = N_{21},$$

where

$$\Theta \equiv \frac{1}{\pi}\int e^{-r_{1A}}e^{-r_{1B}}\,d^3x_1$$

doesn't really depend on whether we integrate over the coordinates of electron 1 or over those of electron 2 (but recall that $r_{1A} \equiv |\mathbf{x}_1 - \mathbf{x}_A|$ and $r_{1B} \equiv |\mathbf{x}_1 - \mathbf{x}_B|$). Moreover, since $u(r)$ is the eigenfunction of the hydrogen-*atom* Hamiltonian,

$$-(1/2)\nabla^2 - 1/r,$$

with eigenvalue $-1/2$ (recall that the energy of the ground-state hydrogen atom equals $-e^2/2a_0$), we see that $-(1/2)\nabla_1^2 - 1/r_{1A}$ operating on $u(r_{1A})$ yields $-(1/2)u(r_{1A})$; with a similar set of rules when we substitute B for A or 2 for 1. These rules now yield

$$H_{11} = -1 + \mathcal{C} = H_{22},$$

$$H_{12} = -\Theta^2 + \mathcal{E} = H_{21},$$

with the -1 and the $-\Theta^2$ coming from the relevant sum of hydrogen-atom Hamiltonians in equation (29.9) for H_{el}, and with $\mathcal{C}$ and $\mathcal{E}$ resulting from the remaining terms in V:

$$\mathcal{C} \equiv \frac{1}{\pi^2}\int\int d^3x_1\,d^3x_2\,e^{-2r_{1A}}\left(-\frac{1}{r_{1B}} - \frac{1}{r_{2A}} + \frac{1}{r_{12}} + \frac{1}{R}\right)e^{-2r_{2B}},$$

$$\mathcal{E} \equiv \frac{1}{\pi^2}\int\int d^3x_1\,d^3x_2\,e^{-r_{1A}}e^{-r_{1B}}\left(-\frac{1}{r_{1A}} - \frac{1}{r_{2B}} + \frac{1}{r_{12}} + \frac{1}{R}\right)e^{-r_{2A}}e^{-r_{2B}}.$$

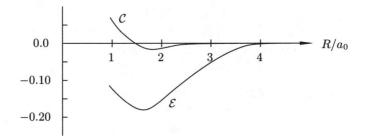

FIGURE 29.1
Values of Coulomb and exchange integrals, $\mathcal{C}$ and $\mathcal{E}$, as a function of the internuclear separation R measured in atomic units.

The quantities $\mathcal{C}$ and $\mathcal{E}$ are called the *Coulomb* and the *exchange* integrals, respectively; they arise from the part of the electrostatic interaction beyond those of the isolated atoms. The Coulomb integral involves the product of the squares of the individual wave functions, $e^{-2r_{1A}}e^{-2r_{2B}}$; the exchange integral involves the mixed terms, $e^{-r_{1A}}e^{-r_{2B}}e^{-r_{1B}}e^{-r_{2A}}$, that represent, in some sense, the effects of the sharing of the two electrons between the two nuclei A and B. Since these expressions, like that for the overlap integral Θ, involve integrations over two centers of attraction, they are somewhat tricky to perform. One thing should be clear, however, with only a little reflection: the results must depend only on the magnitude of $\mathbf{R}$, since the orientation of the molecular axis in physical space cannot affect the energy of the electronic configuration. In any case, analytical results can be obtained by using confocal elliptic coordinates (see problem 9.3 in McQuarrie's book); here we merely quote the results when the internuclear separation equals R:

$$\Theta = e^{-R}\left(1 + R + \frac{1}{3}R^2\right),$$

$$\mathcal{C} = e^{-2R}\left(\frac{1}{R} + \frac{5}{8} - \frac{3}{4}R - \frac{1}{6}R^2\right),$$

with a much more complicated expression for $\mathcal{E}$. Plots of the functions $\mathcal{C}$ and $\mathcal{E}$ appear in Figure 29.1. Notice that only the exchange integral is substantially negative over a significant portion of the spatial range of relevant internuclear spacings. As we will find below, it gives the term that provides the stability of the chemical bond.

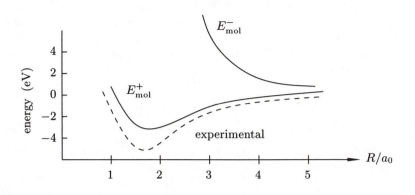

FIGURE 29.2
Comparison of the Heitler-London theory for the H_2 molecule with the electronic energy curve determined experimentally.

Substitution of these results into equation (29.12) now yields a quadratic equation with the possible solutions

$$E_{el}^{\pm} = \frac{H_{11} \pm H_{12}}{1 \pm N_{12}} = -1 + \frac{\mathcal{C} \pm \mathcal{E}}{1 \pm \Theta^2}.$$

The term -1 can be associated with the electronic energy of the separated atoms; thus the part that we should attribute to the molecular bond works out to be

$$E_{mol}^{\pm} = \frac{\mathcal{C} \pm \mathcal{E}}{1 \pm \Theta^2}. \tag{29.13}$$

Since $\Theta^2 \leq 1$ and $\mathcal{E}$ is negative, we see that the plus-sign choice yields the stable state; the minus sign would give an unbound state for any separation R, as illustrated in Figure 29.2.

The Heitler-London treatment predicts the depth of the potential well at R_0 to equal -0.125 in atomic units. To this we must add $\hbar\omega_0/2 = 0.0093$ for the ground vibrational energy, to obtain the ground-state energy level for the molecule of -0.116, or $-3.14\,\text{eV}$. Experiments yield $4.476\,\text{eV}$ for the dissociation energy of the molecule. The experimental curve drawn in Figure 29.2 derives from a compilation of data that also includes analysis of vibrational spectra. The fact that the theoretical value lies everywhere above the true value arises from the general property of the variational method that guarantees $E \geq E_0$. (When making more precise calculations, using, for example, more variable parameters, we must make certain that the governing Hamiltonian includes all the physical effects present in the actual molecule, or else *lower* values could result for the theoretical ground-state energy relative to the experimental data.) More precise calculations for H_2 do exist that yield excellent accord with laboratory measurements of the dissociation energy, the bond length, and the fundamental vibrational frequency ω_0.

EIGENFUNCTIONS

The choice E_{mol}^{+} in the Heitler-London treatment leads to the eigenvector solution

$$c_2 = c_1.$$

For equation (29.10), this corresponds to a symmetric spatial wave function for the two electrons. According to Chapter 27, we must then pair this with the antisymmetric spin ket given by equation (27.9). This state has spin $S = 0$, and is spectroscopically a *singlet* state ($2S + 1 = 1$). Conversely, the choice E_{mol}^{-} leads to the eigenvector solution

$$c_2 = -c_1,$$

which yields an antisymmetric spatial wave function for equation (29.10). The latter may be paired with any of the symmetric spin kets, equations (27.6), (27.7), and (27.8). Thus the unbound molecular state with E_{mol}^{-} as its eigenvalue has $S = 1$ and corresponds to a *triplet* state ($2S + 1 = 3$).

For a molecule, the exchange of a pair of electrons with $S = 0$ gives rise to a net attractive force between the nuclear protons, whereas the exchange of a pair with $S = 1$ gives rise to a net repulsive force. The situation should be contrasted with that of an atom, where LS coupling leads to Hund's rule that states with $S = 1$ lie *lower* in energy than states with $S = 0$. The difference occurs for the following reason. For atoms, where all the positive charges in the system exist at virtually a single point in the nucleus (whose stability against Coulomb repulsion constitutes a problem in nuclear physics, not atomic or molecular physics), it pays to keep the electrons as well separated as possible to minimize their repulsive electrostatic contribution. Thus $S = 1$ states are favored energetically over $S = 0$. In a diatomic molecule, however, we have two centers of positive nuclear charge. Clearly, keeping the electrons far apart cannot help to "glue" these two nuclei together against their natural Coulomb repulsion, and the $S = 1$ ground electronic states lead to a vibrationally unbound H_2 molecule. For binding to take place, we need to stick as much negative charge between the two nuclei as possible. This requires the two valence electrons in our problem to share orbits that have significant occupation of the relatively cramped quarters between the two nuclei, thereby selecting out $S = 0$ as the energetically favored state. Thus the chemists' notation of a chemical bond as a straight line holding together the symbols standing for the nuclear identity of the constituent atoms has more than a figurative meaning.

Mentally, we may picture bringing slowly together from infinity two hydrogen atoms in their ground electronic states to form a hydrogen molecule. If the orbital electrons have oppositely directed spins, the system will form a bound state; if they have parallel spins, Pauli's exclusion principle tends to keep the system from coalescing. In the former case, of course, we still need to get rid of the electronic binding energy if we want to form a permanently bound state. In the discussions that follow, we will show that this requirement cannot be fulfilled by electric dipole emission, because the relevant electronic transitions are forbidden by selection rules, whereas vibrational-rotational transitions are not possible, because the H_2 molecule has no permanent dipole moment (Chapter 30). Thus the actual astrophysical formation of H_2 molecules has to be more complicated than our simple mental image.

EXCITED STATES AND SPECTROSCOPIC NOTATION

Excited electronic-energy levels exist for a molecule just as they do for an atom. In H_2 these are labeled alphabetically, B, C, D, ...; the ground state receives an X, and the lowest unbound state, a b. The electronic wave functions can also have nonzero values for the orbital angular momenta. Even classically, however, the motions of electrons in a diatomic molecule would be clearly different from those in an atom. In an atom, the electric force contributed by the nucleus possesses spherical symmetry; so, to the extent that the coupling to spin can be ignored, the total orbital angular momentum of the electrons (in nonfilled subshells) is conserved. In a diatomic molecule, only the component of the electronic orbital angular momentum Λ about the internuclear axis constitutes a constant of the motion. The other two components of electronic angular momentum feel nonvanishing components of torque from the axisymmetric electric field of the two nuclei, which causes a precession of the vector angular momentum. As the internuclear spacing decreases and the field gains strength, the precession would speed up. Under these circumstances, the vector angular momentum loses its meaning quantum-mechanically, but the projected component Λ remains a good quantum number that takes on only whole integer values. Spectroscopists adopt a molecular notation paralleling the atomic notation, which reads Σ, Π, Δ, ... for $\Lambda = 0, 1, 2, \ldots$.

When $\Lambda \neq 0$, there exists an internal magnetic field induced by the orbital motion of the electrons. This magnetic field causes a precession of the total electronic-spin angular momentum, which destroys its utility as a good quantum number, except for its component Σ along the internuclear axis. (Do not confuse this Σ with the notation for the $\Lambda = 0$ state, any more than you should confuse atomic spin S for the $L = 0$ state.) The values allowed for Σ range from $-S$ to $+S$ in increments of 1.

The quantities Λ and Σ combine to give Ω, the total electronic angular momentum, orbital plus spin. Unlike the relationship between $\mathbf{J}$, $\mathbf{L}$ and $\mathbf{S}$ in an atom, however, Ω, Λ, and Σ refer to a single projection, and can, therefore, be added algebraically rather than vectorially. The interaction of the spin Σ with the magnetic field associated with Λ splits the energies of a given Λ level into $2S + 1$ states, labeled by their values of Ω. This notion applies even when $\Lambda = 0$ and no actual splitting occurs; thus the symbol $^3\Sigma$ denotes a triplet state in which $S = 1$ and $\Lambda = 0$.

Symmetry of the wave function also helps the classification of molecular states. If the electronic wave function of a nondegenerate state (a Σ state) changes sign when reflected about any plane passing through both nuclei of a diatomic molecule, it is denoted a Σ^- state; if it remains unchanged, a Σ^+ state. If the two nuclei have the same charge, e.g., H_2 but also HD, the electric field in which the electrons move is symmetric about a point midway between the two nuclei. If the electronic wave function changes sign when reflected about this center, the state receives a right-hand subscript u; if it remains unchanged, a g. (The origin of this notation comes from the German for "odd" and "even.") Figure 29.3 gives an application of this notation to the electronic configurations of the H_2 molecule. (Recall, however, that each energy curve has a ladder of vibrational and rotational states associated with it, although the unstable curve $b\,^3\Sigma_u^+$ has no bound states, only continuum ones.) The lower dashed line at zero energy corresponds to two stationary hydrogen atoms, each in its ground state, separated by an infinite distance. In the upper dashed curve, we have the same situation, but one of the atoms resides in the first excited level (10.2 eV).

SELECTION RULES FOR ELECTRONIC DIPOLE TRANSITIONS

The selection rules for electronic dipole transitions of a diatomic molecule satisfy

$$\Delta\Lambda = 0, \pm 1, \qquad \Delta S = 0;$$

furthermore, g states combine only with u states and vice versa; and Σ^+ states cannot combine with Σ^- states. The ΔS rule implies that the H_2 in its ground electronic state found in molecular clouds cannot photodissociate via $X\,^1\Sigma_g^+ \to b\,^3\Sigma_u^+$, which helps to protect the molecule and make it more stable in interstellar space than it might otherwise be. (*Collisional* dissociation can occur by this route, but there has to be an adequate supply of hot electrons produced, e.g., by shockwaves.) Moreover, because the molecule lacks a permanent dipole moment (the center of charge coincides

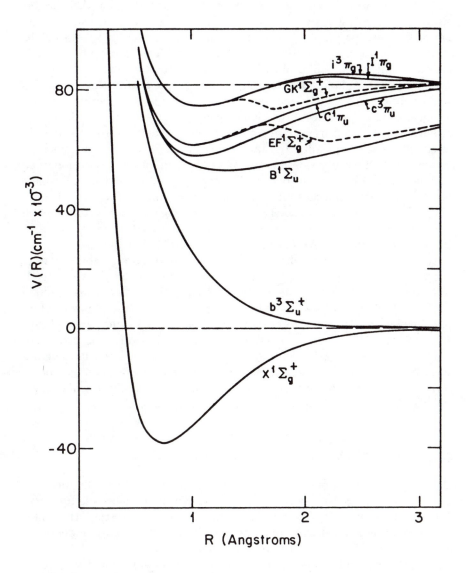

FIGURE 29.3
Some of the lower-lying electronic levels of the H_2 molecule. (From J.M. Shull and S. Beckwith 1982, *Ann. Rev. Astr. Ap.*, **20**, 163.

with the center of mass; consult Chapter 30), two H atoms in their *ground* electronic states (as is overwhelmingly the case in interstellar space) cannot radiatively associate to form an H_2 molecule in the $X\,^1\Sigma_g^+$ state. The formation of H_2 molecules in interstellar clouds must occur via catalytic action on the surfaces of grains. In more dense environments—e.g., in the winds of young and old stars, or in fragmenting clouds after the recombination era of cosmology—H_2 molecules might form via three-body reactions, or via a chain of ion-neutral reactions involving a species such as H^-, or via radiative association from hydrogen atoms in excited electronic states. In interstellar space, shielding by intervening atomic hydrogen gas generally prevents the direct photodestruction of H_2 by bound-free absorption of an extreme ultraviolet photon (review Chapter 10 and the problem in Problem Set 3 dealing with the sharp division between H II and H I regions). The radiative destruction of H_2 therefore involves a two-step process: an allowed photoabsorption from the ground electronic state to a discrete excited level (usually the Lyman or Werner bands, $B\,^1\Sigma_u^+$ or $C\,\Pi_u$), followed 11 percent of the time by a radiative decay to the unstable vibrational continuum of the ground electronic state. The remaining 89 percent of the cases leads to fluorescent radiation as the excited molecule falls and cascades through the bound vibrational states of the ground electronic state. Since the first step involves a number of discrete bound-bound transitions, the molecule can protect itself against complete destruction by *self-shielding*. Molecules near the surface of a large cloud continually dissociate and reform (on grain surfaces) using up the energetic line photons that would otherwise penetrate to the interior to destroy the H_2. (For details, consult D. Hollenbach, M. Werner, and E.E. Salpeter 1971, *Ap. J.*, **163**, 155.)

ORTHO AND PARA FORMS

The H_2 molecule has one other peculiarity that we should briefly mention. It is a *homonuclear* species; i.e., the two nuclei of the molecule are *identical*. In such cases, Pauli's exclusion principle further restricts the states that the system can occupy and the transitions that it may undergo. In particular, for H_2, we can combine the nuclear spins (1/2 each) to form three symmetric spin states (parallel nuclear spins) and one antisymmetric spin state (antiparallel nuclear spins). These correspond to two different forms of hydrogen—*ortho* and *para*—and in a thermodynamically relaxed gas at high temperatures, the ortho form would have an abundance three times as great as the para form. However, radiative transitions between the two forms are forbidden; so relaxation under astrophysical conditions may be quite slow. The existence of nuclear spin may therefore have a profound effect on the statistics of the H_2 molecule, but the energy splittings so introduced are very small. (For a more general overview of the physics of diatomic molecules, see Herzberg's treatise, *Molecular Spectra and Molecular Structure: I, Spectra of Diatomic Molecules.*)

RELATIONSHIP TO OTHER IDEAS

We conclude this chapter with a comment that the idea of the covalent bond had historically an importance that extended beyond the confines of chemistry. In puzzling over the mechanism by which protons and neutrons might be bound together in the nucleus of an atom, Heisenberg got the idea that the binding force might involve nucleons exchanging a particle in a fashion analogous to how a molecular bond is formed by atoms sharing electrons. Unfortunately, his literal idea went nowhere, because he knew he couldn't use electrons in the nuclear context. Yukawa developed the idea along more fertile grounds. Yukawa deduced that the short-range nature of the strong nuclear force implies that the exchanged particle must be much more massive than an electron. The existence of the square of the Compton wavenumber $k_C \equiv mc/\hbar$ in field equations associated with gauge bosons of finite mass m [see equation (25.4)] implies that their static-potential solutions in the presence of a point source of appropriate "charge" have an exponentially attenuated range, varying as

$$\frac{1}{r} \exp(-k_C r),$$

rather than having the infinite range that characterizes the $1/r$ Coulomb solutions associated with gauge bosons of zero rest mass, such as (virtual) photons.

For the appropriate gauge boson to have a Compton length $\hbar/mc$ comparable to nuclear dimensions, 10^{-13} cm, the particle must have a mass m intermediate between a proton and an electron. Yukawa termed the hypothetical particle the *meson* (the "medium one"), and he predicted that it should have rest energy between 100 and 200 MeV. He also realized that it should have a spin of zero, because, for the resulting force to be attractive between like particles (for nuclear binding, protons and neutrons can be considered identical), the exchanged particle must have *even* integer spin (0, 2, etc.). The exchange of bosons with odd integer spin (1, 3, etc.) would result in repulsive forces between like particles.

In any case, the initial confusion of Yukawa's meson with the muon (a heavier version of the electron) and its eventual identification in cosmic-ray showers as the π meson (the pion, with a rest energy of 140 MeV) constitutes a famous story in the history of particle physics. We know today that protons and neutrons have internal substructure, and that the force that mediates their interactions—the exchange of mesons—represents mere "leftovers" from a more basic interaction involving quarks (quantum chromodynamics). Thus the π meson does not, after all, constitute the fundamental carrier of the strong force. Nevertheless, what physicists should never forget is that the original motivation for the now pervasive idea that quantum forces are mediated by the exchange of gauge bosons originated by analogy with the chemistry of the covalent bond.

30

Spectroscopy of Diatomic Molecules

Reference: McQuarrie, *Quantum Chemistry*, pp. 472–484.

In this chapter we discuss the implications of rotational and vibrational structure for the spectroscopy of diatomic molecules. The possibility of combinations of electronic, vibrational, and rotational transitions adds a degree of richness to molecular spectra not present for atoms. As we saw in the last chapter, because the molecular binding has nothing like a central force field, electronic transitions in molecules are much more complicated than those in their atomic counterparts. As we will see in this chapter, however, we have relative simplicity in their purely rotational and/or vibrational transitions.

ELECTRIC DIPOLE TRANSITIONS IN DIATOMIC MOLECULES

To include the possibility of motions of the nuclei, we must reappraise the time-dependent perturbation results derived in Chapter 22 for electronic degrees of freedom alone. In the dipole approximation, the perturbation Hamiltonian H_d equals the energy of an electric dipole $\mathbf{d}$ placed in a radiation electric field $\mathbf{E}$ [see equation (24.29)]:

$$H_d = -\mathbf{E} \cdot \mathbf{d}. \tag{30.1}$$

For a diatomic molecule, we write the total electric dipole moment as a sum of electronic and nuclear parts,

$$\mathbf{d} \equiv \mathbf{d}_{el} + \mathbf{d}_{nuc}, \tag{30.2}$$

where we compute $\mathbf{d}_{el}$ and $\mathbf{d}_{nuc}$ from the positions of the electrons and two nuclei relative to the center of mass. Let $\mathbf{X} \equiv \mathbf{x} - \mathbf{X}_{CM}$ measure the displacement from the center of mass. The electronic dipole moment then equals

$$\mathbf{d}_{\mathrm{el}} = \sum_{a=1}^{N} \mathbf{d}_a, \tag{30.3}$$

where $\mathbf{d}_a \equiv -e\mathbf{X}_a$ is the dipole moment of the a-th electron.

By analogy, the nuclear contribution to the total dipole moment reads

$$\mathbf{d}_{\mathrm{nuc}} = Z_A e\mathbf{X}_A + Z_B e\mathbf{X}_B. \tag{30.4}$$

Since $\mathbf{X}_A = (M_B/M)\mathbf{R}$ and $\mathbf{X}_B = -(M_A/M)\mathbf{R}$, with $\mathbf{R}$ being the relative vector displacement of nucleus A from nucleus B, we notice that $\mathbf{d}_{\mathrm{nuc}}$ vanishes for a *homonuclear* molecule (where $A = B$).

The dipole transition matrix element now becomes

$$\langle \Psi_f | H_{\mathrm{d}} | \Psi_i \rangle = -\langle \Phi_f | \mathbf{E}_\omega \cdot (\mathbf{d}_{\mathrm{el}} + \mathbf{d}_{\mathrm{nuc}}) | \Phi_i \rangle, \tag{30.5}$$

where the time-independent part of the total wave function Φ reads (see Chapter 29)

$$\Phi = \frac{1}{R} Z_{\mathrm{vib}}(R) Y_{Jm}(\theta, \phi) \varphi_{\mathrm{trans}} \varphi_{\mathrm{el}} \Sigma, \tag{30.6}$$

with Σ being the electronic-spin wave function and (R, θ, ϕ) describing the relative displacement of the two nuclei $\mathbf{R}$ in spherical polar coordinates. (The factorization $\varphi_{\mathrm{el}}\Sigma$ is schematic when we have more than two electrons.) In order for expression (30.5) not to vanish, the wave function $\varphi_{\mathrm{trans}} \propto \exp(i\mathbf{k}_{\mathrm{mol}} \cdot \mathbf{X}_{\mathrm{CM}})$ must be the same for the initial and final states. In other words, $\mathbf{k}_{\mathrm{mol}}$ remains the same during the radiative transition, so that photons do not couple to the translational degrees of freedom—at least, not in the approximations made to arrive here. Without loss of generality in what follows, we may therefore set $\varphi_{\mathrm{trans}} = 1$; i.e., we perform the calculations in the rest frame of the molecule where $\mathbf{k}_{\mathrm{mol}} = 0$.

PURE ROTATIONAL SPECTRA

Suppose we consider transitions (typically in the millimeter-wave regime) in which the initial and final electronic and vibrational states remain the same. Equation (30.5) then becomes

$$\langle \Psi_f | H_{\mathrm{d}} | \Psi_i \rangle = \int_0^{2\pi} d\phi \int_0^{\pi} Y_{Jm}^*(\theta, \phi) \langle Z_{\mathrm{vib}} | H_{\mathrm{D}} | Z_{\mathrm{vib}} \rangle Y_{J'm'}(\theta, \phi) \sin\theta \, d\theta, \tag{30.7}$$

where we have denoted the rotational quantum numbers of the initial state by primes and where

$$\langle Z_{\mathrm{vib}} | H_{\mathrm{D}} | Z_{\mathrm{vib}} \rangle \equiv -\int_0^{\infty} \mathbf{E}_\omega \cdot \mathbf{D}(R, \theta, \phi) |Z_{\mathrm{vib}}(R)|^2 \, dR. \tag{30.8}$$

In equation (30.8), $\mathbf{D}(R, \theta, \phi)$ gives the expectation value of the electric dipole moment $\mathbf{d}$ after we integrate over all electronic coordinates in the center of mass frame (including a summation over spin):

$$\mathbf{D}(R, \theta, \phi) \equiv \int (\mathbf{d}_{el} + \mathbf{d}_{nuc}) |\varphi_{el}(\mathbf{X}_1, \mathbf{X}_2, \ldots, \mathbf{X}_N; R)|^2 \, d^3 X_1 d^3 X_2 \ldots d^3 X_N.$$

(30.9)

We call $\mathbf{D}$ the *permanent* dipole moment of the molecule because usually we are interested in the ground-state electronic level for purely rotational transitions. Equation (30.7) then states that for pure rotational transitions to have a nonvanishing probability, the molecule must possess a nonzero permanent dipole moment. This differs from allowed (electronic) radiative processes in atoms which require only that the dipole moment *change* during the transition, i.e., that $\langle \varphi_f | \mathbf{d} | \varphi_i \rangle \neq 0$.

Although we have written $\mathbf{D}(R, \theta, \phi)$ as generally dependent on θ and ϕ, the actual dependence is special. The orientation of

$$\mathbf{D}_{nuc} = \mathbf{d}_{nuc} = \frac{e}{M}(Z_A M_B - Z_B M_A)\mathbf{R}$$

clearly lies along the direction of the line of centers of the two nuclei; the magnitude of $\mathbf{D}_{nuc}$ depends only on R. Similarly, since the electron distribution $|\varphi_{el}|^2$ possesses reflection symmetry with respect to any plane that passes through the line of centers of the two nuclei, integration of $\mathbf{d}_{el}$ in equation (30.9) over electronic coordinates must yield a $\mathbf{D}_{el}$ that has no component perpendicular to $\mathbf{R}$, and whose magnitude again depends only on R. Thus the variation of $\mathbf{D}$ with θ and ϕ enters solely through its orientation in space and not through any intrinsic functional dependence; i.e., we may write

$$\mathbf{D} = D(R)\frac{\mathbf{R}}{R}.$$

(30.10)

For a homonuclear molecule, we have already commented that $\mathbf{d}_{nuc} = 0$. On the other hand, parity considerations prevent a nonzero value for $\mathbf{D}_{el}$ in such a molecule. Less technically, we may say that the center of electronic and nuclear charge lies at the center of mass of the system for a homonuclear molecule like H_2, which then has no *allowed* rotational transitions by which it can be detected. The natural abundance of the species could overcome even this disadvantage were it not for the relatively large temperature equivalent of the level spacings for this light molecule ($2JB$, where $B = 87\,\mathrm{K}$), which makes excitation very difficult under normal interstellar conditions (excluding shocks). For these reasons, radio astronomers use the diatomic molecule CO as a surrogate tracer for molecular clouds in which H_2 is believed theoretically to be by far the dominant species. The molecule CO proves so robust a structure that it survives in as distinctly harsh environments as interstellar space and the atmospheres of cool stars.

For example, spectra of sunspots exhibit its signature. Moreover, CO is much more abundant than molecules other than H_2 or possibly H_2O. Water vapor in the Earth's atmosphere unfortunately generally interferes with observation of the lines of celestial H_2O (except in stellar and interstellar maser sources). The ubiquity of CO arises because C and O lie along the main line of stellar nucleosynthesis (successive alpha captures starting with ^{4}He) and constitute the most abundant elements in the Galaxy after H and He (the latter of which forms no compounds). Carbon and oxygen atoms have such a strong affinity for one another that, to rough approximation in favorable environments, the gas-phase elements first lock each other up in the form of CO, only then leaving the more abundant species (usually O) to form other compounds, such as H_2O or SiO. Moreover, unlike H_2O or SiO, gaseous CO has relatively little tendency to freeze out on interstellar grains.

To compute the electric-dipole-transition probability, consider first the case $\mathbf{E}_\omega$ lying in the $\hat{\mathbf{z}}$ direction, and use equation (30.10) to write

$$\mathbf{E}_\omega \cdot \mathbf{D} = |\mathbf{E}_\omega| D(R) \cos\theta, \qquad (30.11)$$

where θ is the angle between $\mathbf{R}$ and the z-axis (see Figure 28.4).

The rigid-rotor model amounts to the approximation of evaluating D at $R = R_0$, the equilibrium separation for the two nuclei:

$$D(R) \approx D(R_0) \equiv D_0. \qquad (30.12)$$

We refer to D_0 as the *reduced dipole moment*, a quantity often denoted as μ_0 in the astrophysical literature, but we will reserve the symbol μ for the reduced mass of the two nuclei.

The transition matrix element (30.7) now becomes

$$\langle \Psi_f | H_\mathrm{d} | \Psi_i \rangle = -|\mathbf{E}_\omega| D_0 \int_0^{2\pi} d\phi \int_0^\pi \cos\theta\, Y_{Jm}^*(\theta,\phi) Y_{J'm'}(\theta,\phi) \sin\theta\, d\theta.$$

From this expression, the dipole selection rule for purely rotational transitions can be derived in the usual way as

$$\Delta J = \pm 1, \qquad \Delta m = 0,$$

together with the basic requirement that $D_0 \neq 0$. Had we chosen $\mathbf{E}_\omega$ to lie in the $\hat{\mathbf{x}} \pm i\hat{\mathbf{y}}$ (circular polarization) directions, we would have gotten the usual alternative, $\Delta m = \pm 1$.

Consider the de-excitation from rotational level $J' = J + 1$ to level J, with the spontaneous emission of a photon of energy $\hbar\omega = 2(J+1)B$, where B equals the rotational constant of the molecule. From *any* sublevel m' of J', we obtain the total transition probability to *all* m sublevels of J (satisfying the selection rules) by summing over the allowed values of

m. We introduce a transition dipole moment $|\mathbf{D}_{J+1,J}|$ (often denoted $|\mu|$) defined in terms of the permanent dipole moment D_0 via the formula:

$$|\mathbf{D}_{J+1,J}|^2 \equiv \sum_{m=-J}^{J} D_0^2 \left| \oint Y_{Jm}^* \cos\theta Y_{J+1,m'} \, d\Omega \right|^2 = \frac{J+1}{2J+3} D_0^2. \quad (30.13)$$

Equation (23.17) now allows us to express the rate of spontaneous emission by the molecular electric dipole as

$$A_{J+1,J} = \frac{4\omega^3}{3\hbar c^3} |\mathbf{D}_{if}|^2 = \frac{4(J+1)\omega^3}{3(2J+3)\hbar c^3} D_0^2. \quad (30.14)$$

EXAMPLES OF DIAGNOSTICS OF MOLECULAR CLOUDS

For the $^{12}C^{16}O$ molecule, the rotational constant B has a temperature equivalent of 2.77 K; thus the $J = 1$ to $J = 0$ transition occurs at a wavelength equal to 2.6 mm (a frequency of $\nu = 115$ GHz). Because the molecule in its ground electronic and vibrational state has a small permanent dipole moment, the $J = 1$ to $J = 0$ transition of CO has a relatively modest Einstein A value, $A_{10} = 6 \times 10^{-8}$ s^{-1}, and it takes only a relatively small critical density (of H_2) to maintain collisionally a thermal population of this molecule in the first rotationally excited state. If $\langle \sigma v \rangle$ equals the rate coefficient for collisional excitation by a thermal distribution of H_2 molecules, where σ represents the velocity-dependent cross section, radio astronomers define the critical density for a molecular transition characterized by an Einstein A as $A/\langle \sigma v \rangle$. This estimate for the characteristic density from which the transition will be observed typically overestimates the true situation for $^{12}C^{16}O$ because of the phenomenon of *radiative trapping*, i.e., radiative population of excited states because of finite optical-depth effects. For clouds highly opaque to CO line radiation, an "escape probability" formalism yields that the effective downward transition rate A_{eff} from the upper level roughly equals the true Einstein A divided by the optical depth in the core of the line (see Chapter 9). This effect, combined with the high abundance of $^{12}C^{16}O$, typically allows the $J = 1$ level to be thermally populated even at densities of only a few hundred H_2 molecules per cm^{-3}. Since the $J = 1$ to 0 transition of this molecule is optically thick, angularly resolved measurements of the intensity of the line ("brightness temperature") yield good estimates of the kinetic temperature of molecular clouds. Measured in this way, the kinetic temperature of molecular clouds in our Galaxy averages about 10 K, but individual clouds may have dense cores where the temperatures can get appreciably hotter (to 50 K and above). The rarer isotopes of CO (e.g., $^{13}C^{16}O$ and $^{12}C^{18}O$) can give density information when the lines are optically thin.

The $J = 1$ to 0 transition of the $^{12}C^{32}S$ molecule has a frequency $\nu = 48.9\,\text{GHz}$ ($\lambda = 6.1\,\text{mm}$) and an Einstein $A_{10} = 1.8 \times 10^{-6}\,\text{s}^{-1}$. Its critical density (for collisional excitation) is substantially larger than that for $^{12}C^{16}O$, and it is frequently used as a probe of the transition region between envelopes and cores. Another useful diagnostic provided by the CS molecule is the $J = 5$ to 4 transition, which has a frequency $\nu = 245\,\text{GHz}$ ($\lambda = 1.2\,\text{mm}$) and an Einstein $A_{54} = 2.9 \times 10^{-4}\,\text{s}^{-1}$. If we adopt a mean collisional cross section of $\sigma \sim 10^{-15}\,\text{cm}^2$ and a hydrogen thermal velocity of $v_T \sim 3 \times 10^4\,\text{cm s}^{-1}$ ($T \sim 25\,\text{K}$), we easily calculate the critical density for excitation of the last transition as $n_{H_2} \sim 10^7\,\text{cm}^{-3}$; consequently, this line can be observed only from very dense cores of molecular clouds, near the sites of recent or impending star formation.

VIBRATIONAL-ROTATIONAL SPECTRA

We next consider vibrational-rotational transitions in which the vibrational and rotational quantum numbers may change, but the electronic state remains the same. In the harmonic-oscillator approximation, equation (30.5) becomes

$$\langle \Psi_f | H_d | \Psi_i \rangle = -|\mathbf{E}_\omega| \int_0^{2\pi} d\phi \int_0^\pi Y_{Jm}(\theta, \phi) \langle v | D | v' \rangle \cos\theta \, Y_{J'm'}(\theta, \phi) \sin\theta \, d\theta. \tag{30.15}$$

In equation (30.15), we have again made use of equation (30.11), and we write $\langle v | D | v' \rangle$ as referring to the dipole matrix element for transitions between vibrational quantum numbers v' and v:

$$\langle v | D | v' \rangle \equiv -N_v N_{v'} \int_{-\infty}^\infty H_v(x/x_0) D(R) H_{v'}(x/x_0) e^{-x^2/x_0^2} \, dx, \tag{30.16}$$

where $x \equiv R - R_0$, x_0 is the harmonic-oscillator scale length $x_0 \equiv (\hbar/\mu\omega_0)^{1/2}$, H_v is the Hermite polynomial of order v, and N_v is the associated normalization constant. Consistent with the expansion in $x \equiv R - R_0$ needed to obtain the harmonic-oscillator model, we also expand the dipole moment $D(R) = D(R_0 + x)$:

$$D(R) = D_0 + \left(\frac{dD}{dR}\right)_0 x + \cdots, \tag{30.17}$$

where the subscript 0 denotes that quantities are to be evaluated at $R = R_0$. The constant term D_0 yields the pure rotational transitions already discussed before if $v' = v$. To include the possibility of vibrational transitions, where $v' \neq v$, we need to examine the terms that follow the leading one,

D_0. If we take the next term linear in x, substitute it into equation (30.16), and use the recursion relation

$$\xi H_v(\xi) = v H_{v-1}(\xi) + \frac{1}{2} H_{v+1}(\xi),$$

we obtain, for $v' \neq v$:

$$\langle v|D|v'\rangle = x_0^2 N_v N_{v'} \left(\frac{dD}{dR}\right)_0 \int_{-\infty}^{\infty} \left[v H_{v-1}(\xi) + \frac{1}{2} H_{v+1}(\xi)\right] H_{v'}(\xi) e^{-\xi^2} d\xi.$$

Using the orthogonality relations for Hermite polynomials now demonstrates that the selection rule for vibrational transitions in the *fundamental* mode has the form

$$\Delta v = \pm 1,$$

together with the requirement that dD/dR evaluated at $R = R_0$ does not equal zero.

If we take higher-order terms in the expansion (30.17), those proportional to x^2, x^3, etc., we would get the additional possibility of making transitions with $\Delta v = \pm 2$ (first overtone), $\Delta v = \pm 3$ (second overtone), etc. Since differentiating D at R_0 behaves in order of magnitude like dividing by R_0, the transition strengths ($\propto$ the square of the matrix element) would decrease in the process roughly as additional powers of $x_0^2/R_0^2 = \hbar/\mu\omega_0 R_0^2 \sim (m_e/M)^{1/2}$. For self-consistency in such an expansion, we should also consider departures from the harmonic-oscillator approximation (see Problem Set 6). (The higher multipoles in the radiative perturbations should make smaller contributions to the transition rates unless D and its derivative exactly vanish at R_0.)

Within the context of the harmonic-oscillator approximation, all vibrational transitions with $\Delta v = \pm 1$ (or ± 2 for the first overtone, etc.) would be characterized by the emission or absorption of a photon with a single energy $\hbar\omega_0$ (or $2\hbar\omega_0$ for the first overtone, etc.) since the allowed energy levels satisfy $E_{\text{vib}} = (v + 1/2)\hbar\omega_0$. Experimentally, this expectation fails to hold for two reasons: (a) as mentioned already, the binding potential in real molecules departs from a quadratic well, with the consequence that vibrational energy levels with successively larger values of v are spaced by progressively smaller amounts than $\hbar\omega_0$; (b) vibrational transitions are usually accompanied also by changes in the rotational states.

LAMBDA DOUBLING AND ADDITIONAL SELECTION RULES

For a vibrational transition which satisfies $\Delta v = \pm 1$, the selection rule on ΔJ would appear at first sight to be identical to that for pure rotational transitions alone, namely, $\Delta J = \pm 1$. In fact, this proves correct only for $\Lambda =$

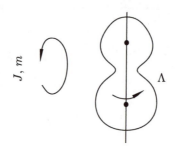

FIGURE 30.1
The component of electronic angular momentum Λ along the internuclear axis constitutes a good quantum number in a diatomic molecule.

0, where Λ represents the quantum number associated with the (conserved) component of electronic angular momentum along the internuclear axis (refer to Figure 30.1 and back to Chapter 29).

The value of J alone determines the parity of a vibrational-rotational level since vibrational excitation does not change the reflection symmetry of the molecular configuration. For $\Lambda = 0$, then, the parity change required for dipole transitions demands $\Delta J = \pm 1$, as we discovered by explicit calculation three sections ago. For $\Lambda \neq 0$, however, each rotational level corresponds to two states (the phenomenon of "Λ doubling"), characterized by the two different sign choices possible for Λ, i.e., by whether the electrons "orbit" with a net angular momentum lined up parallel or antiparallel to $\mathbf{R}$. With everything being equal, these two states have opposite parities, and transitions from a state of one electronic parity to another permit the new possibility $\Delta J = 0$.

The mathematics proceeds as follows. When the electronic wave function changes, the term $\mathbf{d}_{\mathrm{nuc}}$, which does not involve electronic coordinates, drops out of the integral for $\langle \Psi_f | \mathbf{d} | \Psi_i \rangle$ because of the orthogonality between the initial and final electronic wave functions. Moreover, for a vibrational transition accompanied by a change in electronic parity from $\varphi_{\mathrm{el},i}$ to $\varphi_{\mathrm{el},f}$ (possible for $\Lambda \neq 0$), an integration over all electronic coordinates allows

$$\mathbf{D}_{if} \equiv \langle \varphi_{\mathrm{el},f} | \mathbf{d}_{\mathrm{el}} | \varphi_{\mathrm{el},i} \rangle$$

to have no net component along $\mathbf{R}$; i.e., $\mathbf{D}_{if}$ has no dependence on the angle θ between $\mathbf{R}$ and any other fixed vector (such as $\mathbf{E}_\omega$). Hence when we further integrate over the molecular orientation, we get that $\langle Y_{Jm}(\theta, \phi) | \mathbf{D}_{if}(R) | Y_{J'm'}(\theta, \phi) \rangle$ equals zero unless $J' = J$. The additional possibility $\Delta J = 0$ for vibrational transitions arises in the electric dipole approximation without violating parity considerations, therefore, because we have a change of parity in the electronic state replacing one in the rotational state.

APPEARANCE OF VIBRATIONAL-ROTATIONAL SPECTRA

If we adopt both approximations of rigid rotator and harmonic oscillator, we may write the combined vibrational-rotational energy levels as

$$E_{vJ} = (v + 1/2)\hbar\omega_0 + J(J+1)B, \qquad (30.18)$$

where B is the rotational constant of the molecule. The observed (radian) frequency $\omega = |E_f - E_i|/\hbar$ of a transition with $v' = v \pm 1$ then reads

$$\omega = \omega_0 + 2(J+1)B/\hbar \qquad R \text{ branch,}$$

if $J' = J+1$ with $J = 0, 1, 2, \ldots$; and

$$\omega = \omega_0 - 2JB/\hbar \qquad P \text{ branch,}$$

if $J' = J-1$ with $J = 1, 2, 3, \ldots$. There exists no line exactly at ω_0 if $\Lambda = 0$, since $\Delta J = 0$ is then forbidden. The infrared spectrum therefore consists of a *band* of lines due to rotational transitions equally spaced (almost) about the frequency ω_0 due to the vibrational transition in the fundamental mode, as illustrated in Figure 30.2 for astronomical observations of ^{12}CO and ^{13}CO absorption in a protostar.

Notice that the observed lines of the R branch approach closer, but those of the P branch spread apart as ω increasingly departs from ω_0 (i.e., for larger values of J). This disagreement with the simple theory arises because centrifugal stretching modifies the rigid rotator model, and larger values of J distort the molecule more. A related effect involves the anharmonicity of the actual potential well. Expansion of $E_{el}(R)$ beyond the quadratic term in $(R - R_0)^2$ leads to the prediction that, in addition to a dominant series of lines at the fundamental frequency ω_0, there should also occur weaker *overtones* at near-integral multiples of ω_0. The following (theoretically derived) modification for the vibrational-rotational energy levels yields a summary of these effects:

$$E_{vJ} = \left[1 - A\left(v + \frac{1}{2}\right)\right]\left(v + \frac{1}{2}\right)\hbar\omega_0 + [1 - CJ(J+1)]J(J+1)B_v. \quad (30.19)$$

Experimental measurements for the molecule ^{12}C^{16}O yield the rotation constant in temperature units, $B_v/k \approx 2.8\,\mathrm{K}$ weakly dependent on v; the vibrational constant in the same units, $\hbar\omega_0/k = 3100\,\mathrm{K}$; the fractional anharmonicity constant, $A = 0.0062$; and the fractional centrifugal constant,

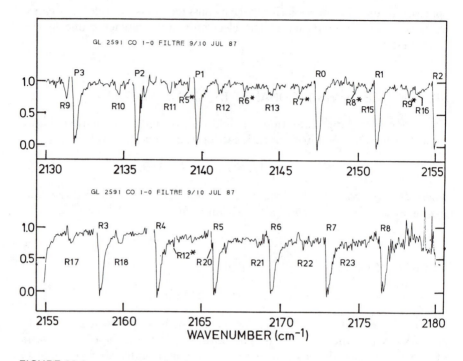

FIGURE 30.2
Vibrational transitions in the fundamental $v = 0$ to 1, with rotational
fine-structure ($\Delta J = \pm 1$ and J as high as $J = 23$ in the R branch), can be seen
in this absorption-line spectrum of CO, where the spectrum of an astronomical
source GL 2591 has been divided by that of a reference star α Lyrae. Labels
above the spectrum refer to ^{12}CO; below, to ^{13}CO. Lines of the $v = 1$ to 2
transition in ^{12}CO carry an asterisk. (From G.F. Mitchell, C. Curry,
J.P. Maillard, and M. Allen 1989, *Ap. J.*, **341**, 1020.)

$C < 5 \times 10^{-4}$. Vibrational-rotational transitions of the CO molecule in
the fundamental $\Delta v = 1$ (e.g., $v = 0$ to 1 or $v = 1$ to 2 for absorption)
occurs at a wavelength $\approx 4.6\,\mu\mathrm{m}$, whereas the first overtone $\Delta v = 2$ (e.g.,
$v = 2$ to 0 or $v = 3$ to 1 for emission) occurs at $2.3\,\mu\mathrm{m}$. These correspond
to windows of partial transmission through the Earth's atmosphere; consequently, the observation of these infrared bands provides one technique
for studying the structure and kinematics of cool stellar atmospheres and
the inner accretion disks and neutral winds from young stellar objects [see,
e.g., the discovery paper by N.Z. Scoville, D.N.B. Hall, S.G. Kleinmann,
and S.T. Ridgeway, 1979, *Ap. J. (Letters)*, **232**, L121].

For the H_2 molecule, $B_v/k \approx 87\,\mathrm{K}$; $\hbar\omega_0/k = 6300\,\mathrm{K}$; $A = 0.027$; and
$C = 7.6 \times 10^{-4}$. Quadrupole vibrational-rotational transitions from the
hydrogen molecule (having the selection rules $\Delta J = 0, \pm 2$ with 0 to 0 being forbidden) have been detected from regions in interstellar space that

have recently undergone shock excitation. For the unusual magnetohydro-
dynamics associated with the shock process, refer to the seminal papers:
B.T.Draine, 1980, *Ap. J.*, **241**, 1021; D.F. Chernoff, D.J. Hollenbach, and
C.F. McKee, 1982, *Ap. J. (Letters)*, **259**, L97.

POPULATION OF ROTATIONAL AND VIBRATIONAL LEVELS IN LTE

The relative strengths of the vibrational-rotational lines that form with
given v's but different J's under optically thin conditions usually satisfy
Boltzmann statistics. (The ^{12}CO lines displayed in Figure 30.2 are not optically thin.) In other words, at densities and temperatures where vibrational
levels can be excited, it usually forms a good approximation to assume that
the rotational states are populated in accordance to Boltzmann's law,

$$n_J/n_0 = (2J+1)\exp[-J(J+1)B/kT],$$

with a level partition function that can be approximated, when $T \gg B/k$,
by replacing the sum over discrete values of J with an integration over
continuous values,

$$\sum_{J=0}^{\infty}(2J+1)\exp[-J(J+1)B/kT]$$

$$\approx \int_{0}^{\infty}(2J+1)\exp[-J(J+1)B/kT]\,dJ = kT/B.$$

$$\frac{n_J(v)}{n(v)} \approx \frac{(2J+1)B}{kT}\exp[-J(J+1)B/kT] \qquad \text{for} \qquad J = 0,1,2,\dots.$$

In the above we have ignored the small v dependence of B and exploited
the fact that $d[J(J+1)] = (2J+1)dJ$, and we have written $n(v)$ to denote
the total number density of molecules in the vibrational level v summed
(integrated) over all rotational levels J. For the fundamental, the rate of
absorption from level v to $v+1$, or rate of spontaneous emission from level
v to $v-1$, with different initial J's is proportional to $n_J(v)/n(v)$ under
optically thin conditions, since the Einstein coefficients for the different J's
differ only to the extent that the frequencies for the vibrational-rotational
transitions are not identical (see Figure 30.3).

In LTE, the mean contribution per molecule to the internal energy
contained in rotations reads

$$\bar{\epsilon}_{\text{rot}} = \sum_{J=0}^{\infty} J(J+1)B\frac{n_J(v)}{n(v)}$$

$$\approx \int_{0}^{\infty} J(J+1)B\exp[-J(J+1)B/kT]\frac{(2J+1)B}{kT}\,dJ$$

$$= kT.$$

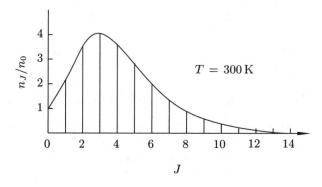

FIGURE 30.3
For optically thin emission, the radiative intensities of lines involving different J's for a given Δv will satisfy Boltzmann statistics if the rotational level populations exist in LTE at a common temperature T.

In other words, for temperatures high enough for us to ignore the discrete nature of the rotational level spacings, equipartition of energy yields $kT/2$ on average for each of two independent axes of rotation, regardless of the population of vibrational levels.

Under many astrophysical conditions of density and temperature, the vibrational levels have less tendency to exist in LTE. For the sake of completeness, however, we record below the relevant formulae if LTE should apply. Within the context of the harmonic-oscillator approximation, the vibrational partition function reads

$$\sum_{v=0}^{\infty} e^{-(v+1/2)\hbar\omega_0/kT} = e^{-\hbar\omega_0/2kT} \sum_{v=0}^{\infty} \left(e^{-\hbar\omega_0/kT} \right)^v = \frac{e^{-\hbar\omega_0/2kT}}{1 - e^{-\hbar\omega_0/kT}}.$$

In LTE, then, the fractional abundance of molecules in any vibrational level v equals

$$\frac{n(v)}{n_{\text{tot}}} = \left(1 - e^{-\hbar\omega_0/kT} \right) e^{-v\hbar\omega_0/kT}, \qquad \text{for} \qquad v = 0, 1, 2, \ldots.$$

The mean contribution per molecule to the internal energy of vibrations then becomes

$$\bar{\epsilon}_{\text{vib}} = \sum_{v=0}^{\infty} \left(v + \frac{1}{2} \right) \hbar\omega_0 \frac{n(v)}{n_{\text{tot}}} = \frac{\hbar\omega_0}{e^{\hbar\omega_0/kT} - 1} + \frac{1}{2}\hbar\omega_0,$$

where we have evaluated the sum $\sum_{v=0}^{\infty} ve^{-vx}$, with $x \equiv \hbar\omega_0/kT$, by taking the negative derivative of $\sum_{v=0}^{\infty} e^{-vx} = (1 - e^{-x})^{-1}$ with respect to x.

Apart from the zero-point contribution, $\hbar\omega_0/2$, $\bar{\epsilon}_{\mathrm{vib}}$ has the expression first derived by Planck in his epochal investigation of the mean thermal energy for a collection of quantized oscillators in thermodynamic equilibrium with respect to the emission and absorption of blackbody radiation of resonant frequency ω_0. At high temperatures, $T \gg \hbar\omega_0/k$, $\bar{\epsilon}_{\mathrm{vib}}$ acquires the classical (Rayleigh-Jeans) expression:

$$\bar{\epsilon}_{\mathrm{vib}} = kT,$$

i.e., on average, a classical oscillator possesses $kT/2$ each in vibrational potential and kinetic energy. However, real molecules tend to dissociate before reaching temperatures much in excess of $\hbar\omega_0/k$. At low temperatures, $T \ll \hbar\omega_0/k$, we have $\bar{\epsilon}_{\mathrm{vib}} \approx \hbar\omega_0/2 = $ constant, independent of T. Thus vibrations (above the zero-point motions) cannot be appreciably excited in the diatomic molecules that make up, for example, the air at room temperature, whereas rotations would contribute to the specific heat capacity only one unit of Boltzmann's constant k, instead of the value $3k/2$ (for three independent axes of rotation) expected on the basis of classical theory (nonnuclear model of atoms and molecules). Since the speed of sound depends on a ratio of specific heat capacities, quantum theory has macroscopic manifestations even in the rate at which our voices carry!

ELECTRONIC-VIBRATIONAL-ROTATIONAL TRANSITIONS

As we saw in Chapter 29, electronic transitions in diatomic molecules satisfy less restrictive vibrational selection rules than those that apply to pure vibrational-rotational spectra. When the electronic configuration changes, the term $\mathbf{d}_{\mathrm{nuc}}$ drops out as before. The angular dependence of $\mathbf{d}_{\mathrm{el}}$ on the molecular orientation θ and ϕ then gives the selection rules: $\Delta J = 0, \pm 1$ and $\Delta m = 0, \pm 1$, with $\Delta J = 0$ not allowed by parity considerations if $\Lambda = 0$ in both the initial and final states. The transition $J = 0$ to $J = 0$ is forbidden independent of Λ.

The dependence of $\mathbf{d}_{\mathrm{el}}$ on azimuthal electronic structure yields the usual selection rule concerning the projection of the angular momentum vector (in this case along the internuclear axis): $\Delta\Lambda = 0, \pm 1$. Notice, in particular, that allowed electronic transitions do *not* require the molecule to possess a *permanent* electric dipole moment. Using this principle, Carruthers made the first detection in 1970 of the H_2 molecule in an ultraviolet rocket experiment that measured the H_2 interstellar absorption lines formed in front of bright hot stars. While potentially a very powerful method for directly measuring the column density of H_2, this technique suffers from the need to choose line of sights toward relatively unobscured OB stars.

If the transition satisfies each of the above selection rules, then the dipole matrix element remains proportional to an integral of the form

$$\mathbf{D}_{if} \propto \int Z^*_{\mathrm{vib},f}(R) Z_{\mathrm{vib},i}(R)\, dR. \qquad (30.20)$$

The square of the vibrational overlap integral, equation (30.20), is called the *Franck-Condon factor*. It has a nonzero value for any Δv; i.e., unlike purely vibrational-rotational transitions, electronic-vibrational-rotational transitions ("vibronic transitions" in short) place no restrictions on Δv. Nevertheless, this does not imply that all values of Δv have an equal probability of occurrence. Figure 30.4 gives a pictorial representation of the relevant physics.

According to the WKB solution (see Problem Set 6), the wave function for a particle (of reduced mass μ) trapped in a potential well possesses $v + 1$ maxima between the classical turning points with the largest amplitudes of the standing wave occurring near the turning points (because the classical particle spends most time turning around there). In between the turning points, the probability density $|Z_{\mathrm{vib}}|^2$ has therefore the appearance illustrated by the light grey shading. Because $\mu \gg m_e$, nuclei hardly move during the interval that it takes electrons to make electronic transitions. Vertical lines depict schematically the latter transitions (upward arrows for radiative absorptions). The overlap integral (30.20) contains large contributions from those arrows that connect one high-value of the vibrational wave function $Z^*_{\mathrm{vib},f}$ with another high-value $Z_{\mathrm{vib},i}$. In other words, we expect electronic transitions that connect vibronic states with large Franck-Condon factors to have greater radiative intensities (either in emission or absorption) than those that connect states with small Franck-Condon factors (see Figure 30.5). Thus detailed analysis of the relative intensities of vibronic lines yields considerable information about the shapes of the curves $E_{\mathrm{el}}(R)$.

BAND STRUCTURE AND P, Q, AND R BRANCHES

For given values of the electronic quantum numbers $n' \to n$, the different possible choices for the changes of the vibrational quantum number Δv lead to vibrational fine structure. For each choice of $v' \to v$, the different possible choices for $\Delta J = 0, \pm 1$ lead to even finer rotational structure. By convention, for emission spectra, the choice $\Delta J = +1$ is called the P branch; $\Delta J = 0$, the Q branch; $\Delta J = -1$, the R branch. In absorption spectra, the signs are reversed. The important thing for spectroscopists is that the R branch starts off at higher frequencies (smaller wavelengths) than the Q branch, which, in turn, begins at higher frequencies (smaller wavelengths) than the P branch. Thus the combination of vibrational plus

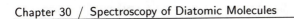

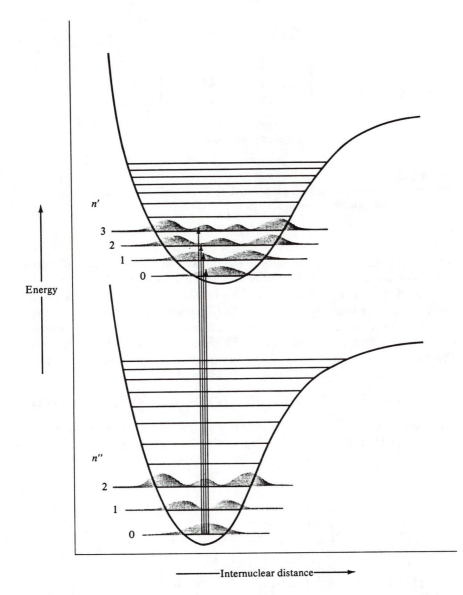

FIGURE 30.4
The physics underlying Franck-Condon factors.

intensities of lines
for previous figure

FIGURE 30.5
Relative intensities (transition probabilities) of vibronic lines follow
Franck-Condon considerations.

rotational changes gives the optical (or ultraviolet or near infrared) spectrum of a diatomic molecule a *band structure* (see Figure 30.6).

Since the moments of inertia of the molecule differ for the initial and final states in an electronic transition, the frequency spacing between successive lines in each rotational branch does not remain approximately constant or even monotonic. For example, if the initial rotation constant $B' > B$ the final rotation constant in an absorption process, the frequency of lines in the R branch ($J' = J + 1$) may be calculated as

$$\nu_R = \nu_{\text{el, vib}} + h^{-1}[2B + (3B - B')J - (B' - B)J^2]. \tag{30.21}$$

Since the terms proportional to J and J^2 have, respectively, positive and negative coefficients, ν_R marches higher at first with increasing J, reaches

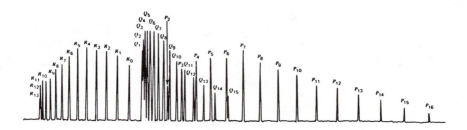

FIGURE 30.6
Electronic transitions accompanied by changes of vibrational and rotational quantum numbers lead to diatomic molecules (here AlH) having band structures for their optical and ultraviolet spectra. (From W.A. Bingel 1969, *Theory of Molecular Spectra.*)

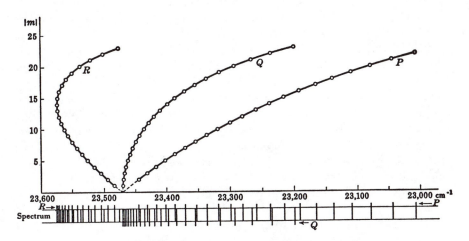

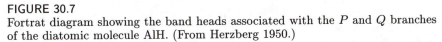

FIGURE 30.7
Fortrat diagram showing the band heads associated with the P and Q branches
of the diatomic molecule AlH. (From Herzberg 1950.)

a maximum for $J = J_{\text{head}}$, and then reverses for larger values of J. In
contrast, the corresponding formulae for the Q $(J' = J)$ and P $(J' = J-1$
for $J \geq 1$) branches read

$$\nu_Q = \nu_{\text{el, vib}} - h^{-1}(B' - B)J(J+1), \qquad (30.22)$$

$$\nu_P = \nu_{\text{el, vib}} - h^{-1}[(B' + B)J + (B' - B)J^2]. \qquad (30.23)$$

As a result, each branch has a sharply defined upper edge in frequency,
called the *band head*, which has a well-known value for each well-studied
molecule (see the Fortrat diagram depicted in Figure 30.7).

LAMBDA DOUBLING AND RADIO ELECTRONIC TRANSITIONS

Because of Λ doubling, electronic transitions in diatomic molecules can
yield not only lines in the ultraviolet, optical, and infrared, but even in
the radio. The phenomenon depends on effects ignored in the Born–
Oppenheimer decoupling of the degrees of freedom contained in the motions
of the electrons and nuclei. The interaction between the rotation of the nuclei and the electron orbital angular momentum (along different axes) yields
a small splitting in the actual energy levels, with electronic transitions between the doubled Λ levels (electrons with opposite senses of rotation about
the internuclear axis), with no changes in J or v, corresponding typically

to tens of cm wavelengths. For example, Shklovskii calculated in the early 1950s that the transition wavelengths of the Λ splitting of the ground states of the astrophysically important hydrides OH, CH, SiH should fall at 18.3, 9.45, and 12.5 cm, respectively. Laboratory work by Townes's group pinned down the precise radio frequencies of the OH transitions, and in a famous experiment performed in 1963, Barrett, Meeks, and Weinreb detected the molecule in absorption in front of a strong source of radio continuum radiation, the supernova remnant Cas A. Although optical astronomers had known for many years that molecules such as CH and CN exist in interstellar clouds, the phenomenon was regarded somehow as a fluke. Thus the discovery of a different species by a radio-astronomical technique heralded the modern era of molecular astronomy. At the present time, some 70 molecules—mostly organic—have been found and identified in interstellar space.

The radio lines of OH remain especially interesting. Hyperfine interaction (between the magnetic moments of the nuclei and the spin of the unpaired electron) splits the Λ doublet of the ground state, nominally at $\lambda = 18.3$ cm, into four lines with frequencies equal to 1,612, 1,665, 1,667, and 1,720 MHz. The two strongest lines (1,667 and 1,665 MHz) have Einstein A's of 2.66×10^{-11} s^{-1} and 2.47×10^{-11} s^{-1}, small only because of the low frequencies of the transitions. By comparison, the 21-cm spin-flip transition of atomic hydrogen (see Problem Set 5) has a much smaller Einstein A equal to 2.85×10^{-15} s^{-1}, because, in addition to the low frequency, the line involves a magnetic dipole transition and therefore has a probability reduced roughly by a factor α^2 compared to electric dipole transitions.

The OH radio-frequency lines have proven astrophysically important for three reasons. First, the small Einstein A values imply that collisional excitation of the molecule can occur in fairly low-density regions, making the detection of OH emission (and absorption) the most widely distributed radio line in the Galaxy next to H$\textsc{i}$. (This does not mean that OH is found only in low-density regions. It can also radiate in the dense cores of molecular clouds, but the emission will be weaker than from molecules with comparable abundances and with transitions having larger A values.) Second, OH has an unpaired electron, giving it a relatively large Lande g factor for the Zeeman effect. Moreover, for a given g factor, because the Zeeman splitting in frequency $\Delta\nu$ is proportional to the magnetic field strength, independent of the frequency of the transition, the relative splitting $\Delta\nu/\nu$ is higher for a low-frequency transition like those of OH than for a high-frequency transition associated with some other hypothetical molecule. This makes the Zeeman contribution to line broadening easier to measure (e.g., by differential polarization measurements) against other, more powerful, sources of line broadening. (For a review, see C. Heiles 1987, in *NATO/ASI Physical Processes in Interstellar Clouds*, ed. G.E. Morfill and M. Scholer (Reidel), p. 429.) Third, astronomers have found OH *maser* emission associated with stars at both extremes of stellar evolution, at birth and near death. Two types are known: type I mase strongest in the "main lines" at 1,665 and 1,667 MHz; type II, in the "satellite lines" at 1,612 and 1,720 MHz. These bright sources of radio-line emission form inviting targets for VLBI investigations and have already yielded much astrophysical information, not only about regions of maser emission, but also about such disparate subjects as Galactic structure and stellar evolution. (For a review, consult M.J. Reid and J.M. Moran, 1981, *Ann. Rev. Astr. Ap.*, **19**, 231.)

PROBLEM SETS

Problem Set 1

1. Neutrinos and antineutrinos (each occupying one spin state of the two possible for spin-1/2 particles, and each having zero chemical potential) are believed to have been in thermodynamic equilibrium with matter and radiation in the early universe (see Problem Set 2).

(a) Show that each species (electron-, muon-, tau-neutrino, etc.) contributes an energy density equal to

$$\mathcal{E} = 8\pi \int_0^\infty \frac{(h\nu^3/c^3)}{e^{h\nu/kT} + 1} \, d\nu.$$

Introduce the new variable $x \equiv h\nu/kT$ and write the above as

$$\mathcal{E} = \frac{8\pi}{h^3 c^3} (kT)^4 I_3,$$

where I_j is the integral,

$$I_j \equiv \int_0^\infty \frac{x^j \, dx}{e^x + 1}.$$

By expanding the denominator,

$$(e^x + 1)^{-1} = \sum_{n=1}^\infty (-)^{n-1} e^{-nx},$$

express the integral as

$$I_j = (j!) \sum_{n=1}^\infty \frac{(-)^{n-1}}{n^{j+1}}.$$

Write the sum as one containing all positive values,

$$\zeta(j + 1) \equiv \sum_{n=1}^\infty \frac{1}{n^{j+1}},$$

where $\zeta(j+1)$ is the Riemann zeta function of argument $j+1$, minus twice another sum over even n:

$$\frac{1}{2^{j+1}} + \frac{1}{4^{j+1}} + \frac{1}{6^{j+1}} + \cdots = \frac{1}{2^{j+1}} \zeta(j + 1).$$

369

Consequently, demonstrate

$$I_j = \left(1 - \frac{1}{2^j}\right)(j!)\zeta(j+1).$$

With $\zeta(4) = \pi^4/90$ from complex variable theory, we have

$$I_3 = \left(\frac{7}{8}\right)\left(\frac{\pi^4}{15}\right),$$

which, apart from the factor of $(7/8)$, is the usual result for the blackbody integral. Thus each species of neutrinos and antineutrinos contributes an energy density

$$\mathcal{E} = \frac{7}{8}aT^4,$$

where a is the standard radiation constant.

(b) Define the energy per unit wavelength $\mathcal{E}_\lambda$ by $\mathcal{E}_\lambda\,d\lambda \equiv -\mathcal{E}_\nu\,d\nu$, where $\lambda \equiv c/\nu$, and show that

$$\mathcal{E}_\nu \propto \frac{x^3}{e^x \pm 1}, \qquad \mathcal{E}_\lambda \propto \frac{x^5}{e^x \pm 1},$$

with $x \equiv h\nu/kT \equiv hc/\lambda kT$. In the above, the plus sign applies to neutrinos; the minus sign to photons. With variations of ν or λ, show that $\mathcal{E}_\lambda$ reaches a maximum (Wien's displacement law) where

$$\frac{5}{x}\left(e^x \pm 1\right) - e^x = 0;$$

whereas $\mathcal{E}_\nu$ reaches a maximum where

$$\frac{3}{x}\left(e^x \pm 1\right) - e^x = 0.$$

Clearly, the numerical value of x will differ for the two cases, which explains why theoretical astrophysicists regard neither $\mathcal{E}_\nu$ nor $\mathcal{E}_\lambda$ as appropriate quantities to plot, but instead prefer $\nu\mathcal{E}_\nu = \lambda\mathcal{E}_\lambda$. Show that $\nu\mathcal{E}_\nu = \lambda\mathcal{E}_\lambda$ peak at an intermediate value of x where

$$\frac{4}{x}\left(e^x \pm 1\right) - e^x = 0.$$

Compute x numerically for the two choices of sign, and compare the numerical value with that usually given for Wien's displacement law for photons. (You may wish to consult Problem 2.)

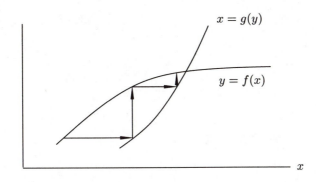

FIGURE P1.1
Successive iterates of $y = f(x)$, $x = g(y)$, for a convergent case.

---

2. We wish to demonstrate the claim made in the text of Chapter 3 that any given iteration scheme generally has only a 50-50 chance of converging. To see this, consider the simplest scheme to find the solution to the simultaneous equations:

$$y = f(x), \qquad x = g(y),$$

where f and g are real functions of real variables x and y. A naive procedure might involve (1) arbitrarily guessing a numerical value for x, (2) plugging x into f to get $y = f(x)$, (3) taking the resulting numerical value of y and plugging into g to get a new value of $x = g(y)$, (4) going back to 2 and repeating until one achieves convergence. Graphically, successive iterations might look as in Figure P1.1. As a specific example, consider the simple case: $f(x) \equiv \sqrt{x}$ and $g(y) \equiv y$. The resulting equation,

$$x = \sqrt{x},$$

has, of course, the roots $x = 0$ and 1. Verify (by repeated square roots on a calculator if need be) that straight iterations starting with any initial guess larger than 0 will produce the root $x = 1$, but not the root $x = 0$. Moreover, in any given problem, it appears just as likely that the labels of the f and g curves could be reversed. As indicated in Figure P1.2, straight iterations via $y = f(x)$ and $x = g(y)$ would then lead to a divergent procedure. (Try it out on the specific example: $y = x$ and $x = y^2$. Now any initial guess $|x| < 1$ leads to the root $x = 0$, but any guess $|x| > 1$ leads

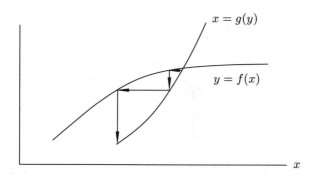

FIGURE P1.2
Successive iterates of $y = f(x)$, $x = g(y)$, for a divergent case.

to divergent successive iterates.) Clearly, then, unless we have additional information about the behaviors of the relevant functions, straightforward iteration procedures generally have (at best) only a 50-50 probability of converging to any desired solution.

(a) To improve our chances, notice that the ability to reverse the path of a divergent case would produce convergence. In other words, if straight iteration of $y = f(x)$, $x = g(y)$ diverges, iteration of $y = g^{-1}(x)$, $x = f^{-1}(y)$ will converge. (Thus $y = x$, $x = y^2$ diverges if we're trying to find the root $x = y = 1$; but $y = \sqrt{x}$, $x = y$ converges. The converse holds if we're trying to find the root $x = y = 0$.) However, finding the inverse functions f^{-1} and g^{-1}, even graphically, may not constitute a practical possibility, e.g., if the relevant "functions" do not have analytical expressions, but represent numerical results from some computational procedure too complicated in practice for us to produce full tabular data. In this case, we need another technique for "going backwards."

The introduction of a relaxation parameter θ represents one such technique. Instead of the straightforward iteration procedure

$$y_n = f(x_n), \qquad x_{n+1} = g(y_n),$$

where the subscript n refers to the n-th iterate, consider taking the linear combination

$$y_n = \theta f(x_n) + (1 - \theta)y_{n-1},$$

before substituting into g to find a new x:

$$x_{n+1} = g(y_n).$$

Choosing a weight θ equal to a fraction between 0 and 1 would correspond just to going slower in the "forward direction" of a straight iteration (under-relaxation); choosing $\theta > 1$ corresponds to over-relaxation; whereas choosing $\theta < 0$ corresponds to "going backwards" (i.e., we like previous iterates better than present ones). Demonstrate explicitly that "going backwards" works for finding the root $x = y = 1$ in the specific example $y = x$, $x = y^2$ if we choose $\theta = -1$ and if we start with $0 < x < 2$. In particular, record the successive iterates for x when you start with $x = 1.5$ or $x = 0.5$. Play around with other negative values of θ (e.g., $\theta = -1/2$ and $\theta = -2$) and comment on the tradeoffs between rate of convergence and range of initial guesses which will lead to a solution.

For extra credit: Discuss how you might try empirically to determine a good choice for θ in any given problem. In particular, consider how you might construct an absolute measure of the error involved in satisfying the simultaneous equations, and how to drive that error systematically to smaller and smaller values at a near to maximal rate.

(b)   Complete linearization represents another technique, one that is somewhat more involved but gives more rapid convergence when it works. Suppose we believe (x_*, y_*) to represent a good guess for the simultaneous solution of the governing equations, although neither $\delta f_* \equiv f(x_*) - y_*$ nor $\delta g_* \equiv g(y_*) - x_*$ exactly equals zero. To obtain an improved estimate for the true roots (x, y), where $y = f(x)$ and $x = g(y)$, expand in a Taylor series about the known values:

$$y = f(x_*) + (x - x_*)f'(x_*), \qquad x = g(y_*) + (y - y_*)g'(x_*).$$

Solving the above as a linear set of simultaneous equations for the corrections, $\delta x \equiv x - x_*$ and $\delta y \equiv y - y_*$, we obtain

$$\delta x = -\frac{g_*'\delta f_* + \delta g_*}{f_*'g_*' - 1}, \qquad \delta y = -\frac{f_*'\delta g_* + \delta f_*}{f_*'g_*' - 1}.$$

After each iteration step, we add the corrections δx and δy to x_* and y_*, call the resulting values the new guesses x_* and y_*, and proceed to the next iteration step. The procedure stops when we have reduced both δf_* and δg_* sufficiently close to zero. For the case $x = g(y) = y$ and $g' = 1$ with $\delta g_* = 0$, verify that the above set becomes Newton's formula for finding (by repeated iteration) the root of the equation $h(x) \equiv f(x) - x = 0$:

$$x = x_* - \frac{h(x_*)}{h'(x_*)}.$$

For extra credit: Generalize the above considerations to the case when x and y are vectors and f and g are vector functions of the same dimension ("arrays" in computer jargon). The importance of this observation rests with the fact that finite-difference techniques can always cast differential and integral equations as matrix equations. If the problem is nonlinear, with the elements of the coefficient matrix dependent on the vector variables themselves, then relaxation techniques or complete linearization (standard packages exist to solve linear matrix problems) may represent the only convenient methods of attacking the problem numerically.

---

3. For an LTE grey atmosphere where $T(\tau)$ is given by equation (4.17) of the text, compute the dimensionless energy distribution $\tilde{\nu}\tilde{F}_{\tilde{\nu}} \equiv \nu F_\nu(0)/\sigma T_{\text{eff}}^4$ (with $\sigma = 2\pi^5 k^4/15c^2 h^3$) as a function of the dimensionless frequency $\tilde{\nu} \equiv h\nu/kT_{\text{eff}}$. (Astronomers refer to νF_ν as the energy distribution, although it would be more accurately termed the "energy *flux* distribution.") Compare your result with the equivalent blackbody $\pi\nu B_\nu(T_{\text{eff}})/\sigma T_{\text{eff}}^4$ and comment appropriately.

To simplify your computational work for this problem, perform one integration by parts to obtain

$$E_2(\tau) = e^{-\tau} - \tau E_1(\tau).$$

Now, notice that Abramowitz and Stegun (1965) give the following approximate formulae. Accurate to 2 parts in 10^7, for $0 \leq \tau < 1$:

$$E_1(\tau) = -\ln\tau + \sum_{n=0}^{5} a_n \tau^n,$$

where $a_0 = -0.57721566$, $a_1 = 0.99999193$, $a_2 = -0.24991055$, $a_3 = 0.05519968$, $a_4 = -0.00976004$, $a_5 = 0.00107857$. Accurate to 2 parts in 10^8, for $1 < \tau < \infty$:

$$E_1(\tau) = \tau^{-1} e^{-\tau} \left[\frac{\sum_{n=0}^{4} c_n \tau^n}{\sum_{n=0}^{4} d_n \tau^n} \right],$$

where $c_0 = 0.2677737343$, $c_1 = 8.6347608925$, $c_2 = 18.0590169730$, $c_3 = 8.5733287401$, $c_4 = 1.00000000$; $d_0 = 3.9584969228$, $d_1 = 21.0996530827$, $d_2 = 25.6329561486$, $d_3 = 9.5733223454$, $d_4 = 1.00000000$. At $\tau = 1$, you may take the average of the two above formulae.

Notice also that the integral in equation (4.20) of the text can be broken into two pieces:

$$F_\nu(0) = 2\pi(f_L + f_0)$$

where $f_L \equiv \int_{\tau_L}^{\infty} E_2(\tau) B_\nu[T(\tau)] d\tau$ and $f_0 \equiv \int_0^{\tau_L} E_2(\tau) B_\nu[T(\tau)] d\tau,$

with τ_L being some large optical depth, say 20. (Justify this choice.) For $\tau \geq \tau_L \gg 1$, show that a sufficient order of approximation is $E_2(\tau) \approx e^{-\tau}/\tau$, $B_\nu \approx (2\nu^2/c^2)kT$, and $T \approx (3/4)^{1/4}T_{\text{eff}}\tau^{1/4}$. Now, perform the integration for f_L analytically and for f_0 numerically.

The observed continuous spectrum of the Sun departs from a perfect blackbody in a manner opposite to the sense you will find above. This fact provided the motivation for astronomers in the early part of the twentieth century to study the frequency dependence of the continuous opacity in the solar atmosphere, leading ultimately to the discovery of the role of the H^- ion discussed in the text.

Problem Set 2

1. Consider an ideal Fermi-Dirac gas with no internal degrees of freedom other than spin $s = 1/2$ (e.g., the free electrons in a star). With the number and internal energy densities defined by $n \equiv N/V$ and $\mathcal{E} \equiv E/V$, use the expression for the grand potential derived in the text to obtain the following limiting thermodynamic relationships.

Nondegenerate, nonrelativistic:
When $e^{-\mu/kT} \gg 1$ and $mc^2/kT \gg 1$,

$$\mathcal{E} = 3P/2,$$
$$P = nkT,$$
$$\mu = kT \ln[n\lambda_T^3/2], \qquad \text{where} \qquad \lambda_T = h/(2\pi mkT)^{1/2},$$
$$S = -Nk \ln[n\lambda_T^3/2].$$

Nondegenerate, ultrarelativistic:
When $e^{-\mu/kT} \gg 1$, and $mc^2/kT \ll 1$,

$$\mathcal{E} = 3P,$$
$$P = nkT,$$
$$\mu = kT \ln[n\lambda_T^3/2], \qquad \text{where} \qquad \lambda_T = (8\pi)^{-1/3}(hc/kT),$$
$$S = -Nk \ln[n\lambda_T^3/2].$$

Notice that nondegeneracy requires the number of particles in a thermal deBroglie volume, $n\lambda_T^3$, to be very small compared to unity, so that μ/kT is a large *negative* number. Notice also that adiabatic processes which keep S constant (given N) satisfy $P \propto n^{5/3}$ under nonrelativistic conditions, and $P \propto n^{4/3}$ under ultrarelativistic conditions. Explain these behaviors on the basis of the fundamental law of thermodynamics

$$dE = TdS - PdV,$$

combined with $E = 3PV/2$ or $E = 3PV$.

Completely degenerate, nonrelativistic:
When $e^{-\mu/kT} \ll 1$ [so that the occupation number may be approximated as the Heaviside step function $\Theta(\mu - \epsilon)$] and $mc^2/kT \gg 1$,

$$\mathcal{E} = 3P/2,$$

$$P = (8\pi h^2/15m)(3n/8\pi)^{5/3},$$
$$\mu = (h^2/2m)(3n/8\pi)^{2/3},$$
$$S = 0.$$

Completely degenerate, ultrarelativistic:

$$\mathcal{E} = 3P,$$
$$P = (2\pi hc/3)(3n/8\pi)^{4/3},$$
$$\mu = hc(3n/8\pi)^{1/3},$$
$$S = 0.$$

Explain the connection between the result (to the lowest order of approximation) $S = 0$ and the "third law of thermodynamics." Comment on the connection between the results $P \propto n^{5/3}$ and $P \propto n^{4/3}$ obtained for the cases of complete nonrelativistic and ultrarelativistic degeneracies, and the same proportionalities when the gas is nondegenerate but limited to adiabatic variations. (Hint: Did we need $P = nkT$ to derive the latter?) Finally, compute the value $(3n/8\pi)^{-1/3}$, which equals in order of magnitude the mean distance between two fermions of the same spin, when the formulae for the completely degenerate nonrelativistic and ultrarelativistic cases have the same pressure P. (In other words, what is the characteristic transition density between nonrelativistic and ultrarelativistic degeneracy?) What is the name in physics for the combination h/mc? What is its relationship to the uncertainty principle?

---

2. In the early universe, until a temperature of $T \approx 10^{10}\,\mathrm{K}$ (when $kT \approx 2m_ec^2$), electron-positron pairs are copiously produced and maintained in equilibrium with the cosmic background photons (blackbody radiation) via the reaction

$$e^- + e^+ \rightleftharpoons 2\gamma,$$

where γ represents gamma rays. At even higher temperatures, the electrons and positrons would even have been in equilibrium with respect to weak-interaction processes of the type

$$e^- + e^+ \rightleftharpoons \nu + \bar{\nu},$$

where ν and $\bar{\nu}$ refer here to (electron) neutrinos and antineutrinos.

When considering the production and destruction of particles with rest masses, we find it convenient to redefine the zero point of energies with the vacuum state, rather than with the free state where particles have zero momentum. This adds the rest energy mc^2 to all formulae derived so far for particle energies ϵ and chemical potentials μ. Convince yourself that nothing else we have done depends on this modification. Show that the generalized law of mass action for the above reactions requires

$$\mu_- + \mu_+ = 0,$$

where $\mu_\pm$ is to be determined from the requirement

$$n_\pm = \frac{2}{h^3} \int_0^\infty \frac{4\pi p^2\, dp}{e^{(\epsilon - \mu_\pm)/kT} + 1},$$

with the present zero-point convention giving

$$\epsilon = (p^2 c^2 + m^2 c^4)^{1/2}.$$

In conventional big-bang theory, there exists a net electron number $n_- - n_+ = n_0$, but it is fractionally small $(n_0/n_- \sim 10^{-9})$ during the epoch of concern. To the extent that $n_+ = n_-$, the symmetry of the expressions for their (Fermi-Dirac) integrals implies $\mu_+ = \mu_-$, which combines with $\mu_+ + \mu_- = 0$, to give $\mu_+ = \mu_- = 0$.

For $kT \gg mc^2$, with $\epsilon \approx pc$, show that

$$n_e = I_2 \lambda_{Te}^{-3}, \qquad \text{with} \qquad I_2 \equiv \int_0^\infty \frac{x^2\, dx}{e^x + 1} = \frac{3}{4}(2!)\zeta(3) = 1.803,$$

where the evaluation of the integral I_2 can be obtained by the method of Problem Set 1, and where λ_{Te} has its ultrarelativistic expression (see Problem 1 above). (Notice that degeneracy effects here only change a factor of 2 to a factor of 1.803.) For $kT \ll mc^2$, show that

$$n_e = 2\lambda_{Te}^{-3} e^{-m_e c^2/kT},$$

where λ_{Te} has its usual nonrelativistic expression. Assume the cross sections for mutual annihilation of electrons and positrons to be comparable to the Thomson cross section $\sigma_T = 8\pi r_e^2/3$, where $r_e \equiv e^2/m_e c^2$, and show that the characteristic time scale $(n_e c \sigma_T)^{-1}$ to maintain thermal equilibrium, say, at $T = 10^{11}$ K, is much shorter than the age of the universe $(t \sim 0.3 \text{ s})$. Describe why the assumption of pair-equilibrium will fail for much smaller T.

For $T > 10^{10}$ K, when electrons, neutrinos, and their antiparticles are plentiful, protons p^+ and neutrons n have equilibrium thermodynamic concentrations governed by weak interactions of the type (ν = electron neutrino below):

$$p^+ + e^- \rightleftharpoons n + \nu,$$

plus other similar reactions. Apply the law of mass action to the above reaction and show

$$n_n/n_p = e^{-Q/kT},$$

where Q is the rest energy difference between a free neutron and a free proton: $Q = (m_n - m_p)c^2$. Calculate the ratio n_n/n_p at a temperature $T = 10^{10}$ K. Once the temperature drops significantly below 10^{10} K because of the expansion of the universe, most photons lack the energy to create electron-positron pairs; the electrons and positrons quickly recombine (leaving a residue of 1 part in 10^9 of the original excess of electrons). The neutrinos and antineutrinos also become largely decoupled from matter. Neutron capture reactions will begin that take about three minutes to complete. (This value depends on the number of neutrino species that contribute to the relativistic energy content of the early universe from Problem 1 of Problem Set 1. In turn, this relativistic energy content determines the rate of expansion; see Shu 1982, pp. 386–390, for an elementary discussion. Thus big-bang nucleosynthesis of the helium-4 considered in this problem can help set astrophysical constraints on the number of allowable neutrino types in particle physics.) In any case, since free neutrons will decay on a time scale of about 10 minutes, roughly 70 percent of the neutrons present at $T = 10^{10}$ K will end up in atomic nuclei, mostly helium-4. Adopt the above numbers and compute the primordial helium abundance (as a mass fraction of the total) that results from standard big-bang nucleosynthesis.

---

3. Consider the thermal ionization process in a pure hydrogen plasma: $H^+ + e^- \rightleftharpoons H$, with the ionization potential I from the ground state equaling 13.6 eV.

(a) For simplicity, ignore the excited bound states, and show that Saha's equation can be written

$$\frac{n_e n_+}{n_0} = \frac{g_e g_+}{g_0} \left(\frac{\lambda_{T0}^3}{\lambda_T^3 + \lambda_{Te}^3} \right) e^{-I/kT},$$

where the g's represent statistical weights and the λ_T's represent (nonrelativistic) de Broglie thermal wavelengths. Since $g_e = g_+ = g_0 = 2$, $m_+ \approx m_0$, $n_e = n_+$ (overall charge neutrality), with $n_+ + n_0 = \text{constant} \equiv n_H$ (proton conservation), show that we may write Saha's equation as

$$\frac{\alpha^2}{1 - \alpha} = \frac{g_e}{n_H \lambda_{Te}^3} e^{-I/kT},$$

where $\alpha \equiv n_+/n_H = n_e/n_H$, $1 - \alpha = n_0/n_H$, and $\lambda_{Te} = h/(2\pi m_e kT)^{1/2}$.

(b) Compute the fractional ionization α at the center of the Sun where $T = 1.5 \times 10^7$ K and $n_H = 1.0 \times 10^{26}$ cm^{-3}. Does your answer contradict the usual statement made in elementary textbooks that high *temperatures* keep the solar interior nearly completely ionized? As it turns out, the solar interior *is* nearly fully ionized, but the reason resides more in the effects of *pressure* ionization than of *thermal* ionization. [See the discussion in Shu 1982, Problem 5.5. Alternatively, compute n_{max} from equation (7.30) and comment physically.]

---

4. An important relation required for the linear pulsation theory of variable stars concerns the fractional changes in pressure P and in density ρ (or specific volume $v \equiv \rho^{-1}$):

$$\frac{dP}{P} + \Gamma_1 \frac{dv}{v} = 0.$$

For adiabatic compressions of a perfect monatomic gas, we all know $\Gamma_1 = \gamma = c_P/c_v = 5/3$. In the envelope of a star, however, real compressions behave in a more complicated fashion because of the effects of thermal ionization and variations in radiation pressure.

(a) As an example of the general calculation for the adiabatic index Γ_1 under dynamic pulsations, imagine a small enclosed volume V at temperature T of non-degenerate, partially ionized, hydrogen gas mixed with a thermal radiation field. (We may consider the volume essentially closed because the envelopes of stars—where the effects of pulsations have the most weight—are optically thick and prevent much radiative diffusion of the trapped heat.) The total local pressure P equals the sum of the radiation and gas pressures,

$$P = P_r + P_g = \frac{1}{3}aT^4 + \frac{NkT}{V},$$

where $N = nV$ represents the total number of particles in volume V and can be obtained from the sum of the electrons, protons, and neutral H atoms,

$$N = N_e + N_+ + N_0 = (1+\alpha)N_H.$$

In the above, we have borrowed the notation of the previous problem, so α satisfies the (simplified) Saha's equation

$$\frac{\alpha^2}{1-\alpha} = b\frac{V}{N_H}T^{3/2}e^{-I/kT},$$

where

$$b \equiv 2h^{-3}(2\pi m_e k)^{3/2} = \text{constant}.$$

Argue that the internal energy in volume V, E, which equals the sum of the energy of ionization and the thermal energies of particles and photons, has the expression

$$E = (1 - \alpha)N_{\rm H}(-I) + \frac{3}{2}P_{\rm g}V + 3P_{\rm r}V.$$

Write $\beta \equiv P_{\rm g}/P$, $\mu \equiv \alpha(1 - \alpha)/[2 + \alpha(1 - \alpha)]$, $\lambda \equiv (3/2) + (I/kT)$, and show that

$$\frac{dP}{P} = \beta\frac{dN}{N} + (4 - 3\beta)\frac{dT}{T} - \beta\frac{dV}{V},$$

$$\frac{dN}{N} = \frac{d\alpha}{1 + \alpha} = \mu\left(\frac{dV}{V} + \lambda\frac{dT}{T}\right),$$

$$\frac{2}{3}\frac{dE}{PV} = \frac{2}{3}\frac{N_{\rm H}I}{PV}(1 + \alpha)\frac{dN}{N} + \beta\frac{dP_{\rm g}}{P_{\rm g}} + 2(1 - \beta)\frac{dP_{\rm r}}{P_{\rm r}} + [\beta + 2(1 - \beta)]\frac{dV}{V}.$$

Because we can assume negligible diffusion of heat on the time scale of a dynamical pulsation, changes tend to occur adiabatically:

$$T\,dS = dE + P\,dV = 0.$$

Notice that we have counted ionization energy in E; otherwise, we would still need explicitly to account for changes in the neutral hydrogen content due to ionization and subtract here $I\,d[(1-\alpha)N_{\rm H}]$ from the thermal energy contained in the gas plus radiation field. Use $dE = -P\,dV$ to eliminate dE; write $(dP_{\rm g}/P_{\rm g}) = (dN/N) + (dT/T) - (dV/V)$, $(dP_{\rm r}/P_{\rm r}) = 4(dT/T)$; and reduce the governing set of equations to

$$\begin{bmatrix} \beta & (4 - 3\beta) & -\beta \\ 1 & -\mu\lambda & -\mu \\ 2\beta\lambda & (24 - 21\beta) & 2(4 - 3\beta) \end{bmatrix}\begin{pmatrix} dN/N \\ dT/T \\ dV/V \end{pmatrix} = \begin{pmatrix} dP/P \\ 0 \\ 0 \end{pmatrix}.$$

(b) Regard the right-hand side as known and solve the above for dV/V, dT/T, and dN/N as

$$\frac{dV}{V} = -\frac{\mathcal{N}_V}{\mathcal{D}}\frac{dP}{P}, \qquad \frac{dT}{T} = \frac{\mathcal{N}_T}{\mathcal{D}}\frac{dP}{P}, \qquad \frac{dN}{N} = -\frac{\mathcal{N}_N}{\mathcal{D}}\frac{dP}{P},$$

where Cramer's rule (with $-\mathcal{D}$ as the determinant of the coefficient matrix) shows

$$\mathcal{N}_V = (24 - 21\beta) + 2\mu\beta\lambda^2,$$

$$\mathcal{N}_T = 2[(4 - 3\beta) + \mu\beta\lambda],$$

$$\mathcal{N}_N = \mu[(24 - 21\beta) - 2\lambda(4 - 3\beta)] = \mu(\mathcal{N}_V - \lambda\mathcal{N}_T),$$

$$\mathcal{D} = -\beta\mathcal{N}_N + (4 - 3\beta)\mathcal{N}_T + \beta\mathcal{N}_V = \beta(1 - \mu)\mathcal{N}_V + \frac{1}{2}\mathcal{N}_T^2.$$

Define the adiabatic indices for dynamical changes, Γ_1, Γ_2, and Γ_3, by

$$\frac{dP}{P} + \Gamma_1 \frac{dV}{V} = 0, \qquad \frac{dP}{P} - \frac{\Gamma_2}{\Gamma_2 - 1}\frac{dT}{T} = 0, \qquad \frac{dT}{T} + (\Gamma_3 - 1)\frac{dV}{V} = 0,$$

and obtain the following expressions for Γ_1, Γ_2, and Γ_3:

$$\Gamma_1 = \frac{\mathcal{D}}{\mathcal{N}_V}, \qquad \Gamma_2 = \frac{1}{1 - \mathcal{N}_T/\mathcal{D}}, \qquad \Gamma_3 = 1 + \frac{\mathcal{N}_T}{\mathcal{N}_V}.$$

(c)   What forms do Γ_1, Γ_2, and Γ_3 take when the fractional ionization $\alpha = 0$ or 1 (i.e., when $\mu = 0$)? For $\mu = 0$, compute the limiting values when $\beta = 0$ (all radiation) and 1 (all gas). Do these limiting values make sense? Can Γ_1, Γ_2, or Γ_3 ever be less than 4/3?

Hint: Consider the case when $\beta = 1$ but $\alpha \neq 0$ or 1. For example, consider a situation where $\alpha = 1/2 \Rightarrow \mu = 1/9$, while λ is of order, say, 15 (consult Chapter 7). Because partial ionization can still take place when $I \gg kT$, we have the possibility of $\mathcal{N}_T \ll \mathcal{N}_V$ and $\mathcal{D}$. In these circumstances, what does Γ_2, and $\Gamma_3 \approx 1$ imply heuristically about the increase in the number of degrees of freedom in a gas when we can liberate the electrons from the atoms? In terms of the relative change of temperature dT/T compared to relative changes of pressure, dP/P, and volume, dV/V? How does partial ionization help to thermostat a gas? (Try to use *physical* descriptions rather than mathematical ones in answering these questions.)

Problem Set 3

1. Convince yourself that you understand the derivation of equation (10.37) of the text for the ionization structure of an H II region:

$$\exp\left[-\tau_S F(z)\right] = \frac{3z^2}{\tau_S}(1-f)^2 f^{-1}, \tag{1}$$

where

$$F(z) \equiv \int_0^z f(z')dz'. \tag{2}$$

Calculate the numerical value of α from equation (10.12) for a value of $T = 10,000\,\mathrm{K}$. Recompute now for yourself the radius R_S of a Stromgren sphere for an O5 star when $n_0 = 100\,\mathrm{cm}^{-3}$. Given the photoabsorption cross section $\alpha_1 = 7.9 \times 10^{-18}\,\mathrm{cm}^{-2}$ at the Lyman edge, verify that the parameter

$$\tau_S \equiv R_S n_0 \alpha_1 \tag{3}$$

is large compared to unity. For $\tau_S \gg 1$, find the approximate solution $f(z)$ to equation (1) for $z < 1$ and $z > 1$, giving particular attention to how the transition is made near $z = 1$.

Hint: Use the method of matched asymptotic expansions (essentially an expansion in inverse powers of the large parameter τ_S). Thus, for $z < 1$, define $f \equiv \tau_S^{-1}\psi$ and $F \equiv \tau_S^{-1}\Psi$, with $d\Psi/dz = \psi$ being an order unity function, and show that to lowest order in large τ_S:

$$e^{-\Psi}\frac{d\Psi}{dz} = 3z^2.$$

Integrate this to show

$$e^{-\Psi} = 1 - z^3, \qquad \psi = \frac{3z^2}{1-z^3}. \tag{4}$$

What is the physical interpretation for the first of the above relations in terms of using up ultraviolet photons? The second, in terms of the fraction of neutral atoms f inside the H II region? Clearly, since f cannot exceed 1, show that the solution (4) must break down when we approach within a fractional distance of order τ_S^{-1} of $z = 1$.

For $z > 1$, define $f = 1 - x_e$, where $x_e \ll 1$ is the electron (or proton) fraction, and show that an approximate solution for x_e reads

$$x_e^2 = \frac{x_{e0}^2}{z^2}e^{-\tau_S(z-1)}, \tag{5}$$

where

$$x_{e0}^2 = \frac{\tau_S}{3} \exp\left[-\tau_S \int_0^1 f(z)\,dz\right],$$

is a constant (of order unity) obtained by integrating from $z = 0$ to 1 the true solution for f (instead of just the first term in an asymptotic expansion as we have implicitly carried out above). Interpret the z^{-2} and $e^{-\tau_S(z-1)}$ parts of the dropoff physically. Clearly, apart from a geometric dilution, the ionized fraction x_e becomes exponentially small as z becomes larger than 1 by a fractional amount greater than τ_S^{-1}. Argue, therefore, that the transition from $f \approx 0$ to $f \approx 1$ must be made in a dimensional distance $\sim \tau_S^{-1} R_S$. Give the physical interpretation for this distance.

To investigate the structure of the transition layer in more detail, stretch the coordinate variable by defining

$$\zeta \equiv \tau_S(z - 1). \tag{6}$$

Introduce the variable Ψ as in the first part of this problem, and show that for ζ of order unity, we may write

$$\frac{\tau_S}{3}\frac{d\Psi}{d\zeta}e^{-\Psi} = \left(1 - \frac{d\Psi}{d\zeta}\right)^2.$$

To get rid of the τ_S dependence, transform to the new variable

$$\Psi_* \equiv \Psi + \ln(6/\tau_S), \tag{7}$$

and solve the resulting relation as a quadratic equation for $f = d\Psi_*/d\zeta$ to obtain

$$\frac{d\Psi_*}{d\zeta} = 1 - e^{-\Psi_*}\left[(1 + 2e^{\Psi_*})^{1/2} - 1\right]. \tag{8}$$

Justify the choice of sign that led to equation (8) on the basis that we want $f = d\Psi_*/d\zeta \leq 1$.

With the substitution of variables $u = (1 + 2e^{\Psi_*})^{1/2}$ so that $u\,du = e^{\Psi_*}\,d\Psi_*$, equation (8) can be integrated analytically to yield

$$\zeta - \zeta_0 = 2\left\{\ln\left[(1 + 2e^{\Psi_*})^{1/2} - 1\right] - \frac{1}{(1 + 2e^{\Psi_*})^{1/2} - 1}\right\}, \tag{9}$$

where ζ_0 is an integration constant. Demonstrate that the solution possesses the behavior

$$e^{-\Psi_*} = \frac{1}{2}(\zeta_0 - \zeta) \qquad \text{as} \qquad \zeta \to -\infty. \tag{10}$$

But the outer limit of the inner solution (4) satisfies

$$e^{-\Psi} \to 3(1 - z) \qquad \text{as} \qquad z \to 1^-;$$

hence argue that ζ_0 must equal zero if we are to have a match with the inner limit of the transition-region solution (10), when ζ and Ψ_* are given by equations (6) and (7). With the identification $\zeta_0 = 0$, equation (10) implies that, as we head toward the interior of the H II region from the middle of the transition zone, the fraction of neutrals f asymptotically drops as the inverse of the scaled distance ζ: $f = d\Psi_*/d\zeta \to -\zeta^{-1}$ as $\zeta \to -\infty$.

On the other hand, as $\zeta \to +\infty$, show that equation (9) has the asymptotic behavior

$$\Psi_* \to \zeta, \tag{11}$$

which matches the exponential decay predicted by equation (5). Use equation (9), with $\zeta_0 = 0$, to tabulate ζ against Ψ_*, starting at large negative values and proceeding to large positive values. Plot up the result for the fraction of neutrals $f = d\Psi_*/d\zeta$ [given in terms of Ψ_* by the right-hand side of equation (8)] as a function of ζ. Comment on the sharpness with which ionization-bounded H II regions turn into H I regions. (In actuality, young luminous stars arise in *molecular* clouds, so most OB stars tend to be surrounded by *photodissociation* regions [PDRs] as well as photoionization [H II] regions.)

---

2. Consider the angular phase distribution and polarization properties for Thomson scattering. Let the free electron reside at the origin of a coordinate system with the Z axis along the original direction of incident radiation with propagation vector $\hat{\mathbf{k}}'$. Let the direction $\hat{\mathbf{k}}$ to the observer make a polar angle Θ with respect to $\hat{\mathbf{k}}'$. Let the $Y = z$ axis lie along $\mathbf{E}'$, with θ being the angle between the direction of $\mathbf{E}'$ and $\hat{\mathbf{k}}$. The projection of $\hat{\mathbf{k}}$ onto the $z = Y$ axis yields

$$\cos\theta = \sin\Theta\sin\Phi, \tag{12}$$

where Φ is the azimuthal angle in the (X, Y, Z) coordinate system. The dipole oscillations induced in a free electron satisfies

$$\ddot{\mathbf{d}} = \frac{e^2}{m_e}\mathbf{E}'.$$

At a distance r along $\hat{\mathbf{k}}$ from the electron, this produces a scattered electromagnetic radiation field,

$$\mathbf{B} = \frac{1}{c^2 r}\ddot{\mathbf{d}} \times \hat{\mathbf{k}}, \qquad \mathbf{E} = \mathbf{B} \times \hat{\mathbf{k}}.$$

According to the derivation in the text, the angular pattern of scattered power is given by

$$\frac{dP}{d\Omega} = \frac{|\ddot{\mathbf{d}}|^2}{4\pi c^3}\sin^2\theta,$$

with equation (12) implying

$$\sin^2\theta = 1 - \sin^2\Theta\sin^2\Phi.$$

Unpolarized incident light would have a uniform distribution of $\mathbf{E}'$ with respect to Φ. If we average the above equation with respect to Φ, we obtain

$$\langle \sin^2 \theta \rangle = 1 - \frac{1}{2} \sin^2 \Theta = \frac{1}{2}(1 + \cos^2 \Theta).$$

Show now that the differential scattering cross section for Thomson scattering can be written as

$$\sigma(\Omega) = \sigma_T \phi(\Theta), \tag{13}$$

where $\sigma_T = (8\pi/3)r_e^2$ and

$$\phi(\Theta) = \frac{3}{16\pi}(1 + \cos^2 \Theta) \tag{14}$$

is the angular phase function for Thomson scattering. Verify that ϕ satisfies the normalization condition:

$$\int_0^{2\pi} d\Phi \int_0^{\pi} \phi(\Theta) \sin \Theta \, d\Theta = 1.$$

Notice that this scattering phase function has forward-backward symmetry. Argue that the same phase function would hold for Rayleigh scattering (or resonant line scattering), but the cross section would no longer be grey in frequency.

For extra credit: Deduce the scattering matrix $\overset{\leftrightarrow}{\mathbf{R}}$ associated with the Stokes parameters I_ν^+, I_ν^-, U_ν, and V_ν.

Hint: Rename the axes before comparing with equation (12.40).

---

3. In this problem, we derive the electromagnetic fields associated with the Lienard-Wiechert potentials. Jackson (1962, Chapter 14) demonstrates that the manipulations are actually somewhat easier if we work with equation (13.22) rather than directly with equation (16.1); however, we shall follow the latter route here and show that we recover his expression (14.14) for $\mathbf{E}$.

(a) Begin by noting that equation (16.3) dotted with itself yields

$$R^2 = x^2 + \mathbf{r} \cdot \mathbf{r} - 2\mathbf{x} \cdot \mathbf{r}.$$

Take the τ derivative of the above and show that

$$\frac{\partial R}{\partial \tau} = -\frac{\mathbf{R}}{R} \cdot \mathbf{v}. \tag{15}$$

Demonstrate now that the differentiation of $R = c(t - \tau)$ with respect to t, yields the expression

$$\frac{\partial \tau}{\partial t} = \frac{R}{R - \mathbf{R} \cdot \mathbf{v}/c}. \tag{16}$$

(b) Take the gradient of the expression for R^2 to obtain

$$2R\boldsymbol{\nabla}R = 2\mathbf{x} + 2\mathbf{r} \cdot \mathbf{v}\boldsymbol{\nabla}\tau - 2\mathbf{r} - 2\mathbf{x} \cdot \mathbf{v}\boldsymbol{\nabla}\tau.$$

Hint: $\boldsymbol{\nabla}$ operating as a gradient on any scalar product $\mathbf{x} \cdot \mathbf{r}$ yields $\mathbf{r}$ (when we hold $\mathbf{r}$ fixed) plus $\mathbf{x} \cdot \mathbf{v}\boldsymbol{\nabla}\tau$ (when we hold $\mathbf{x}$ fixed).

On the other hand, with R given by $c(t - \tau)$, we obtain $\boldsymbol{\nabla}R = -c\boldsymbol{\nabla}\tau$. Collecting terms, show that we get

$$\boldsymbol{\nabla}\tau = \frac{-\mathbf{R}}{c(R - \mathbf{R} \cdot \mathbf{v}/c)}. \tag{17}$$

(c) Consider now taking the negative gradient of the first row of equation (16.1):

$$-\boldsymbol{\nabla}\phi = \frac{q}{(R - \mathbf{R} \cdot \mathbf{v}/c)^2}[\boldsymbol{\nabla}R - \boldsymbol{\nabla}(\mathbf{R} \cdot \mathbf{v}/c)].$$

Calculate the term $\boldsymbol{\nabla}R$ as $-c\boldsymbol{\nabla}\tau$, and demonstrate that the term $\boldsymbol{\nabla}(\mathbf{R} \cdot \mathbf{v}/c)$ has the expression $\mathbf{v}/c$ [when we operate on the $\mathbf{x}$ dependence of $\mathbf{R}$ in equation (16.3)] plus $[\partial(\mathbf{R} \cdot \mathbf{v}/c)/\partial\tau]\boldsymbol{\nabla}\tau$ (when we operate on the τ dependence of everything inside the parenthesis). But $\partial(\mathbf{R} \cdot \mathbf{v}/c)/\partial\tau = -v^2/c + \mathbf{R} \cdot \dot{\mathbf{v}}$. Consequently, collect terms together containing $\boldsymbol{\nabla}\tau$ and get

$$\boldsymbol{\nabla}R - \boldsymbol{\nabla}(\mathbf{R} \cdot \mathbf{v}/c) = (-c + v^2/c - \mathbf{R} \cdot \dot{\mathbf{v}}/c)\boldsymbol{\nabla}\tau - \mathbf{v}/c.$$

Substitute in the expression (17), and arrive at the desired expression,

$$-\boldsymbol{\nabla}\phi = \frac{q}{(R - \mathbf{R} \cdot \mathbf{v}/c)^3}\left[\mathbf{R}\left(1 - \frac{v^2}{c^2} + \mathbf{R} \cdot \frac{\dot{\mathbf{v}}}{c^2}\right) - \frac{\mathbf{v}}{c}\left(R - \mathbf{R} \cdot \frac{\mathbf{v}}{c}\right)\right]. \tag{18}$$

(d) In a similar manner, show that if we differentiate the second row of equation (16.1) with respect to t and multiply by $-1/c$, we obtain

$$-\frac{1}{c}\frac{\partial\mathbf{A}}{\partial t} = \frac{q}{(R - \mathbf{R} \cdot \mathbf{v}/c)^3}$$

$$\times \left[\left(-\mathbf{R} \cdot \frac{\mathbf{v}}{c} + R\frac{v^2}{c^2} - R\mathbf{R} \cdot \frac{\dot{\mathbf{v}}}{c^2}\right)\frac{\mathbf{v}}{c} - R\left(R - \mathbf{R} \cdot \frac{\mathbf{v}}{c}\right)\frac{\dot{\mathbf{v}}}{c^2}\right]. \tag{19}$$

Add equations (18) and (19), and show that we obtain for equation (16.5) the desired result, equation (16.7) of the text.

For extra credit: Go through Jackson's derivation or otherwise convince yourself of the correctness of equation (16.8) of the text.

(e) In the wave zone, we may approximate $R \approx \hat{\mathbf{k}}x$ where x equals the large mean distance to the charge. In this limit, show that equations (16.7) and (16.8) yield a received angular power proportional to

$$\mathcal{A} \equiv \frac{g^2}{|\dot{\mathbf{v}}|^2},$$

where g^2 is given by equation (17.3) of the text. Construct a coordinate system with the velocity $\mathbf{v}$ lying along the z axis, with the acceleration $\dot{\mathbf{v}}$ lying in the x-z plane making an angle i with respect to $\mathbf{v}$, and with the unit vector $\hat{\mathbf{k}}$ having spherical polar angles (θ, ϕ). For $\dot{\mathbf{v}}$ parallel to $\mathbf{v}$, i.e., for $i = 0$, show that

$$\mathcal{A} = (1 - \beta \cos \theta)^{-6} \sin^2 \theta,$$

where $\beta \equiv v/c$. For $\dot{\mathbf{v}}$ perpendicular to $\mathbf{v}$, i.e., for $i = 90°$, show that

$$\mathcal{A} = (1 - \beta \cos \theta)^{-4} - (1 - \beta^2)(1 - \beta \cos \theta)^{-6} \sin^2 \theta \cos^2 \phi.$$

For β very close to 1 so that $\gamma \equiv (1 - \beta^2)^{-1/2} \approx [2(1 - \beta)]^{-1/2}$, show in either case that $\mathcal{A}$ has large values of order γ^{10} or γ^8 (strong forward focusing) when θ lies within an angle of order γ^{-1} of $\theta = 0$ (the direction of $\mathbf{v}$).

Problem Set 4

1. In this problem, we consider the electromagnetic structure in the near zone of an uncharged but magnetized object which rotates at an angular speed ω. Suppose the intrinsic size L of the object is much smaller than the wavelength $\lambda = 2\pi c/\omega$ of the electromagnetic waves that it emits. For distances x of the same order as λ and $\gg L$, the expansion of equation (13.18) for the vector potential must be carried out totally inside the integral. With

$$|\mathbf{x} - \mathbf{x}'| \approx x \left(1 - \hat{\mathbf{k}} \cdot \frac{\mathbf{x}'}{x}\right) \qquad \text{where} \qquad \hat{\mathbf{k}} = \mathbf{x}/x, \tag{1}$$

show that we may write

$$\mathbf{A} = \frac{1}{cx}\left[\int \left(1 + \hat{\mathbf{k}} \cdot \frac{\mathbf{x}'}{x}\right) \mathbf{j}_\mathrm{e}(\mathbf{x}', t - x/c)\, d^3x' \right.$$
$$\left. + \frac{\hat{\mathbf{k}}}{c} \cdot \frac{\partial}{\partial t} \int \mathbf{x}' \mathbf{j}_\mathrm{e}(\mathbf{x}', t - x/c)\, d^3x' + \cdots\right]. \tag{2}$$

Assume that the object has no electric dipole moment, and obtain the approximation

$$\mathbf{A} = \frac{1}{cx}\left[\left(\frac{1}{x} + \frac{1}{c}\frac{\partial}{\partial t}\right) \hat{\mathbf{k}} \cdot \int \mathbf{x}' \mathbf{j}_\mathrm{e}(\mathbf{x}', t - x/c)\, d^3x'\right]. \tag{3}$$

Write $\mathbf{x}'\mathbf{j}_\mathrm{e}$ in terms of its symmetric and antisymmetric parts:

$$\mathbf{x}'\mathbf{j}_\mathrm{e} = \frac{1}{2}(\mathbf{x}'\mathbf{j}_\mathrm{e} + \mathbf{j}_\mathrm{e}\mathbf{x}') + \frac{1}{2}(\mathbf{x}'\mathbf{j}_\mathrm{e} - \mathbf{j}_\mathrm{e}\mathbf{x}').$$

Assume that the object has no electric quadrupole moment, and show $\mathbf{A}$ to be given by

$$\mathbf{A} = \frac{1}{x}\left[\left(\frac{1}{x} + \frac{1}{c}\frac{\partial}{\partial t}\right) \mathbf{M}\right] \times \hat{\mathbf{k}},$$

where $\mathbf{M}$ is the magnetic dipole moment of the object evaluated at time $t - x/c$:

$$\mathbf{M} \equiv \frac{1}{2c}\int \mathbf{x}' \times \mathbf{j}_\mathrm{e}(\mathbf{x}', t - x/c)\, d^3x'. \tag{4}$$

Show that $\mathbf{A}$ can also be written more compactly as

$$\mathbf{A} = \nabla \times \left(\frac{\mathbf{M}}{x}\right). \tag{5}$$

In the very near zone, where $L \ll x \ll \lambda$, equation (4) may be evaluated at the instantaneous time t. In this case, show that $\mathbf{A}$ has the expression

$$\mathbf{A} = \frac{\mathbf{M} \times \mathbf{x}}{|\mathbf{x}|^3}. \tag{6}$$

Show also that the corresponding magnetic field $\mathbf{B} = \nabla \times \mathbf{A}$ in the very near zone is given by

$$\mathbf{B} = \frac{3(\mathbf{M} \cdot \hat{\mathbf{k}})\hat{\mathbf{k}} - \mathbf{M}}{|\mathbf{x}|^3}, \tag{7}$$

which, with $\hat{\mathbf{k}} = \mathbf{x}/|\mathbf{x}|$, is the standard expression for the magnetic field of a (quasistatic) magnetic dipole.

---

2. We apply the considerations of Problem 1 to the problem of the Crab pulsar, widely accepted to be a rapidly spinning, magnetized neutron star. The spin period $2\pi/\omega$, identified with the spacing between pulse arrival times, equals 0.033 s. Assume that the pulsar has a magnetic dipole moment $\mathbf{M}$ inclined at an angle i with respect to the spin axis, so that we may write $\mathbf{M}(t)$ as

$$\mathbf{M}(t) = M_0[\mathbf{e}_3 \cos i + (\mathbf{e}_1 \cos \omega t + \mathbf{e}_2 \sin \omega t) \sin i], \tag{8}$$

where M_0 is the magnitude of the dipole moment, $\mathbf{e}_3$ represents a unit vector in the direction of the spin axis (assumed fixed in space), and $\mathbf{e}_1$ and $\mathbf{e}_2$ form two other orthonormal vectors. According to equation (15.12) of the text, if this rotating dipole resides in a vacuum, it would radiate an electromagnetic wave field given by

$$\mathbf{B} = \frac{1}{c^2 x} \left[(\ddot{\mathbf{M}} \times \hat{\mathbf{k}}) \times \hat{\mathbf{k}} \right], \qquad \mathbf{E} = \mathbf{B} \times \hat{\mathbf{k}}.$$

Show that the total power P emitted as magnetic dipole radiation equals

$$P = \frac{2|\ddot{\mathbf{M}}|^2}{3c^3}. \tag{9}$$

Applied to our problem, demonstrate that the above equation becomes

$$P = \frac{2\omega^4}{3c^3} M_0^2 \sin^2 i.$$

The observed luminosity from the Crab nebula in all forms of emission amounts perhaps to 3×10^{38} erg s^{-1}. What would $M_0 \sin i$ need to be if all of this energy were supplied initially in the form of low-frequency magnetic-dipole radiation? Assume $i = 45°$, and compute the implied strength of the magnetic field at the magnetic pole of a neutron star of radius $R = 15$ km.

Hint: Apply equation (7) at the magnetic pole of the star after verifying that the wavelength $\lambda = 2\pi c/\omega$ is much larger than the size of the neutron star.

How does this field strength compare with what you have heard about magnetized neutron stars? At what distance roughly would the quasistatic dipole magnetic field turn into a radiation field?

Hint: Reconsider the difference in Problem 1 between the quasistatic zone, the near zone, and the far zone (or wave zone). In particular, since the quasistatic magnetic field dominates at small x, but decays as x^{-3}, whereas the radiation **B** decays at large x only as x^{-1}, calculate the distance x where the two formulae give the same answer in order of magnitude. Show that the latter distance satisfies $x \sim c/\omega = \lambda/2\pi$. The light cylinder corresponds to the locus where the axial radius ϖ satisfies $\omega\varpi = c$. What relationship does it bear to the above considerations?

The moment of inertia of a homogeneous sphere of mass m and radius R is $I = (2/5)mR^2$, and it would have a kinetic energy K of rotation equal to $I\omega^2/2$. Compute K for the Crab pulsar if it were a homogeneous sphere of mass $m = 1.4M_\odot$. If the power P emitted by the pulsar comes ultimately at the expense of the stored rotational energy, argue for the evolutionary equation

$$\frac{d}{dt}\left(\frac{1}{2}I\omega^2\right) = -P,$$

which gives a spindown rate,

$$\dot{\omega} = -\frac{P\omega}{2K}, \tag{10}$$

independent of the detailed model for the pulsar emission process (e.g., magnetic dipole radiation). The measured spindown rate of the Crab pulsar yields $\omega/\dot{\omega} \approx 2{,}500\,\text{yr}$. How does this compare with the theoretical model with the empirically measured value of the luminosity of the Crab nebula (assumed equal to P) and calculated value of K? Discuss qualitatively how the extent of any disagreement might be used to constrain the structure of neutron star models, e.g., the influence of the equation of state for neutron star matter on more realistic (lower) estimates for the moment of inertia I.

If we adopt magnetic dipole radiation as the model for pulsar emission, show that the implied spindown rate satisfies

$$\dot{\omega} = -\left(\frac{2M_0^2 \sin^2 i}{3c^3 I}\right)\omega^3.$$

Discuss how the above relation might be used as a test of the theory.

Hint: Show that the model implies a braking index $n \equiv \omega\ddot{\omega}/\dot{\omega}^2 = 3$. The observed index $n \approx 2.5$. Better (because it gives a longer time line), integrate the above equation to show that the actual age t of the pulsar according to this model is less than $1/2$ of its nominal spindown time $\omega/\dot{\omega} \approx 2{,}500\,\mathrm{yr}$ ($t = \omega/2\dot{\omega}$ only if the pulsar were born spinning infinitely fast). When did the Chinese observe a supernova in this part of the sky?

For extra credit: In what ways can you criticize the above theory or make it better? Some possible points you might consider include: (1) how to convert the low-frequency radiation input into the observed forms of energy in the Crab nebula and pulsar, (2) whether the magnetic dipole radiation can get out in the first place if the pulsar magnetosphere does not constitute a perfect vacuum, and (3) whether the magnetic field near the surface of the star can be well-approximated by a dipole.

Hint: Papers published by Ostriker and Gunn, Goldreich and Julian, Ruderman and coworkers, Arons and coworkers, Blandford and coworkers, etc., might prove useful to look at if you wish to pursue this topic seriously.

---

3. In this problem we consider how radio observations of free-free emission provide diagnostics of the physical conditions of H II regions. (Analogous considerations apply to X-ray observations of hot gas in galaxy clusters.) Assume an H II region to have a uniform electron temperature T and density n_e which we would like to determine by observational means. Since the free-free emission associated with the thermal distribution of electrons occurs under conditions of LTE, satisfy yourself that equation (3.18) of the text yields the solution for the equation of transfer:

$$I_\nu = I_\nu(0)e^{-\tau_\nu} + B_\nu(T)\left(1 - e^{-\tau_\nu}\right). \tag{11}$$

For radio observations spanning, say, $\lambda \sim 100\,\mathrm{cm}$ to $1\,\mathrm{mm}$, show that $h\nu \ll kT$ for all likely values of T. Thus it represents a good approximation to replace $B_\nu(T)$ by its Rayleigh-Jeans limit:

$$\mathrm{Rayleigh - Jeans}: \qquad B_\nu(T) = \frac{2\nu^2}{c^2}kT.$$

Motivated by this simplification, radio astronomers then like to express the specific intensity I_ν in terms of a *brightness temperature* T_b defined through

$$T_b \equiv \frac{c^2}{2\nu^2 k}I_\nu. \tag{12}$$

Show now that equation (11) can be rewritten as

$$T_b = T_b(0)e^{-\tau_\nu} + T\left(1 - e^{-\tau_\nu}\right), \tag{13}$$

which applies not only to free-free radiation, but wherever (a) we may ignore the effects of scattering, and (b) we may assume that the source function has an LTE value at a uniform temperature T throughout the region being observed. (Notice that we have not yet assumed uniformity of density.)

Applied to free-free emission, τ_ν has the form:

$$\tau_\nu = \int \rho \kappa_\nu^{\mathrm{ff}} ds,$$

where the integral is taken along the line of sight through the H II region and $\rho \kappa_\nu^{\mathrm{ff}}$ is given by equation (15.29) of the text. Assume for simplicity a pure hydrogen plasma, expand the exponential in the correction for stimulated emission, $1 - e^{-h\nu/kT}$, for small $h\nu/kT$, and show that

$$\rho \kappa_\nu^{\mathrm{ff}} = C n_{\mathrm{e}}^2 T^{-3/2} \nu^{-2} \bar{g}_\nu^{\mathrm{ff}},$$

where C is a constant coefficient,

$$C \equiv \left(\frac{2m_{\mathrm{e}}}{3\pi k} \right)^{1/2} \left[\frac{4\pi e^6}{3m_{\mathrm{e}}^2 ck} \right].$$

The Gaunt factor $\bar{g}_\nu^{\mathrm{ff}}$ in the radio regime reads

$$\bar{g}_\nu^{\mathrm{ff}} = \frac{\sqrt{3}}{2\pi} \left[\ln \left(\frac{8k^3 T^3}{\pi^2 e^4 m_{\mathrm{e}} \nu^2} \right) - 5\gamma \right],$$

where $\gamma = 0.5772\ldots$ is Euler's constant. Compute $\bar{g}_\nu^{\mathrm{ff}}$ for $\nu = 10^9$ Hz and $T = 10^4$ K, and show that, unlike the optical case, $\bar{g}_\nu^{\mathrm{ff}}$ should not be approximated by unity here. Notice also that Planck's constant h has dropped out of all equations, so that the considerations are purely classical.

Define the *emission measure* as the integral

$$\mathrm{EM} \equiv \int n_{\mathrm{e}}^2 ds,$$

and show that τ_ν can be expressed as

$$\tau_\nu = (\mathrm{EM}) C T^{-3/2} \nu^{-2} \bar{g}_\nu^{\mathrm{ff}}. \tag{14}$$

At low frequencies, $\tau_\nu \gg 1$, whereas at high frequencies, $\tau_\nu \ll 1$. With no background source, show that this implies $T_{\mathrm{b}} \approx T$ at low frequencies, while $T_{\mathrm{b}} \approx T\tau_\nu$ at high frequencies.

For a spherical H II region with radius R_{S}, show that the observed flux (measured in Janskys = 10^{-26} watts m^{-2} Hz^{-1} by radio astronomers) $F_\nu = \pi I_\nu (R_{\mathrm{S}}^2/r^2)$ where r is the distance to the source. The size R_{S} can be

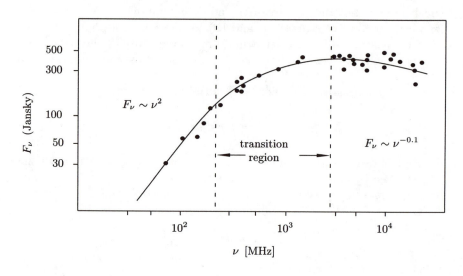

FIGURE P4.1
Observations of free-free emission from the Orion nebula.

determined if the source is angularly resolved and its distance known. Show
now that

$$F_\nu = \frac{2\pi k}{c^2} \left(\frac{R_S}{r} \right)^2 \nu^2 T_b \qquad (15)$$

will be proportional to ν^2 at low frequencies and to $\bar{g}_\nu^{ff}$ (a nearly flat function $\propto \nu^{-0.1}$) at high (radio) frequencies. Describe qualitatively how this
information could be used to deduce T and EM if the spectrum on both
sides of the turnover frequency ν_c (where $\tau_\nu = 1$) can be measured. (A
better way in the radio to obtain the electron temperature is to measure
the strength of the H109α recombination line.)

Figure P4.1 contains observations of an H II region with size $R_S \approx 0.6$ pc
in the Orion nebula, which lies at a distance $r \approx 500$ pc. Compute approximate values for T and n_e from this data. Quoted values in the literature
are $T \approx 8,000$ K, and $n_e \approx 2,000$ cm^{-3}. Check to see at what frequency
$\tau_\nu = 1$ for your results. Your derived temperature will not agree with the
value 8,000 K, because the low-frequency measurements have larger effective beam sizes than the high-frequency measurements; thus corrections
need to be applied to obtain the true underlying ν^2 dependence at low
ν (see the discussion in Osterbrock 1989, pp. 128–130). This fact should
provide a warning against the naive use of free-free radiation to deduce

the electron temperatures of H II regions, and partially explains why radio astronomers prefer to use recombination lines for this purpose.

Is the size $R_S = 0.6$ pc consistent with Stromgren's theory of H II regions (see Problem Set 3, but note the density difference) if the exciting source is an O5 star? There's actually some indication that the Orion H II region is *density-bounded* (runs out of gas) rather than *ionization-bounded* (runs out of ultraviolet photons).

Problem Set 5

1. In this problem we compute the Einstein A for allowed transitions of the hydrogen atom, ignoring fine-structure and hyperfine effects (e.g., spin-orbit coupling, relativistic corrections, etc.).

(a) For the transition,

$$(n', \ell', m') \rightarrow (n, \ell, m),$$

with $n' > n$, we need to compute the square of the electric-dipole matrix element, $|\langle i|\mathbf{x}|f\rangle|^2$, where the hydrogen wave function associated with the quantum state (n, ℓ, m) is given by

$$|n\ell m\rangle = R_{n\ell}(r)Y_{\ell m}(\theta, \phi).$$

Standard textbooks (e.g., Merzbacher 1961, Chapter 10) give the expression:

$$R_{n\ell}(r) = \left\{ \frac{(n - \ell - 1)!}{2n[(n + \ell)!]^3} \left(\frac{2}{na_0} \right)^3 \right\}^{1/2} e^{-r/nr_0} \left(\frac{2r}{nr_0} \right)^{\ell} L_{n+1}^{2\ell+1}(2r/nr_0), \tag{1}$$

where $r_0 \equiv \hbar^2/m_e e^2$ is the Bohr radius, $L_{n+1}^{2\ell+1}(x)$ is an associated Laguerre polynomial, and $R_{n\ell}(r)$ satisfies the normalization condition:

$$\int_0^\infty R_{n\ell}^2(r)r^2\, dr = 1.$$

Now, write $\mathbf{x}$ as

$$\mathbf{x} = \frac{1}{2}\left[(x + iy)(\mathbf{e}_x - i\mathbf{e}_y) + (x - iy)(\mathbf{e}_x + i\mathbf{e}_y)\right] + z\mathbf{e}_z,$$

together with the expressions (see Chapter 24):

$$x \pm iy = (8\pi/3)^{1/2}rY_{1,\pm 1}, \qquad z = (4\pi/3)^{1/2}rY_{10},$$

and show that (also derivable as a consequence of the Wigner-Eckart theorem):

$$|\langle n'\ell'm'|\mathbf{x}|n\ell m\rangle|^2 = r_0^2 \mathcal{Y}_{\ell'm'\ell m} R_{n'\ell'n\ell}^2,$$

where $\mathcal{Y}_{\ell'm'\ell m}$ and $\mathcal{R}_{n'\ell'n\ell}$ are the dimensionless quantities:

$$\mathcal{Y}_{\ell'm'\ell m} \equiv \frac{4\pi}{3}\left[\left|\oint Y_{\ell'm'}^* Y_{11} Y_{\ell m}\,d\Omega\right|^2 + \left|\oint Y_{\ell'm'}^* Y_{10} Y_{\ell m}\,d\Omega\right|^2\right.$$

$$\left. + \left|\oint Y_{\ell'm'}^* Y_{1,-1} Y_{\ell m}\,d\Omega\right|^2\right]; \tag{2}$$

$$\mathcal{R}_{n'\ell'n\ell} \equiv \frac{1}{r_0}\int_0^\infty R_{n'\ell'} R_{n\ell}\,r^3\,dr. \tag{3}$$

(b) The discussion in Chapter 27 on Clebsch-Gordon coefficients allows us to write:

$$Y_{1,\pm 1} Y_{\ell m} = \left(\frac{3}{8\pi}\right)^{1/2}\left[a_\pm(\ell,m) Y_{\ell+1,m\pm 1} - b_\pm(\ell,m) Y_{\ell-1,m\pm 1}\right],$$

$$Y_{10} Y_{\ell m} = \left(\frac{3}{4\pi}\right)^{1/2}\left[a_0(\ell,m) Y_{\ell+1,m} + b_0(\ell,m) Y_{\ell-1,m}\right],$$

where

$$a_\pm(\ell,m) \equiv \left[\frac{(\ell\pm m+1)(\ell\pm m+2)}{(2\ell+1)(2\ell+3)}\right]^{1/2},$$

$$a_0(\ell,m) \equiv \left[\frac{(\ell+m+1)(\ell-m+1)}{(2\ell+1)(2\ell+3)}\right]^{1/2},$$

$$b_\pm(\ell,m) \equiv \left[\frac{(\ell\mp m)(\ell\mp m-1)}{(2\ell+1)(2\ell-1)}\right]^{1/2},$$

$$b_0(\ell,m) \equiv \left[\frac{(\ell+m)(\ell-m)}{(2\ell+1)(2\ell-1)}\right]^{1/2}.$$

For given final state ℓ and m, show that we get

$$\oint Y_{\ell'm'}^* Y_{1,\pm 1} Y_{\ell m}\,d\Omega = \left(\frac{3}{8\pi}\right)^{1/2} a_\pm(\ell,m)$$
$$\text{if } \ell' = \ell+1 \text{ and } m' = m\pm 1,$$

$$\oint Y_{\ell'm'}^* Y_{1,\pm 1} Y_{\ell m}\,d\Omega = -\left(\frac{3}{8\pi}\right)^{1/2} b_\pm(\ell,m)$$
$$\text{if } \ell' = \ell-1 \text{ and } m' = m\pm 1,$$

$$\oint Y_{\ell'm'}^* Y_{10} Y_{\ell m}\,d\Omega = \left(\frac{3}{4\pi}\right)^{1/2} a_0(\ell,m)$$
$$\text{if } \ell' = \ell+1 \text{ and } m' = m,$$

$$\oint Y_{\ell'm'}^* Y_{10} Y_{\ell m}\,d\Omega = \left(\frac{3}{4\pi}\right)^{1/2} b_0(\ell,m)$$
$$\text{if } \ell' = \ell-1 \text{ and } m' = m;$$

with the integrals equaling zero otherwise. Thus show that equation (2) implies

$$\mathcal{Y}_{\ell+1,m+1,\ell,m} = \frac{1}{2}a_+^2(\ell,m),$$

$$\mathcal{Y}_{\ell+1,m,\ell,m} = a_0^2(\ell,m), \tag{4}$$

$$\mathcal{Y}_{\ell+1,m-1,\ell,m} = \frac{1}{2}a_-^2(\ell,m);$$

while

$$\mathcal{Y}_{\ell-1,m+1,\ell,m} = \frac{1}{2}b_+^2(\ell,m),$$

$$\mathcal{Y}_{\ell-1,m,\ell,m} = b_0^2(\ell,m), \tag{5}$$

$$\mathcal{Y}_{\ell-1,m-1,\ell,m} = \frac{1}{2}b_-^2(\ell,m),$$

with the last three quantities equaling zero if $\ell = 0$.

(c) For given ℓ', show that the sum of the squares of the matrix elements, and therefore the transition rates, to all allowed values of ℓ and m are the same for different values of m'. In other words, prove that the sums

$$a_+^2(\ell'-1,m'-1) + 2a_0^2(\ell'-1,m') + a_-^2(\ell'-1,m'+1)$$

and

$$b_+^2(\ell'+1,m'-1) + 2b_0^2(\ell'+1,m') + b_-^2(\ell'+1,m'+1)$$

do not depend on the value of m'. Thus, without regard to the radial structure, the rates of depopulation and population (by the general relationships among the Einstein's coefficients) do not depend on the m'-values of an atom. Argue physically that this must be the case if there exists nothing special about the particular choice of the axis for the z-projection of the angular momentum.

(d) For hydrogenic atoms, it is possible to do the radial integrals for $\mathcal{R}_{n'\ell'n\ell}$ analytically. Here, however, we specialize to the simple case of Lyman-alpha transitions:

$$(2,1,m') \rightarrow (1,0,0) \qquad \text{with} \qquad m' = +1, 0, -1.$$

For $\ell = 0$ and $m = 0$, verify explicitly that equation (4) yields

$$\mathcal{Y}_{1,1,0,0} = \mathcal{Y}_{1,0,0,0} = \mathcal{Y}_{1,-1,0,0} = 1/3.$$

Prove the same equalities directly by using $Y_{00} = \text{constant} = (4\pi)^{-1/2}$ and the orthonormal properties of the spherical harmonic functions, Y_{11},

Y_{10}, $Y_{1,-1}$. This proof represents a check on the formula for the integral of three Y's in the special case $m = 0$; it also gives an explicit display of the constancy of rates with different m'.

When written out explicitly, the two radial wave functions relevant to Lyman-alpha transitions have the forms

$$R_{1,0}(r) = 2r_0^{-3/2}e^{-r/r_0}, \qquad R_{2,1} = -(24)^{-1/2}r_0^{-3/2}\left(\frac{r}{r_0}\right)e^{-r/2r_0}.$$

Hence show that

$$\mathcal{R}_{2,1,1,0} = -\frac{1}{\sqrt{6}}\int_0^\infty x^4 e^{-3x/2}\,dx = -\frac{1}{\sqrt{6}}\left(\frac{2}{3}\right)^5 4!.$$

We now have

$$|\langle 2,1,m|\mathbf{x}|1,0,0\rangle|^2 = \frac{2^{15}}{3^{10}}r_0^2 \qquad \text{for} \qquad m = +1, 0, -1.$$

For any two possible spin states m'_s and three possible values for m' of the initial $2p$ electron, therefore, show that we have [see equation (23.17)]:

$$A_{2p1s} = \frac{4e^2}{3\hbar}\left(\frac{\omega}{c}\right)^3 |\langle 2,1,m',m'_s|\mathbf{x}|1,0,0,m_s = m'_s\rangle|^2 = \frac{2^{17}}{3^{11}}\left(\frac{\omega}{c}\right)^3\left(\frac{e^2 r_0^2}{\hbar}\right),$$

where ω equals the radian frequency of the Lyman-alpha line:

$$\omega = \frac{3}{8}\frac{m_e e^4}{\hbar^3}.$$

Substitute in the expression $r_0 = \hbar^2/m_e e^2$ and demonstrate that

$$A_{2p1s} = \left(\frac{2}{3}\right)^8 \alpha^5 \left(\frac{m_e c^2}{\hbar}\right),$$

where $\alpha \equiv e^2/\hbar c \approx 1/137$ is the fine-structure constant and $m_e c^2/\hbar$ is the radian Compton frequency of the electron. Compute now the numerical value of A_{2p1s}, as well as the effective rate $A_{21} = 3A_{2p1s}/4$ if we assume statistical equilibrium in the population of $2p$ and $2s$.

---

2. In this problem we compute the hyperfine splitting of the hydrogen atom in its ground electronic state.

(a) According to the discussion in Chapters 24 and 25, the electron possesses a spin magnetic moment,

$$\mathbf{m} = -\frac{g_e e}{2m_e c}\mathbf{s},$$

where s equals the electron-spin angular momentum and $g_e = 2.0023193048$ is the electron g factor given by both theory and experiment. The proton possesses a spin magnetic moment that has an analogous formula

$$\mathbf{M} = \frac{g_p e}{2 m_p c} \mathbf{S},$$

where $\mathbf{S}$ (usually denoted $\mathbf{I}$) equals the proton-spin angular momentum and $g_p = 5.5855$ is the proton g factor obtained by nuclear magnetic resonance experiments. (The neutron has a g factor equal to -3.8270, which indicates that it, like the proton, contains substructure—presumably in the form of quarks.) Associated with the magnetic moment $\mathbf{M}$ of the proton should be a dipole magnetic field (see Problem Set 4):

$$\mathbf{B} = \frac{3(\mathbf{M} \cdot \hat{\mathbf{k}})\hat{\mathbf{k}} - \mathbf{M}}{|\mathbf{x}|^3},$$

which couples with the magnetic moment $\mathbf{m}$ of the electron in a hydrogen atom to give rise to an interaction Hamiltonian,

$$H_m = -\mathbf{m} \cdot \mathbf{B}.$$

The term H_m splits the degeneracy of the ground electronic state, with the energy level dependent on the magnitude of the total-spin angular momentum $\mathbf{F} \equiv \mathbf{S} + \mathbf{s}$ of the system. Without proof (but see Problem Set 6), accept the statement that the spin eigenkets of F^2 and F_z correspond to the triplet states ($F = 1$)

$$\Sigma_{\pm 1}^1 = \alpha_\pm(\mathrm{p})\alpha_\pm(\mathrm{e}), \qquad \Sigma_0^1 = \frac{1}{\sqrt{2}}\left[\alpha_+(\mathrm{p})\alpha_-(\mathrm{e}) + \alpha_-(\mathrm{p})\alpha_+(\mathrm{e})\right],$$

and the singlet state ($F = 0$)

$$\Sigma_0^0 = \frac{1}{\sqrt{2}}\left[\alpha_+(\mathrm{p})\alpha_-(\mathrm{e}) - \alpha_-(\mathrm{p})\alpha_+(\mathrm{e})\right].$$

In the above, the subscript labels the eigenvalue of F_z in units of $\hbar$, whereas the superscript labels the eigenvalue F. Notice, in particular, that we do not attach the superscript labels "sym" and "anti," since the proton p and the electron e constitute distinguishable particles, so we do not consider interchanges of their (internal) coordinates.

Show now that the formal $1/|\mathbf{x}|^3$ dependence of a dipole magnetic field yields a divergent result if we try a straightforward calculation of the energy splittings via (see Chapter 26)

$$E_m = \langle \varphi\Sigma | H_m | \varphi\Sigma \rangle,$$

where φ is the $1s$ orbital wave function (see Problem 1),

$$\varphi = 2r_0^{-3/2} e^{-r/r_0} Y_{00},$$

and Σ represents one of eigenkets describing the combined spin states of the electron and proton. In the above we have denoted $r \equiv |\mathbf{x}|$.

(b)   To get a nondivergent answer, we resort to a trick invented by Fermi. Problem 1 of Problem Set 4 shows that the vector potential $\mathbf{A}$ associated with a magnetic dipole moment $\mathbf{M}$ may be written

$$\mathbf{A} = \boldsymbol{\nabla} \times \left(\frac{\mathbf{M}}{r} \right).$$

Expressing $\mathbf{B} = \boldsymbol{\nabla} \times \mathbf{A}$, show that H_m becomes

$$H_m = -\mathbf{m} \cdot \left\{ \boldsymbol{\nabla} \times \left[\mathbf{M} \times \boldsymbol{\nabla} \left(\frac{1}{r} \right) \right] \right\}$$

$$= (\mathbf{m} \cdot \mathbf{M}) \nabla^2 \left(\frac{1}{r} \right) - [(\mathbf{m} \cdot \boldsymbol{\nabla})(\mathbf{M} \cdot \boldsymbol{\nabla})] \left(\frac{1}{r} \right).$$

Subtract and add $1/3$ of the first term, from the first and to the second terms, respectively, and write H_m as

$$H_m = H_m^{(0)} + H_m^{(2)},$$

where

$$H_m^{(0)} \equiv \frac{2}{3} (\mathbf{m} \cdot \mathbf{M}) \nabla^2 \left(\frac{1}{r} \right),$$

$$H_m^{(2)} = - \left[(\mathbf{m} \cdot \boldsymbol{\nabla})(\mathbf{M} \cdot \boldsymbol{\nabla}) - \frac{1}{3} (\mathbf{m} \cdot \mathbf{M}) \nabla^2 \right] \left(\frac{1}{r} \right).$$

The operator in the square brackets defining $H_m^{(2)}$ transforms as a tensor of rank 2; consequently (by the Wigner-Eckart theorem), its acting on a spherically symmetric function will produce angular variations that contain only spherical harmonic decompositions of the form $Y_{2m}(\theta, \phi)$. Therefore this term—which contains the bad behavior at small r—contributes nothing when we perform the angular integrations:

$$\langle \varphi \Sigma | H_m^{(2)} | \varphi \Sigma \rangle = 0.$$

(c)   Compute now the expression

$$E_m = \langle \varphi \Sigma | H_m^{(0)} | \varphi \Sigma \rangle,$$

noting the identity

$$\nabla^2 \left(\frac{1}{r} \right) = -4\pi\delta(\mathbf{x}),$$

where $\delta(\mathbf{x})$ is the three-dimensional Dirac delta function. In particular show that

$$E_m = \frac{2g_e g_p e^2}{3 m_e m_p c^2 a_0^3} \langle \Sigma | \mathbf{S} \cdot \mathbf{s} | \Sigma \rangle.$$

Use the identity

$$\mathbf{S} \cdot \mathbf{s} = \frac{1}{2}(\mathbf{F} \cdot \mathbf{F} - \mathbf{S} \cdot \mathbf{S} - \mathbf{s} \cdot \mathbf{s})$$

to evaluate

$$\langle \Sigma | \mathbf{S} \cdot \mathbf{s} | \Sigma \rangle = \frac{1}{2}[F(F+1) - S(S+1) - s(s+1)]\hbar^2,$$

where $S = s = 1/2$ for the proton and electron spins. Consequently, show that the upper and lower states have the associated energies

$$E_m = \frac{1}{6}E_0 \quad \text{for} \quad F = 1, \qquad E_m = -\frac{1}{2}E_0 \quad \text{for} \quad F = 0,$$

where E_0 represents the energy scale,

$$E_0 \equiv g_e g_p \alpha^4 \left(\frac{m_e}{m_p} \right) m_e c^2,$$

and $\alpha \equiv e^2/\hbar c$ stands for the fine-structure constant. The small factor, m_e/m_p, on top of the powers of α (two extra in comparison with Bohr levels) gives the effect its name: "hyperfine" splitting.

(d) Compute the rest frequency ν and the wavelength $\lambda = c/\nu$ of the hydrogen spin-flip transition, $F = 1 \to F = 0$:

$$h\nu = \frac{2}{3}E_0.$$

Predicted by van de Hulst in 1944 to be present in the interstellar medium despite its enormously small A value of 2.85×10^{-15} s^{-1} (see below), and detected by Ewen and Purcell in 1951, this transition constitutes, perhaps, the single most important discovery in all of radio astronomy. As measured experimentally by Crampton, Kleppner, and Ramsey, the transition frequency is also one of the most accurately measured quantities in all of modern physics.

---

3. In this problem, we compute the Einstein coefficient A_{10} for spontaneous emission of the 21-cm line.

(a) Show for magnetic dipole transitions that the analogous formula to equation (23.17) reads

$$A_{if} = \frac{4\omega^3}{3\hbar c^3} \left| \langle f | \mathbf{M} + \mathbf{m} | i \rangle \right|^2 ,$$

where in the present application

$$\mathbf{M} = \mu_p \boldsymbol{\sigma}_p, \qquad \mathbf{m} = -\mu_e \boldsymbol{\sigma}_e,$$

with

$$\mu_p \equiv \frac{e g_p \hbar}{4 m_p c}, \qquad \mu_e \equiv \frac{e g_e \hbar}{4 m_e c},$$

and we have used the definitions $\mathbf{S} \equiv \hbar \boldsymbol{\sigma}_p / 2$, $\mathbf{s} \equiv \hbar \boldsymbol{\sigma}_e / 2$ to eliminate $\mathbf{S}$ and $\mathbf{s}$ in $\mathbf{M}$ and $\mathbf{m}$ in favor of the spin matrices $\boldsymbol{\sigma}_p$ and $\boldsymbol{\sigma}_e$. Since the spin matrices do not act on the spatial part of the wave function, the latter (given by the $1s$ orbital for the initial and final states) integrates out of the matrix element. Thus a spontaneous transition from any one of the three upper $(F = 1)$ states to the lower $(F = 0)$ state has the associated rate

$$A_{10} = \frac{4\omega^3}{3\hbar c^3} \left| \langle \Sigma_0^0 | \mu_p \boldsymbol{\sigma}_p - \mu_e \boldsymbol{\sigma}_e | \Sigma^1 \rangle \right|^2 .$$

(b) With $\boldsymbol{\sigma}$ given by

$$\boldsymbol{\sigma} = \hat{\mathbf{x}} \sigma_x + \hat{\mathbf{y}} \sigma_y + \hat{\mathbf{z}} \sigma_z,$$

where

$$\sigma_z \alpha_\pm = \pm \alpha_\pm, \qquad \sigma_x \alpha_\pm = \alpha_\mp, \qquad \sigma_y \alpha_\pm = \pm i \alpha_\mp,$$

show that

$$\langle \Sigma_0^0 | \mu_p \boldsymbol{\sigma}_p - \mu_e \boldsymbol{\sigma}_e | \Sigma_{\pm 1}^1 \rangle = -\frac{1}{\sqrt{2}} (\hat{\mathbf{x}} \pm i \hat{\mathbf{y}})(\mu_p + \mu_e),$$

$$\langle \Sigma_0^0 | \mu_p \boldsymbol{\sigma}_p - \mu_e \boldsymbol{\sigma}_e | \Sigma_0^1 \rangle = \hat{\mathbf{z}}(\mu_p + \mu_e).$$

(c) Demonstrate now that the downward transition rates from all three upper levels are equal and given by

$$A_{10} = \frac{4\omega^3}{3\hbar c^3} (\mu_p + \mu_e)^2.$$

Since $\mu_e \gg \mu_p$, the magnetic moment of the electron dominates the expression, motivating us to write it in the form:

$$A_{10} = \frac{g_e^2}{12} \left(1 + \frac{g_p m_e}{g_e m_p} \right)^2 \left(\frac{r_e \omega}{c} \right) \left(\frac{\hbar \omega}{m_e c^2} \right) \omega,$$

with $r_e \equiv e^2 / m_e c^2$ given as the classical radius of the electron, and $\omega / 2\pi = \nu$ given by the results of Problem 2. Compute now the numerical value of A_{10}, and compare it with the cited value $2.85 \times 10^{-15} \text{ s}^{-1}$.

Problem Set 6

---

1. Let the total-spin operator of a two-electron atom be given by

$$\mathbf{S} = \mathbf{S}_1 + \mathbf{S}_2, \tag{1}$$

where $\mathbf{S}_1$ operates on the spinor $\alpha_\pm(1)$ and $\mathbf{S}_2$, on $\alpha_\pm(2)$.

(a) The total spinor Σ of a two-electron system must be symmetric if the spatial part of the wave function is antisymmetric with respect to the interchange of the coordinates of the two electrons, and antisymmetric if the spatial part of the wave function is symmetric. For a two-electron atom there exist three symmetric combinations:

$$\begin{aligned}
\Sigma_{+1}^{\text{sym}} &= \alpha_+(1)\alpha_+(2), \\
\Sigma_0^{\text{sym}} &= \frac{1}{\sqrt{2}}[\alpha_+(1)\alpha_-(2) + \alpha_-(1)\alpha_+(2)], \\
\Sigma_{-1}^{\text{sym}} &= \alpha_-(1)\alpha_-(2);
\end{aligned} \tag{2}$$

and only one antisymmetric combination:

$$\Sigma_0^{\text{anti}} = \frac{1}{\sqrt{2}}[\alpha_+(1)\alpha_-(2) - \alpha_-(1)\alpha_+(2)]. \tag{3}$$

Demonstrate that S_z operating on any of the above spinors yields as an eigenvalue $\hbar$ times the value, ± 1 or 0, attached as a subscript to the spinor. Show also that S^2 operating on the same spinors yields as an eigenvalue $\hbar^2$ times $S(S+1) = 2$ for the symmetric states, and 0 for the antisymmetric state. In other words, the three symmetric states correspond to $S = 1$ (multiplicity $2S + 1 = 3$ because of the three possible projections -1, 0, $+1$ for S_z); whereas the one antisymmetric state corresponds to $S = 0$ (multiplicity $2S + 1 = 1$ because of only one projection 0 for S_z).

(b) Verify by direct multiplication that the four spinor states constructed above are orthogonal to one another; for example, if we use Dirac's notation to indicate the inner product between two spinors (turning one to a row vector and multiplying into the column vector of the other), we have

$$\langle \Sigma_{+1}^{\text{sym}} | \Sigma_{-1}^{\text{sym}} \rangle \equiv \langle \alpha_+(1)|\alpha_-(1)\rangle \langle \alpha_+(2)|\alpha_-(2)\rangle = 0, \tag{4}$$

since $\langle \alpha_+ | \alpha_- \rangle = 0$ when the spinors have the same argument. Verify also that each spinor has been normalized; i.e., its inner product with itself equals 1. Thus the four spinors of part (a) form an *orthonormal set*.

(c) Consider now the structure of the helium atom with $Z = 2$. If we ignore spin-orbit interactions, the starting Hamiltonian reads

$$H = H_Z(1) + H_Z(2) + \frac{e^2}{r_{12}}, \tag{5}$$

where

$$H_Z(a) = -\frac{\hbar^2}{2m_e} \nabla_a^2 - \frac{Ze^2}{r_a} \quad \text{for} \quad a = 1, 2 \tag{6}$$

gives the effects of the motion and interaction with the nucleus of the a-th electron. Since the two electrons are indistinguishable fermions, the wave function for stationary states must consist of either a singlet,

$$^1\Psi(1, 2) = U^{\text{sym}}(\mathbf{r}_1, \mathbf{r}_2)\Sigma^{\text{anti}}e^{-iEt/\hbar},$$

or a triplet,

$$^3\Psi(1, 2) = U^{\text{anti}}(\mathbf{r}_1, \mathbf{r}_2)\Sigma^{\text{sym}}e^{-iEt/\hbar},$$

where Σ^{anti} represents the single antisymmetric spinor Σ_0^{anti}, but Σ^{sym} could be any one of the three symmetric choices Σ_{-1}^{sym}, Σ_0^{sym}, Σ_{+1}^{sym}. Our ability to factor out the spatial part U completely from the spinor part Σ depends on the Hamiltonian having no explicit dependence on the spin operator.

Consider, in particular, the calculation of the ground electronic state. Since $\Sigma = \Sigma^{\text{anti}}$ for this state, we want to construct $U^{\text{sym}}(1, 2)$ for two electrons sharing the same orbital ($n = 1$ and $\ell = 0$, denoted as $1s$ for short). The *Hartree-Fock method* begins by assuming that this spatial wave function may be decomposed into a product of (identical) single-particle wave functions:

$$U^{\text{sym}}(\mathbf{r}_1, \mathbf{r}_2) = u(\mathbf{r}_1)u(\mathbf{r}_2), \tag{7}$$

where u may have an arbitrary dependence on its argument. Nevertheless, the factorization into separate dependences on $\mathbf{r}_1$ and $\mathbf{r}_2$ clearly cannot exactly satisfy a Schrödinger equation where the basic Hamiltonian depends on the quantity $r_{12} = |\mathbf{r}_1 - \mathbf{r}_2|$. Get around this difficulty by settling for the best choice of u which minimizes $\langle \Psi | H | \Psi \rangle$ (see the discussion in Chapter 29). Using the notation of variational calculus, we require

$$\delta\langle \Psi | H | \Psi \rangle = \delta\langle U | H | U \rangle = \langle \delta U | H | U \rangle + \langle U | H | \delta U | \rangle = 0,$$

where the variation of δU for $U(1, 2) = u(1)u(2)$ is given by

$$\delta U = u(1)\delta u(2) + u(2)\delta u(1).$$

Show that the symmetry with which $\mathbf{r}_1$ and $\mathbf{r}_2$ enter into the equations and the Hermitian character of the governing Hamiltonian allow us to express the extremal constraint as

$$\int \int u^*(1)\delta u^*(2) H u(1) u(2)\, d^3 r_1\, d^3 r_2 = 0. \tag{8}$$

As written, however, we cannot set the integrand of $d^3 r_2$ equal to zero at all values of $\mathbf{r}_2$ since the variations of $\delta u^*(2)$ cannot be taken completely arbitrarily, but must satisfy the integral constraint that $[u(1)+\delta u(1)][u(2)+ \delta u(2)]$ remains normalized, i.e., that

$$\int \int u(1)u(2)u^*(1)\delta u^*(2)\, d^3 r_1\, d^3 r_2 = 0. \tag{9}$$

To accommodate this constraint, multiply equation (9) by a Lagrange multiplier E (which we can prove to be real if we manipulate our variations a little more carefully) and subtract the result from the extremal condition (8). Now we can set the integrand of $d^3 r_2$ equal to zero, obtaining the requirement

$$\left[\int d^3 r_1 u^*(1) H u(1) \right] u(2) = E u(2),$$

where we have made use of the condition

$$\int u(1)u^*(1)\, d^3 r_1 = 1.$$

With the notation $\mathbf{r}_2 = \mathbf{r}$ and $\mathbf{r}_1 = \mathbf{r}'$, substitute in the expression (6) for H and obtain the effective Schrödinger equation

$$H_{\text{eff}} u(\mathbf{r}) = E_{\text{eff}} u(\mathbf{r}), \tag{10}$$

where

$$H_{\text{eff}} \equiv -\frac{\hbar^2}{2m_{\text{e}}}\nabla^2 + V(\mathbf{r}),$$

$$V \equiv -\frac{Ze^2}{r} + W(\mathbf{r}),$$

with

$$W(\mathbf{r}) \equiv \int \frac{e^2}{|\mathbf{r}-\mathbf{r}'|} u^*(\mathbf{r}')u(\mathbf{r}')\, d^3 r',$$

$$E_{\text{eff}} = E - \int u^*(\mathbf{r}') \left[-\frac{\hbar^2}{2m_{\text{e}}}\nabla'^2 - \frac{Ze^2}{r'} \right] u(\mathbf{r}')d^3 r' = \text{constant}.$$

Since equation (10) implies that we may replace

$$\left[-\frac{\hbar^2}{2m_{\text{e}}}\nabla'^2 - \frac{Ze^2}{r'} \right] u(\mathbf{r}')$$

in the integrand by $[E_{\text{eff}} - W(\mathbf{r}')]u(\mathbf{r}')$, we obtain the identification

$$E = 2E_{\text{eff}} - \int \int \frac{e^2 |u(\mathbf{r})|^2 |u(\mathbf{r}')|^2}{|\mathbf{r} - \mathbf{r}'|} \, d^3r \, d^3r'.$$

In this form, we easily identify E as the energy of the two-electron system, and E_{eff} and H_{eff} are the effective energy and Hamiltonian of the corresponding one-particle state. Since the term $2E_{\text{eff}}$ counts, on average, the electrons' mutual potential energy twice, we must subtract this term (the double integral) once in deriving E from $2E_{\text{eff}}$.

The effective potential V in which each electron moves equals the sum of the Coulomb interaction with the nucleus, $-Ze^2/r$, and the average (repulsive) interaction with the other electron, W. The latter will possess spherical symmetry, i.e., $W(\mathbf{r}) = W(r)$, if u corresponds to an s-orbital (angular momentum $\ell = 0$). This derivation via a variational principle provides the mathematical justification for the claim that the Hartree-Fock procedure, which we wrote down intuitively in Chapter 27, yields the best possible central-field approximation. (Q.E.D.)

---

2. The electronic energy as a function of the internuclear spacing R in a diatomic molecule is often approximated as a *Morse potential* minus the value of that potential at infinite separation:

$$E_{\text{el}}(R) = E_0 \left[1 - e^{-(R-R_0)/L} \right]^2 - E_0,$$

where E_0 and R_0 equal constants having, respectively, the units of energy and length.

(a) Show that the Morse potential has a minimum at $R = R_0$, where $E_{\text{el}}(R_0)$ has depth $-E_0$. For small amplitude oscillations in classical mechanics, the curvature of the potential energy at the equilibrium position, i.e., the value of $E_{\text{el}}''(R_0)$, determines the fundamental oscillation frequency ω_0. Show that ω_0 has the value

$$\omega_0 = \left(\frac{2E_0}{\mu L^2} \right)^{1/2}, \tag{11}$$

where μ equals the reduced mass of the two nuclei. Compute the numerical values of L and R_0 in Angstroms for the H_2 molecule, given the experimental data on the fundamental vibrational frequency, dissociation energy, and rotational constant: $\hbar\omega_0 = 0.548\,\text{eV}$, $E_0 - \hbar\omega_0/2 = 4.476\,\text{eV}$, and $B/k = 87\,\text{K}$, where $B = \hbar^2/2\mu R_0^2$. Now, plot $E_{\text{el}}(R)$ (in eV) versus R (in Å), and compare your curve with the shape shown in the text for the ground electronic state of the H_2 molecule.

(b)   Evaluate the centrifugal contribution to the effective potential at the equilibrium position R_0; introduce the dimensionless coordinate

$$x \equiv \left(\frac{\mu\omega_0}{\hbar}\right)^{1/2} (R - R_0);$$

and show that the Schrödinger equation (28.25) for the vibrational part of the wave function Z takes the nondimensional form,

$$\frac{d^2 Z}{dx^2} + \left[\epsilon - \Lambda^2(1 - e^{-x/\Lambda})^2\right] Z = 0. \tag{12}$$

In the above, we have defined [see equation (11)]

$$\Lambda^2 \equiv \frac{2E_0}{\hbar\omega_0} = \left(\frac{\mu\omega_0}{\hbar}\right) L^2, \qquad \epsilon \equiv \frac{2E_{\text{vib}}}{\hbar\omega_0},$$

where

$$E_{\text{vib}} \equiv E_{\text{int}} + E_0 - \frac{J(J+1)\hbar^2}{2\mu R_0^2}$$

in the usual way. The second-order ODE (12) is to be solved by suitably applying boundary conditions at $x = -\Lambda R_0/L$ (i.e., $R = 0$) and as $x \to \infty$ (i.e., $R \to +\infty$). Homogeneous conditions lead to the necessity of regarding ϵ as an eigenvalue of the problem.

(c)   From the data given in part (a), verify the numerical value of $\Lambda^2 = 17.3$ for the ground electronic state of H_2. Interpret the result $\Lambda^2 \gg 1$ physically.

The harmonic-oscillator approximation amounts to regarding Λ as an arbitrarily large parameter. In the limit that $\Lambda \to \infty$, we may make the replacement

$$\Lambda^2(1 - e^{-x/\Lambda})^2 \to x^2. \tag{13}$$

Consistent with this approximation, we may push the left-hand boundary to $x = -\infty$. Since the potential (13) has unlimited height at $x = \pm\infty$, we naturally apply the boundary conditions:

$$Z \to 0 \qquad \text{as} \qquad x \to \pm\infty. \tag{14}$$

The ODE (12) now reads

$$\frac{d^2 Z}{dx^2} + (\epsilon - x^2)Z = 0,$$

which has the standard eigensolutions

$$Z(x) = H_v(x)e^{-x^2/2}, \qquad \epsilon = (2v + 1), \qquad \text{with} \qquad v = 0, 1, 2, \ldots,$$

where $H_v(x)$ equals the Hermite polynomial of order v. Given the definition of ϵ, verify that the vibrational energy has the quantized levels

$$E_{\text{vib}} = \left(v + \frac{1}{2}\right)\hbar\omega_0, \qquad v = 0, 1, 2, \ldots.$$

(d) The approximation of part (c) must break down when Λ has a large but finite value. The left-hand side of equation (13) reaches a maximum height Λ^2 when $x \to +\infty$, whereas the right-hand side has no such limitation. For Λ^2 finite, then, the true eigensolutions to equation (12) with $\epsilon > \Lambda^2$ can reach $x = +\infty$ as oscillatory disturbances rather than as exponentially damped ones. Thus, rather than bound states, the eigensolutions must correspond to *free* states (the "vibrational continuum"). If we were simply to extrapolate the harmonic-oscillator approximation $\epsilon = (2v + 1)$ to the continuum limit $\epsilon = \Lambda^2 \approx 17$, we would predict that the vibrational levels of the H_2 molecule in its ground electronic state become unbound after $v = 8$. In fact, because the spacing of successive energy levels decreases with increasing v, the real molecule has additional bound states. Use the WKB technique to estimate the effect by solving equation (12) subject to the (approximate) boundary condition (14).

Hint: Begin by defining

$$K^2(x; \epsilon) \equiv \epsilon - \Lambda^2(1 - e^{-x/\Lambda})^2, \tag{15a}$$

so that equation (12) reads

$$\frac{d^2 Z}{dx^2} + K^2(x; \epsilon)Z = 0. \tag{15b}$$

For given ϵ satisfying $0 < \epsilon < \Lambda^2$, classical turning points occur at the roots $x_\pm$ where $K^2(x; \epsilon) = 0$, i.e., at

$$x_\pm = -\Lambda \ln\left[1 \mp (\epsilon/\Lambda^2)^{1/2}\right].$$

The solution x_- has a negative value; x_+, a positive one. For x well inside the interval (x_-, x_+), $K^2(x; \epsilon)$ is large and positive, so that equation (15b) has the approximate oscillatory solutions

$$Z \propto \frac{1}{\sqrt{K}} e^{\pm i\phi(x)}, \tag{16a}$$

$$\phi(x) \equiv \int^x K(x'; \epsilon)\, dx', \tag{16b}$$

with a zero-point for the integrated phase $\phi(x)$ that we need not specify for our purposes here. Equation (16a) corresponds to right- and left-propagating waves. For x in a small neighborhood of $x_\pm$, where $K^2(x; \epsilon)$ goes through zero and then becomes negative, the solution (16a) breaks down. For small $(x - x_-)$, if we Taylor-expand about the zero and keep only the linear term, we obtain

$$K^2(x; \epsilon) = A(x - x_-),\qquad(17)$$

where A equals a constant whose value need not concern us here other than that it is large and positive (since $K^2[x; \epsilon]$ is large and negative for $x < x_-$ and is large and positive for $x > x_-$.) If we substitute equation (17) into equation (15b) and redefine the coordinate variable $\xi \equiv A^{1/3}(x - x_-)$, show that we obtain

$$\frac{d^2Z}{d\xi^2} + \xi Z = 0,$$

which is *Airy's differential equation.* (See Abramowitz and Stegun 1965, pp. 446–449.) This has the general solution

$$Z = a\mathrm{Ai}(-\xi) + b\mathrm{Bi}(-\xi),$$

with the Airy function $\mathrm{Ai}(-\xi)$ and the Bairy function $\mathrm{Bi}(-\xi)$ having the properties, respectively, of decaying and growing exponentially as $\xi \rightarrow -\infty$ (i.e., $x < x_-$ by a finite amount). To satisfy the boundary condition (14), we keep therefore only the Airy function $\mathrm{Ai}(-\xi)$. It has the following asymptotic limiting behavior as $\xi \rightarrow +\infty$:

$$\mathrm{Ai}(-\xi) = \pi^{-1/2}\xi^{-1/4} \sin\left(\frac{2}{3}\xi^{3/2} + \frac{\pi}{4}\right)$$
$$= \frac{1}{2i}\pi^{-1/2}\xi^{-1/4} \left[\exp\left(+i\frac{2}{3}\xi^{3/2} + i\frac{\pi}{4}\right) + \exp\left(-i\frac{2}{3}\xi^{3/2} + i\frac{3\pi}{4}\right)\right].$$
$$(17)$$

Since $(2/3)\xi^{3/2}$ can be identified as the limit

$$\int_{x_-}^{x} K(x'; \epsilon)\, dx'$$

for $x \rightarrow x_-$ from the propagating side, we see that the forms (16a) and (18) asymptotically match in the common regime of validity. The solution has the physical interpretation that when a left-propagating wave $e^{-i\phi(x)}$ travels toward the turning point x_-, it reflects (with no change in amplitude) into a right-propagating wave $e^{+i\phi(x)}$ with a change in phase $\Delta\phi = \pi/4 - 3\pi/4 = -\pi/2$. Use a similar calculation to show that when a right-propagating wave $e^{+i\phi(x)}$ travels toward the turning point x_+, it reflects (again, with no change in amplitude) into a left-propagating wave $e^{-i\phi(x)}$, also with a change in phase $\Delta\phi = -\pi/2$.

Begin now arbitrarily with a left-propagating wave at x_+. In propagating from x_+ to x_-, the wave accumulates a phase change (we use the negativewave number to flip the limits of integration)

$$\Delta\phi = \int_{x_-}^{x_+} K(x; \epsilon)\, dx'.\qquad(19a)$$

As the wave reflects from x_-, it suffers a phase change

$$\Delta\phi = -\frac{\pi}{2}. \tag{19b}$$

The reflected wave propagating to the right then travels to x_+, accumulating another phase change

$$\Delta\phi = \int_{x_-}^{x_+} K(x; \epsilon)\, dx. \tag{19c}$$

As this wave then reflects from x_+, it also suffers a phase change

$$\Delta\phi = -\frac{\pi}{2}. \tag{19d}$$

The reflected left-propagating wave will be in phase with the original one (so as to allow a standing pattern and a stationary state) only if the total phase change $(\Delta\phi)_{\text{tot}}$ equals a non-negative integer v times 2π. Summing equations (19a)–(19d), dividing out a factor of 2π, and defining the integral function

$$I(\epsilon) \equiv \frac{1}{\pi} \int_{x_-}^{x_+} K(x; \epsilon)\, dx, \tag{20}$$

show that we have the requirement

$$I(\epsilon) = v + \frac{1}{2}, \qquad v = 0, 1, 2, \ldots, \tag{21}$$

which represents the desired WKB quantum condition for determining the allowable eigenvalues ϵ. (Q.E.D.)

(e) To facilitate the evaluation of $I(\epsilon)$ in equation (20), where equation (15a) defines K, introduce a new variable $y \equiv \Lambda\epsilon^{-1/2}(1 - e^{-x/\Lambda})$, and show that I becomes

$$I(\epsilon) = \frac{\epsilon}{\pi} \int_{-1}^{+1} (1 - y^2)^{1/2} (1 - \beta y)^{-1}\, dy, \qquad \text{where} \qquad \beta \equiv \frac{\sqrt{\epsilon}}{\Lambda}.$$

Another substitution, $z = 1 - \beta y$, puts the integral into a form that can be integrated by elementary means. Show that evaluation at the limits leads to the result

$$I(\epsilon) = \Lambda^2 \left[1 - \left(1 - \frac{\epsilon}{\Lambda^2} \right)^{1/2} \right].$$

Equation (21) now implies

$$\epsilon = 2 \left(v + \frac{1}{2} \right) - \frac{1}{\Lambda^2} \left(v + \frac{1}{2} \right)^2, \qquad v = 0, 1, 2, \ldots.$$

Note that we recover the harmonic-oscillator result, $\epsilon = 2v + 1$, in the limit $\Lambda^2 \to \infty$. For $\Lambda^2 = 17.3$, compute ϵ as a function of v from $v =$

0 to $v = 18$. Comment on the decreasing spacing of successive energy levels as we approach the continuum limit $\epsilon = \Lambda^2 = 17.3$. What is the maximum value of the vibrational quantum number v for bound states? Those results do not agree completely at high v with experimental results (compare with Figure 28.5). Compare also the spectroscopic determination of the anharmonicity constant $A = 0.027$ in H_2 with the theoretical value of $1/2\Lambda^2$; comment on the adequacy of the three-parameter Morse potential representation if we want to reproduce all the information available via high-resolution spectroscopy.

Bibliography

M. Abramowitz and I.A. Stegun. 1965. *Handbook of Mathematical Functions*. Dover.

W.A. Bingel. 1969. *Theory of Molecular Structure*. Wiley.

S. Chandrasekhar. 1950. *Radiative Transfer*. Oxford University Press.

D.D. Clayton. 1968. *Principles of Stellar Evolution and Nucleosynthesis*. McGraw-Hill.

E.V. Condon and G.H. Shortley. 1967. *The Theory of Atomic Spectra*. Cambridge University Press.

P. A. M. Dirac. 1958. *Quantum Mechanics*. Oxford University Press.

H. Goldstein. 1959. *Classical Mechanics*. Addison-Wesley.

G. Herzberg. 1950. *Molecular Spectra and Molecular Structure I. Spectra of Diatomic Molecules*. Van Nostrand.

K. Huang. 1963. *Statistical Mechanics*. Wiley.

J. D. Jackson. 1962. *Classical Electrodynamics*. Wiley.

H. Jeffreys and B. S. Jeffreys. 1962. *Mathematical Physics*. Cambridge University Press.

V. Kourganoff. 1963. *Basic Methods in Transfer Problems*. Dover.

L. D. Landau and E. M. Lifshitz. 1951. *Classical Theory of Fields*. Addison-Wesley.

R. Leighton. 1959. *Principles of Modern Physics*. McGraw-Hill.

D. McQuarrie. 1983. *Quantum Chemistry*. University Science Books.

E. Merzbacher. 1961. *Quantum Mechanics*. Wiley.

D. Mihalas. 1978. *Stellar Atmospheres*. Freeman.

D. Mihalas and B. W. Mihalas. 1984. *Foundations of Radiation Hydrodynamics*. Oxford University Press.

D. Osterbrock. 1989. *Astrophysics of Gaseous Nebulae and Active Galactic Nuclei.* University Science Books.

A.G. Pacholczyk. 1977. *Radio Galaxies.* Pergamon Press.

G. B. Rybicki and A. P. Lightman. 1979. *Radiative Processes in Astrophysics.* Wiley-Interscience.

J. J. Sakurai. 1967. *Advanced Quantum Mechanics.* Addison-Wesley.

L. Schiff. 1955. *Quantum Mechanics.* McGraw-Hill.

M. Schwarzschild. 1958. *Structure and Evolution of the Stars.* Princeton University Press.

F.H. Shu. 1982. *The Physical Universe.* University Science Books.

T. Y. Wu. 1986. *Quantum Mechanics,* World Scientific.

INDEX

Q

R